Modern Hydroforming Technology

Shijian Yuan

Modern Hydroforming Technology

国防工業出版社
National Defense Industry Press

Springer

Shijian Yuan
School of Material Science and Engineering
Harbin Institute of Technology
Harbin, Heilongjiang, China

ISBN 978-981-19-5774-1 ISBN 978-981-19-5775-8 (eBook)
https://doi.org/10.1007/978-981-19-5775-8

Jointly published with National Defense Industry Press
The print edition is not for sale in China (Mainland). Customers from China (Mainland) please order the print book from: National Defense Industry Press.

Translation from the Chinese language edition: “现代液压成形技术(第3版)” by Shijian Yuan, © National Defense Industry Press 2020. Published by National Defense Industry Press. All Rights Reserved.

This Springer imprint is published by the registered company Springer Nature Singapore Pte Ltd.
The registered company address is: 152 Beach Road, #21-01/04 Gateway East, Singapore 189721, Singapore

Contents

1 Introduction ... 1
1.1 Types and Characteristics of Hydroforming Technology ... 1
1.1.1 Definition and Types of Hydroforming ... 1
1.1.2 Technological Characteristics of Tube Hydroforming ... 2
1.1.3 Characteristics of Sheet Hydroforming Technology ... 7
1.1.4 Characteristics of Shell Hydroforming Technology ... 9
1.2 Overview of Hydroforming Technology ... 10
1.2.1 Overview of Tube Hydroforming Technology ... 10
1.2.2 Overview of Sheet Hydroforming Technology ... 14
1.2.3 Overview of Shell Hydroforming Technology ... 16
1.3 Development and Perspective of Hydroforming Technology ... 18
1.3.1 Development and Perspective of Tube Hydroforming ... 18
1.3.2 Development Direction and Perspective of Sheet Hydroforming ... 19
1.4 Recent Developments of Hydroforming ... 20
1.4.1 Recent Developments of Tube Hydroforming ... 20
1.4.2 Recent Developments of Sheet Hydroforming ... 24
1.4.3 Recent Developments of Shell Hydroforming ... 25

2 Hydroforming of Variable Diameter Tubular Parts ... 27
2.1 Process Stages and Application Fields ... 27
2.1.1 Process Stages ... 27
2.1.2 Application Fields ... 29
2.2 Determination of Main Process Parameters ... 29
2.2.1 Initial Yielding Pressure ... 29
2.2.2 Splitting Pressure ... 30
2.2.3 Calibration Pressure ... 30
2.2.4 Axial Feeding Force ... 33
2.2.5 Clamping Force ... 34
2.2.6 Axial Feeding Length ... 34
2.3 Defects and Loading Path ... 35

2.3.1 Type of Defects ... 35
2.3.2 Process Window and Loading Path ... 36
2.3.3 Limit Expansion Ratio ... 38
2.3.4 Control and Use of Wrinkles ... 40
2.4 Thickness Distribution and Influencing Factors ... 44
2.4.1 Thickness Distribution ... 44
2.4.2 Thickness Dividing Circle ... 46
2.5 Tubes for Hydroforming ... 47
2.5.1 Material ... 47
2.5.2 Requirement of Tube Materials ... 48
2.5.3 Types and Specification of Tubes ... 48
2.5.4 Measurement of Mechanical Properties of Tubes ... 50
2.5.5 Testing Methods and Apparatus for Mechanical Properties of Anisotropic Tubes ... 51
2.6 Friction and Lubrication During Hydroforming ... 58
2.7 Hydroforming Process of Typical Variable Diameter Tubular Parts ... 60
2.7.1 Hydroforming of an Aluminum Alloy Variable Tubular Part ... 60
2.7.2 Hydroforming of a Low Carbon Steel Tubular Part ... 62
2.7.3 Hydroforming of a Ω-shaped Joint Part ... 64
2.7.4 Hydroforming of an Irregular Tubular Part with Double Cones ... 65
2.7.5 Hydroforming of a Long Waved Pipe ... 67

3 Hydroforming of Curved Parts with Irregular Cross-Sections ... 71
3.1 Process and Typical Cross-Sections ... 71
3.1.1 Process Description ... 71
3.1.2 Typical Cross-Sections ... 73
3.2 Tube Bending Processes ... 74
3.2.1 Common Bending Processes and Characteristics ... 74
3.2.2 Minimum Bending Radius of Tube ... 76
3.2.3 Cross-Section Distortion and Prevention Approaches ... 77
3.2.4 Calculation of the Bending Moment ... 78
3.2.5 Thickness Variation ... 79
3.2.6 Ultimate Diameter-Thickness Ratio of Tube Bending ... 79
3.3 Hydro-Bending Process of Tubes ... 80
3.3.1 Principle and Characteristics of Hydro-Bending ... 80
3.3.2 Tube Hydro-Bending Without End Constraint ... 82
3.3.3 Tube Hydro-Bending with End Constraint ... 83
3.4 Type of Defects ... 86
3.5 Thickness Distribution of a Square Cross-Section ... 90
3.5.1 Effect of Expansion Ratio on Thickness Distribution ... 93
3.5.2 Effect of Friction Coefficient on Thickness Distribution ... 93

3.5.3 Effect of Parting Surface on Thickness Distribution 94
3.5.4 Effect of Material Mechanical Properties on Thickness Distribution 94
3.6 Principle and Method for Reducing Calibration Pressure 95
3.6.1 Principle of Reducing Calibration Pressure by Concave Preformed Cross-Section 95
3.6.2 The Correlation Between the Tangential Pushing Force and the Concave Depth 96
3.6.3 Calculation of Calibration Pressure for Concave Preformed Cross-Section 99
3.6.4 Shape Optimization Design of Concave Preform 102
3.6.5 Hydro-Pressing Process for the Preform 106
3.7 Hydroforming of Typical Parts with Curved Axis 109
3.7.1 Hydroforming of the Engine Cradle 109
3.7.2 Hydroforming of an Instrument Panel Beam 112
3.7.3 Hydroforming of an Aluminum Alloy Tubular Part 113
3.7.4 Hydroforming of an Aluminum Alloy Subframe Tubular Part 114
3.7.5 Hydroforming of MPV Passenger Car Subframe 120
3.7.6 Hydroforming of a DP590 Trailing Arm 121
3.7.7 Hydroforming of Torsion Beam of a High Strength Steel (780 MPa) 125
3.7.8 Hydroforming of a Crash Box 128

4 Hydroforming of Multi-branch Tubular Parts 131
4.1 Classifications and Processes of Multi-branch Tubes 131
4.2 Defect and Limit Height of Branch 135
4.2.1 Typical Defects 135
4.2.2 Limit Height of Branch 137
4.3 Thickness Distribution of Hydroformed Three-Branch Tubular Parts 138
4.4 Multi-step Hydroforming of Three-Branch Tube of Aluminium Alloys 142
4.4.1 Geometry and Challenges 142
4.4.2 Simulations of Multi-step Hydroforming of Three-Branch Tubular Part 143
4.4.3 Experiment on Multi-step Forming of Three-Branch Tube 144
4.5 Hydroforming of Ultra-Thin Three-Branch Tube of Stainless Steel 148
4.5.1 Geometry and Challenges 148
4.5.2 Finite Element Simulations of Multi-step Forming of Three-Branch Ultra-Thin Three-Branch Tube 149
4.5.3 Experiment on Multi-step Forming of Three-Branch Ultra-Thin Tube 150

4.6 Hydroforming of a Three-Branch Thin-Walled Tube of Ni-Based Superalloys ... 151
4.6.1 The Geometry of Ni-Based Superalloy Three-Branch Tube and Forming Difficulties ... 151
4.6.2 Experimentation of Multi-step Forming of Ni-Based Superalloy Three-Branch Tube ... 152

5 Stress–Strain Analysis for Tube Hydroforming ... 155
5.1 The Stress–Strain States and the Yield Locus During Hydroforming of Tubes ... 155
5.1.1 Initial Filling Stage ... 155
5.1.2 Forming Stage ... 156
5.1.3 Calibration Stage ... 158
5.2 The Stress–Strain States During Hydroforming of Curved Tubes and Three-Branch Tubes ... 159
5.2.1 The Stress–Strain States During Hydroforming of Curved Tubes ... 159
5.2.2 The Stress–Strain States During Hydroforming of Y-shaped Tubes ... 162
5.3 The Stress Locus of Tube Hydroforming ... 163
5.4 Stress-State of Corner Zone and Splitting Mechanism ... 165
5.4.1 Stress Analysis of Corner Zone ... 165
5.4.2 Mechanism of Splitting at the Corner Zone ... 166

6 Equipment and Tooling of Hydroforming ... 171
6.1 Construction and Functions of Hydroforming Equipment ... 171
6.1.1 Construction of Hydroforming Equipment ... 171
6.1.2 Sub-Systems of Hydroforming Equipment ... 172
6.2 Parameters of Hydroforming Equipment ... 177
6.2.1 Definitions of Main Parameters ... 177
6.2.2 Determination of Main Parameters ... 179
6.2.3 Recommendations of Standard and Parameters of Hydroforming Equipment ... 180
6.3 Typical Structures and Characteristics of the Die-Clamping Press ... 181
6.3.1 The Long-Stroke Die-Clamping Press ... 181
6.3.2 The Short-Stroke Die-Clamping Press ... 183
6.4 Large-Scale Hydroforming Equipment ... 185
6.4.1 Servo-Control of Ultra-High Pressure and Multi-axial Displacement ... 185
6.4.2 Computer Numerical Control System and Control Software ... 186
6.4.3 Series of Hydroforming Equipment Developed by HIT ... 187
6.4.4 The Layout of Hydroforming Production Line ... 188
6.5 Tooling of Hydroforming ... 191

6.5.1 The Structure and Material of Forming Dies 191
6.5.2 Effects of Thickness of Die Sidewall and Bottom Plate 192

7 Sheet Hydroforming Process 195
7.1 Procedure and Characteristics of Process 195
7.1.1 Forming Procedure 195
7.1.2 Characteristics of Sheet Hydroforming 197
7.1.3 Application Fields of Sheet Hydroforming 198
7.2 The Calculation of Main Process Parameters 199
7.2.1 The Critical Pressure of Liquid Cavity 199
7.2.2 The Forming Force 202
7.2.3 Blankholding Force 203
7.3 The Limit Drawing Ratio and Type of Defects 204
7.3.1 The Limit Drawing Ratio 204
7.3.2 The Type of Defects 205
7.3.3 The Stresses During Sheet Hydroforming Defects 208
7.4 Dimensional Accuracy and Thickness Distribution of Hydroformed Parts 210
7.4.1 Dimensional Accuracy 210
7.4.2 The Thickness Distribution 211
7.4.3 Springback Variation 213
7.5 The Equipment and Tooling of Sheet Hydroforming 215
7.5.1 The Construction of Sheet Hydroforming Equipment 215
7.5.2 The Characteristics and Main Parameters of Sheet Hydroforming Equipment 216
7.5.3 The Structure and Materials of Die-Sets 218
7.6 The Sheet Hydroforming with Radial Pressure 220
7.6.1 Process Procedure 220
7.6.2 Effect of Radial Pressure on the Forming Limit 221
7.6.3 Effect of Radial Pressure on the Thickness Distribution 222
7.7 The Sheet Hydroforming with Pre-bulging 223
7.7.1 Process Procedure 223
7.7.2 Thickness Distribution of Sheet Hydroforming with Pre-bulging 224
7.7.3 The Strengthening of Sheet Hydroforming with Pre-bulging 225
7.8 The Sheet Hydroforming of Typical Parts 227
7.8.1 Parabolic Cross-Section Shaped Parts 227
7.8.2 Box-Shaped Panel with a Single Curvature 229
7.8.3 Box-Shaped Panel with Double Curvatures 236
7.8.4 The Hydroforming of 2219 Aluminium Alloy Sheet 240

8 Die-Less Hydroforming of Shells 247
8.1 Structure of Shells and Manufacturing Process 247
8.2 Die-Less Hydroforming for Spherical Vessels 248

8.2.1 Forming Principle and Advantages 248
8.2.2 Forming Pressure 248
8.2.3 Polyhedral Structure Before Hydroforming 249
8.3 Die-Less Hydroforming of Ellipsoidal Shells 251
8.3.1 Die-Less Hydroforming of Ellipsoidal Shell 251
8.3.2 Relationship Between Stress and Axial Length Ratio of an Ellipsoidal Shell 253
8.3.3 Hydroforming Pressure for Ellipsoidal Shells 255
8.3.4 Hydroforming Experiment of Ellipsoidal Shell 256
8.3.5 Deformation and Wrinkling Behavior in Hydroforming of Ellipsoidal Shell 257
8.4 Die-Less Hydroforming of Ellipsoidal Shells with Double Generating Lines 262
8.4.1 Principle of Hydroforming of Ellipsoidal Shell with Double Generating Lines 262
8.4.2 Structural Design of Ellipsoidal Shell with Double Generating Lines 263
8.4.3 Hydroforming Experiment of Ellipsoidal Shell with Double Generating Lines 266
8.4.4 Variation of Curvature Radius During Hydroforming of Ellipsoidal Shell with Double Generating Lines 269
8.4.5 Variation of Shell Volume During Hydroforming of Ellipsoidal Shell with Double Generating Lines 272
8.4.6 Variation of Stress During Hydroforming of Ellipsoidal Shell with Double Generating Lines 275
8.5 Die-Less Hydroforming of Prolate Ellipsoidal Shells with Double Generating Lines 277
8.5.1 Hydroforming Experiment of Prolate Ellipsoidal Shell with Double Generating Lines 278
8.5.2 Volume Variation of Prolate Ellipsoidal Shell with Double Generating Lines in Hydroforming 281
8.5.3 Thickness Variation of Prolate Ellipsoidal Shell with Double Generating Lines in Hydroforming 282
8.6 Die-Less Hydroforming of Toroidal Shells 282
8.6.1 Die-Less Hydroforming Process of Toroidal Shells 282
8.6.2 Stress Distribution and Forming Pressure of Toroidal Shells 283
8.6.3 Die-Less Hydroforming Experiment of Toroidal Shells 284
8.6.4 Analysis on Wrinkling During Toroidal Shell Forming 286
8.6.5 Effect of Initial Structures of Toroidal Shell on Hydroforming 287

References 293

Chapter 1
Introduction

1.1 Types and Characteristics of Hydroforming Technology

1.1.1 Definition and Types of Hydroforming

Hydroforming is a kind of metal forming technology, in which the liquid is used as the force transmission medium to deform a sheet or tube into desired shapes. It is also known as the hydraulic forming.

Hydroforming can be divided into water pressure forming and oil pressure forming, based on the selected liquid medium. In water pressure forming, pure water or the emulsion consisting of water and a certain proportion of emulsified oil is used as the force-transmitting medium. While the hydraulic oil is used as the force-transmitting medium in oil pressure forming. In addition, according to the shape of raw blank, hydroforming can be also classified into three types [1], i.e. tube hydroforming [2], sheet hydroforming [3] and shell hydroforming [4].

The liquid pressures required by both sheet and shell hydroforming are relatively low, while that of tube hydroforming is higher. Therefore, the tube hydroforming is also called internal high pressure forming (IHPF). Regarding the sheet hydroforming, the hydraulic oil is often used as the forming medium, and the maximum forming pressure is generally lower than 100 MPa; Regarding the shell hydroforming, pure water is used, and the maximum forming pressure is generally lower than 50 MPa. However, the forming medium used in tube hydroforming is usually emulsion, and a maximum forming pressure up to 400 MPa can be reached in certain industrial productions.

Modern hydroforming technology has been developed from the mid-1980s. There are two the main characteristics of modern hydroforming. Firstly, only a die or punch is needed because the liquid medium can act as the punch or die correspondingly, thus half of the die cost and processing time can be saved. When the liquid medium is used as the punch, many complex-shaped parts which beyond the capability of rigid punch forming, can be manufactured successfully. For the shell hydroforming,

S. Yuan, *Modern Hydroforming Technology*,
https://doi.org/10.1007/978-981-19-5775-8_1

neither punch nor die is used, hence this process is also called as the die-less hydroforming. The second characteristic is the ability of being instantly controlled when the liquid is used as the forming medium. The pressure can be controlled precisely by the hydraulic closed-loop servo and computer control systems based on the user-defined loading curve. Hence, the process parameters are able to be controlled within the set value regions, and adjusted with the proceeding of process time. The flexibility of process is significantly improved. In addition to above main characteristics, each of the three different hydroforming technologies, tube, sheet and shell hydroforming, also exhibits its unique characteristics, and will be introduced in detail in the following content.

1.1.2 Technological Characteristics of Tube Hydroforming

The main characteristic of tube hydroforming technology is that hollow parts with irregular cross-sections and two-dimensional (2D) or three-dimensional (3D) axial curve can be manufactured integrally [5]. Irregular shapes, such as rectangular, trapezoid, oval, and/or other special-shaped closed sections, can be formed from a circular cross-sectioned tube initially, as schematically shown in Fig. 1.1. While using the traditional manufacturing processes to form such hollow parts, at least two segmented pieces are firstly stamped, and then welded into an integrated part. To reduce welding distortion, the spot welding is normally used. Obviously, confined closed cross-sections are unable to be obtained under this routine. Furthermore, the cross-section shapes of the stamped parts are relatively simple, which are unable to satisfy the structural design requirements.

Tube hydroforming is an advanced manufacturing technology to meet the lightweight development of vehicle structures, typically automobile and aircraft. Material and structure lightweight are two main approaches to achieve the lightweight objective. Regarding the material lightweight approach, some light materials, i.e. aluminum alloys, magnesium alloys, titanium alloys and composite materials, can

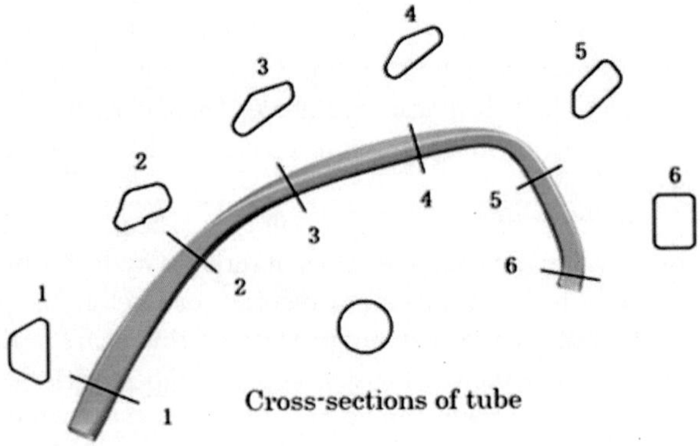

Fig. 1.1 Variable sections hollow parts

be used. For the structural lightweight approach, hollow structures with varied cross-sections, thin-walled shells with varied thickness, integral structures can be adopted. According to the statistical data, for a certain weight reduction target in the aerospace industry, the contribution of lightweight is materials to weight reduction occupies about 2/3, and the rest 1/3 is achieved by the structural weight reduction [6]. In comparison, structure lightweight is the dominant way to reduce weight of vehicles for the automobile industry alternatively. Once the applied material is fixed, the reduction of vehicle weight is accomplished by designing a reasonable lightweight structure. For the structures sustaining bending and torsional load in-service, the use of hollow structures with varied cross sections can not only reduce the mass but also fully explore the material strength. Several methods of weight reduction by the structural approach are illustrated using some cases given as follows [7].

(1) For the case of sustaining varied external loads, under the premise of guaranteeing strength, reasonable variable cross sections can be designed to adapt to different loads and reduce the structure mass. As shown in Fig. 1.2, for a cantilever beam, there are three different designs according to the characteristics of bending moment (Fig. 1.2a). The strength of all these three structures are able to satisfy the in-service requirements. Figure 1.2b shows the structure with equal strength, in which, same strength of each cross-section is obtained. Such a structure owns the highest load-bearing efficiency with lowest weight, however, the change of section shape conforms to a quadratic curve, which brings the greatest difficulty in manufacturing. Figure 1.2c gives the scheme of structure with uniform cross-section, which is the simplest shape to be machined. In this case, the strength at the sections S_2 and S_3 are obviously abundant and the usage of materials is the largest, which is about the twice of that using an equal strength beam. Figure 1.2d gives the design of variable cross section structure. Three different sections are designed to make the sections S_2, S_3 and S_1 have the same strength. Compared to the design of equal cross-section structure, 1/3 of the material can be saved using the variable cross-section structure, and its manufacture is much easier. Generally, equal strength structure is complex-shaped and difficult to be manufactured. Therefore, the variable cross-section hollow structure is a reasonable choice of structure design, which not only satisfies the strength requirement, but also achieves the goal of saving materials, reducing weight and simplifying the processing.

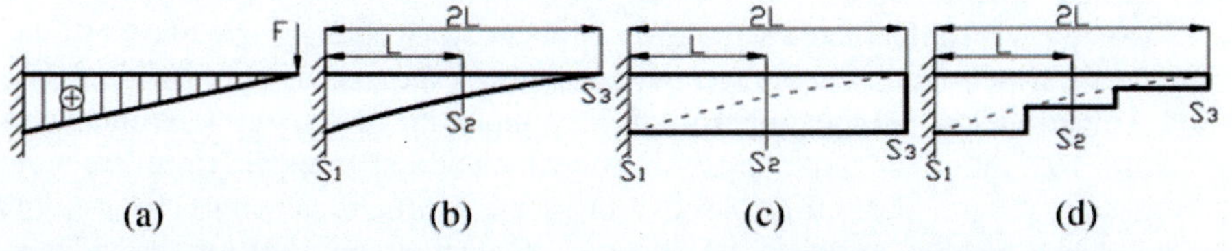

Fig. 1.2 Cantilever beams with different structures. **a** Bending moment diagram; **b** quadratic curved structure; **c** structure with uniform cross-sectioned structure; **d** variable cross sectioned structure

Table 1.1 Bending modulus of different cross-section shapes

Cross-section type	Initial tube	Square cross-section	Rectangular cross-section
Schematic diagram	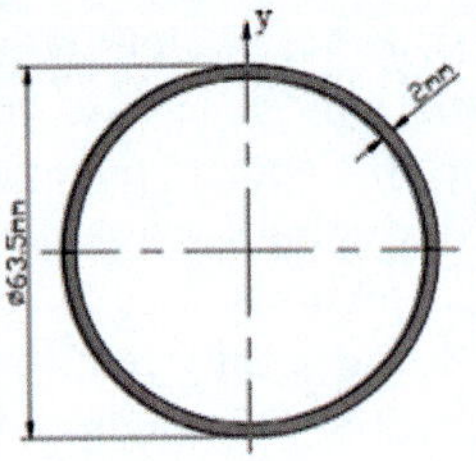	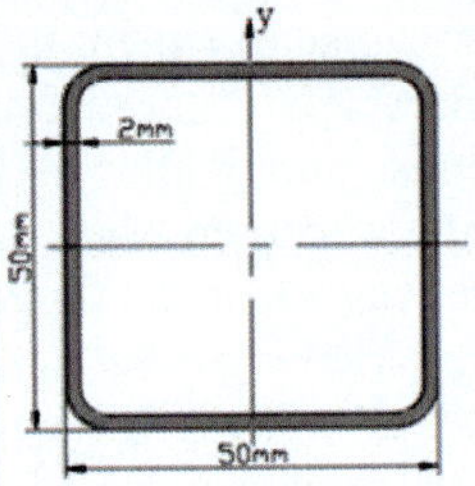	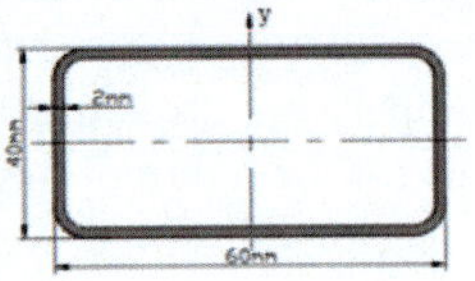
Bending modulus ratio	1	1.27	1.66

(2) The stiffness and strength of structure can be improved by using reasonable cross-section shapes due to the different modulus of various shapes once the masses are same. For instance, once a circular tube with a 63.5 mm diameter and 2 mm thickness is formed into different rectangular cross-sections, its bending modulus changes significantly, as shown in Table 1.1. When the circular tube is formed into a square cross section, a 27% increase of bending modulus is obtained. The bending modulus along the Y axis has an increase of 66% when the rectangular section with a ratio of 1.5:1 is formed. Therefore, the improvement of bending strength can be easily achieved by changing the cross-section shape only.

(3) Hollow structures are used to replace solid structures. Figure 1.3 shows the comparison between a solid beam and a hollow beam. Under the same strength condition, the use of hollow structure with variable sections (Fig. 1.3d) is able to reduce the mass to 16.8 kg from 46.2 kg of a solid structure (Fig. 1.3c), which is a 64% weight reduction. Using machining to manufacture such hollow structures, the mass of the initial rod material is 168.3 kg if machining is used, which results in a 90% of material waste. While using tube hydroforming alternatively, the initial tube mass is only 17.2 kg. Therefore, the adoption of tube hydroforming parts can, not only reduce the mass of structure, but also can save material.

(4) Structure with closed cross-sections is used to replace welded structure. For the structures manufactured by traditional spot or lap welding, there exists welded flange which increases the weight. In addition, the existence of welding spot results in a low fatigue properties. For example, Fig. 1.4 shows the schematics of a 76.2 mm × 63.5 mm cross-section which adopts three different structural designs. Figure 1.4a, b are welded overlapped structures using TIG welding and spot welding at flange, respectively. In comparison, Fig. 1.4c provides an integral structure with closed cross-sections. Comparing the weight reduction, a 9% weight reduction is achieved using the integral structure to replace the

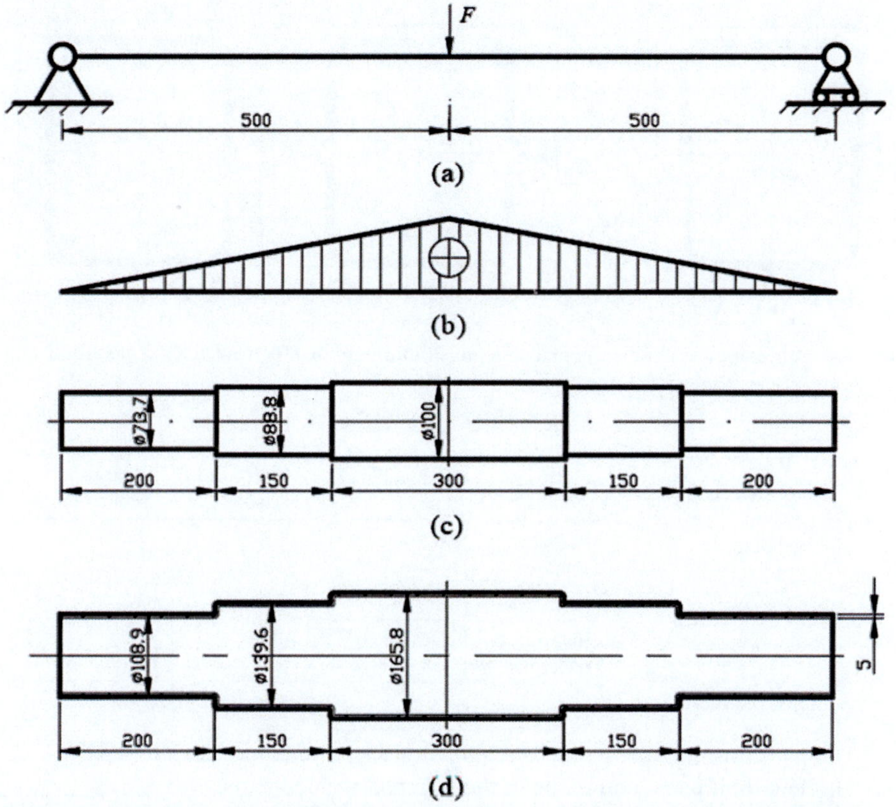

Fig. 1.3 Comparison of solid beam and hollow beam (unit: mm). **a** Simply supported beam; **b** bending moment diagram; **c** solid beam; **d** hollow beam

TIG welded overlapped structure, and a greater weight reduction 21% is found for replacing the spot welded structure at flange.

Considering the capability of hydroforming to manufacture closed cross-sectioned hollow parts with various shapes and dimensions of cross-sections along the part axis. The goal of cooperatively achieving reducing weight, saving material and simplifying processing routine can be achieved, with the combination of the above mentioned lightweight structural designs. Reasoning from the processing, the main advantages of hydroforming compared to conventional stamping-welding process can be summarized as follows [1, 5, 8–11]:

(1) *Weight reduction and materials saving*. The above practical cases have comprehensively discussed the methods of using hydroforming parts to achieve the weight reduction. The comparisons of part weights between hydroforming parts and stamped-welded parts in automobiles are given in Table 1.2. In general, for frame and beam structural parts, a 20–40% weight reduction can be achieved

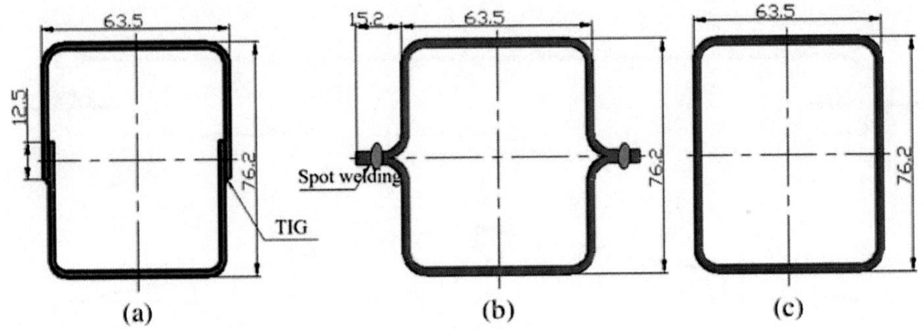

Fig. 1.4 Schematics of close sectioned structures (unit: mm). **a** TIG welded lap structure; **b** spot welded structure; **c** integral closed sectioned structure

Table 1.2 Weight comparison between the stamped parts and hydroformed parts

Part name	Stamped parts /kg	Hydroformed parts/kg	Weight reduction (%)
Radiator support	16.5	11.5	24
Subframe	12	7.9	34
Instrument panel beam	2.72	1.36	50

by hydroforming parts, and the reduction magnitude increases to 40–50% for hollow shaft parts, compared to the stamped-welded parts.

(2) *Fewer parts and lower die cost*. Compared to the necessity of multiple sets of forming dies in conventional stamping processes, normally, only one set of dies is needed in tube hydroforming. In addition, the number of sub-frame parts can be reduced to 1 compared to 6 of using stamping parts.
(3) *Reduced amount of subsequent machining and assembly welding*. Taking the hydroforming of radiator support as an example, the heat dissipation area increases 43%. In the meantime, the number of welding joints is reduced from initial 174 to 20, and the number of assembling operation is reduced from 13 to 6, which significantly increases the production efficiency by 66%.
(4) *Higher strength and stiffness*. A significant improvement of fatigue strength is obtained. For the radiator support part, the increase of strength are 39% and 50%, along the vertical and horizontal direction, respectively.
(5) *Material utilization*. The material utilization is 90–95%, while that of stamped parts is only 60–70%.
(6) *Lower production cost*. According to the statistical data of the already used parts by a German company, compared to the conventional stamped parts, the cost of hydroformed parts is reduced by 15–20%, and the die cost is reduced by 20–30%.

However, there still exists several disadvantages for hydroforming:

(1) Large capacity hydraulic press is needed to be used as the die clamping press due to the high internal pressure. For example, regarding a tube with an 100 mm inner diameter and 2.5 m length to be hydroformed, the clamping force is 25,000 kN for a 100 MPa internal pressure, and the value could go up to 50,000 kN if the internal pressure goes up to 200 MPa.
(2) High pressure source and the closed-loop instant control system are very complex and expensive.
(3) The forming defects and wall thickness distribution are dependent on the loading path, considering the cost of trial-manufacturing is very high. Comprehensive numerical simulation needs to be fully explored to optimize process parameters.

1.1.3 Characteristics of Sheet Hydroforming Technology

Sheet hydroforming can be classified into two types: hydro-deep drawing and liquid-punch deep drawing. For hydraulic deep drawing, the liquid medium is used to replace the female die cavity, while for liquid-punch, the liquid medium is used as the punch alternatively.

The main advantages of hydraulic deep drawing are increased forming limit and reduced forming operations [12, 13]. For example, in order to form the conical part shown in Fig. 1.5a, six forming operations needed using conventional deep drawing process, together with six sets of dies (Fig. 1.5b), in addition, annealing treatment has to be performed between two operations. In comparison, this conical part can be successfully formed using the hydraulic deep drawing in only single operation, which significantly simplifies the process and saves the die cost. Figure 1.6 shows the comparison of limit drawing ratio of several processes, the limit draw ratio of single operation deep drawing can reach 2, while the ratio is increased to 2.9 using hydraulic deep drawing process.

The main advantage of liquid-punch deep drawing is the capability of forming complex shaped part with a deep cavity within one operation. Due to the calibration using high pressure liquid at the late stage of forming, the sheet is fully attached to the die cavity, contributing to forming complex space profiles with relatively small transition corner radius.

The main disadvantages of sheet hydroforming technology are summarized into two aspects:

(1) Low production efficiency. Constraint by leading time of filling and discharging of liquid, the cycle time of sheet hydroforming is longer than that of conventional deep drawing. Therefore, the application fields of sheet hydroforming focus on the complex parts of small volume luxury passenger cars or high strength steel parts. While in the military industry, sheet hydroforming is suitable for forming complex shaped parts of aluminium alloys and other hard-to-form materials.

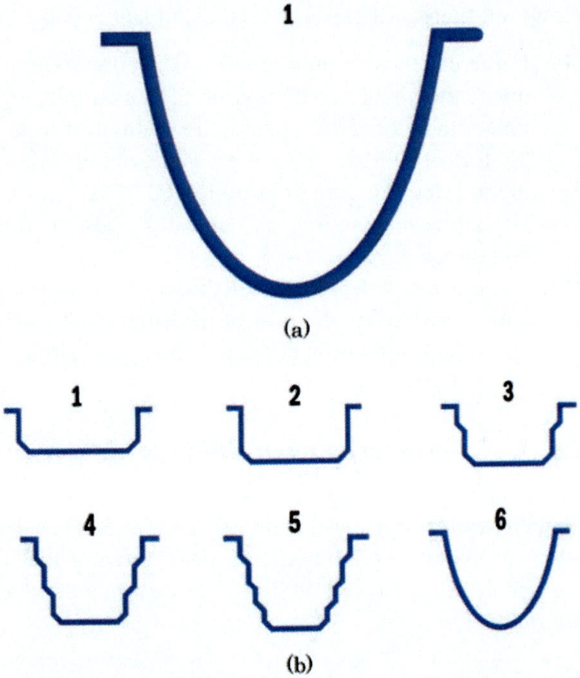

Fig. 1.5 Comparison of forming processes between sheet hydroforming and conventional deep drawing. **a** Sheet hydroforming; **b** conventional deep drawing

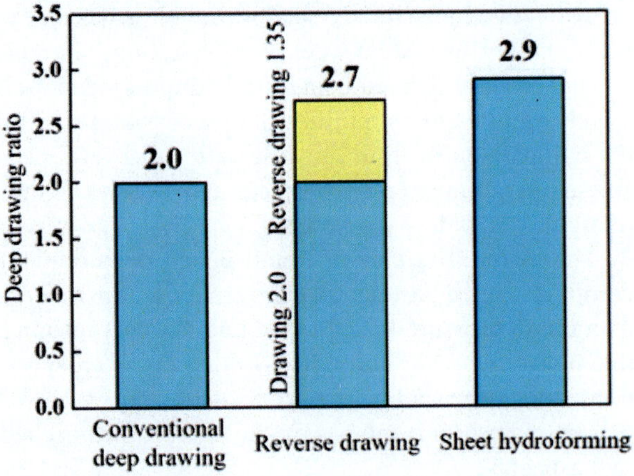

Fig. 1.6 The comparison of limit drawing ratio of several processes

(2) Requirement of large capacity equipment. In terms of forming parts with same dimensions, due to the effect of counter force exerted by the liquid, the required forming load of sheet hydroforming is higher that of using conventional deep drawing.

1.1.4 Characteristics of Shell Hydroforming Technology

In shell hydroforming, a closed polyhedral shell with a specific shape is usually used as the preformed billet. After the shell is filled with liquid, the internal pressure is increased with the continuous filling of liquid. Subsequently, the shell can be bulged once the internal pressure is increased to a certain magnitude, and gradually deformed to the target shapes, such as spherical, ellipsoid, and toroidal shape.

The main advantages of shell hydroforming are [14]:

(1) Forming dies and press are not used. From the view of mechanics, the principle of this technology is to utilize the self-force equilibrium system of the integral closed shell being loaded by the inner pressure. In this condition, such an integral closed shell is not only a deformed body but also the object with a balanced force system. Therefore, the avoidance of using press and dies to form large scale shell structures is achieved.
(2) The thickness and diameter of shell are easy to change. For a required shell with a given diameter and thickness, it can be directly bulged to final part once a reasonable closed polyhedron shell preform is designed. While using the conventional die-pressing process, a set of dies are needed for the forming of spherical shell with a certain diameter, and several sets of dies have to be used for a specific sized ellipsoidal shell.
(3) High dimensional accuracy of products. Conventional process of manufacturing shell structure is forming first then welding, resulting in the welding distortion at the early stage. Considering the shell hydroforming process selects the routine of first welding then forming, the welding distortion is avoided. The forming in shell hydroforming essentially is calibration of the shape, which results in the high dimensional accuracy of final shells.
(4) Lower cost and shortened manufacture cycle time. The time of die-pressing and die cost are saved due to the avoidance of using press and dies.

The main disadvantages of shell hydroforming are:

(1) As the routine of shell hydroforming technology is "first welding and then forming", the welding seams on the closed polyhedral shell sustain certain plastic deformation during forming. Splitting can be induced potentially if there exists initial imperfects in the welding seams, which would result in the failure of whole shell. Such a problem is more serious for thick plates and low alloy steels. Therefore, the control of welding quality is critical for shell forming.
(2) For large scale shells, the construction of support base during forming faces great difficulty and normally brings high cost. For example, to form a spherical shell with a diameter of 12.3 m and a volume of 1000 m^3, it is necessary to build the infrastructure to support 1000 t water and the self-weight of the shell.

1.2 Overview of Hydroforming Technology

1.2.1 Overview of Tube Hydroforming Technology

As early as the 1950s, the tube hydro-bulging process has been used in the production of T-shaped copper alloy tubular parts of the pipeline system, and the joint structures of a bicycle frame. At that time, the forming pressure used was lower than 25 MPa, and only simple parts with low dimensional accuracy can be produced. The main problem was due to the poor controllability of process parameters. The essential differences between modern tube hydroforming and early tube hydro-bulging are [11]:

(1) For modern hydroforming, the forming pressure is very high, which is generally up to 400 MPa, even as high as 1000 MPa in certain industrial productions.
(2) The process parameters are controllable. The internal pressure and axial displacement are closed-loop controlled by computer according to the given loading curve. The precision of controlling ultra-high pressure reaches 0.2–0.5 MPa, and that of displacement control reaches 0.05 mm.
(3) The complex-shaped parts with high precision can be formed. Complex parts with 3D axis and irregular cross sections can be integrally formed, which are mainly used in the manufacture of automobile and aircraft parts. For tube hydroforming, two technological breakthroughs have promoted its significant development. One is the dynamic sealing of ultra-high-pressure water medium, which achieves the stable sealing at ultra-high pressures greater than 400 MPa for long-time production conditions. The other is the computer close-loop control, using which, not only the designed loading path can be monitored precisely, but also a quick response and feedback of system is obtained, which can guarantee the production rhythm within about 30 s.

Moreover, driven by the increasingly demands of weight reduction of automobile structures to save fuel, reduce emissions and increase safety of the whole car body, the tube hydroforming has been developed significantly. In the early 1980s, the universities, research institutes and related companies in both Germany and the United States have systematically carried out the research on fundamentals and processing for applications of tube hydroforming extensively. In Europe, research were mainly conducted at universities such as Paderborn and Stuttgart, while the development of equipment and parts was mainly carried out by the companies of Schuler, SPS and AP&T. In the United States, the research and application of tube hydroforming were mainly carried out by the Ohio state university, general motors (GM), Vari-Form and Hydrodynamic companies. Professor Dohmann from the University of Paderborn in Germany is the first scholar to begin the systematic research of modern tube hydroforming technology in the world. Several representative papers have been published in Journal of Materials Processing Technology (JMPT), which have become the classical research in the field of tube hydroforming [5, 8, 9]. The tube hydroforming technology is usually used in automobile, aviation, aerospace, bicycle and pipeline industries. [10, 15–19]. Particularly, the tube hydroforming has been most widely used in the automobile industry to manufacture various types of parts of an automobile, which are (1) parts of chassis, e.g. engine cradle, sub-frame, back shaft, carling and bumper etc. (2) car body members, e.g. panel support beam, radiator support, seat frame, roof side rail and top beam. (3) engine and drive systems, e.g. manifold, exhaust pipes, cam shaft and drive shaft. (4) steering and suspension systems, e.g. control arm and steering lever. Regarding the application of tube hydroforming in aircraft, there are hollow frame beam, hollow crankshaft of engine and tubular parts with irregular cross-sections, etc. In aerospace applications, tubular joint of rocket power system and irregular cross-sectioned air intake channels can be manufactured using tube hydroforming. In 1990, the instrument panel bracket for the small commercial vehicle of Chrysler was produced by Vari-Form of the United States using the tube hydroforming. In 1994, Ford Motor company for the first time achieved the mass production of sub-frame used in the Contour and Mystique models using tube hydroforming in North America. In 1993, Mercedes-Benzes established its tube hydroforming plant to produce automobile chassis and various structural parts. Later, the hydroformed tubular parts were adopted by several European car makers such as Volkswagen and BMW.

Until now, many fundamental problems of tube hydroforming have been systematically investigated through theoretical analysis, numerical simulations and experimentation by researchers from different countries. These fundamental problems mainly concentrate on the correlations between failure and loading path, the forming zone and limit, the thickness distribution, the measurement of tube properties, the establishment of forming limit diagram (FLD), anisotropy effect, the friction behavior at high pressure, and preform optimization etc. [20–30].

The fundamental theory and key technology have been investigated systematically by a group from Harbin Institute of Technology (HIT), China. Regarding the hydroforming of variable diameters tubular parts [31–45], the mechanism of defect occurrence, thickness distribution and the boundary circle of thickness were firstly given. The method of using "useful wrinkles" as preform was proposed, with the required geometric and mechanics conditions of obtaining "useful wrinkles" systematically established. The variable diameters tubular parts were formed successfully using the low carbon steel, stainless steel and aluminum alloy. Regarding the hydroforming of curved tubular parts [7, 46–55], the complex irregular cross-section shapes were simplified into three typical sections: rectangular, trapezoid and prolate. For this sub-kind of tube hydroforming, the corner filling behavior, thickness distribution, defect types and solutions of overcoming defects of typical cross-sections were investigated. In addition, a method of using concave preform to reduce the calibration pressure was proposed, which has been successfully used in the manufacture of automotive structural parts. For the hydroforming of multi-branch tubular parts [56–58], focusing on the Y-shaped tubular part, effects of correlations between internal pressure and displacement, axial feeding ratio, structure form and retracting speed of middle punch on forming were systematically investigated. Oblique three-branch tubular parts of thin-wall stainless steel and large diameter aluminium alloys were successfully manufactured. To thoroughly reveal the plastic deformation mechanism of three typical hydroforming processes, the stress–strain characteristics and stress locus of typical points for different locations during different stages of hydroforming were given [59–64].

The dominant feature of variable diameters tubular parts is varied cross-section diameter along the axis. Such kind of parts are commonly used in automobile exhaust pipes, pipeline system of aircrafts and rockets. From the view of structure, the variable diameters tubular parts can be divided into symmetrical and asymmetrical ones, as shown in Fig. 1.7. The asymmetry could be classified into up-down asymmetry, left–right asymmetry or complete asymmetry. During the hydroforming of these parts, the expansion ratio is an important evaluation index to represent the technical level and difficulty. The expansion ratio refers to the change ratio of perimeter from the original tube to the maximum cross-section of the part, which depends on the tube material, length of forming zone, lubrication and loading path [23]. For the hydroforming of symmetrical variable diameters tubular part, the forming length is 2 times of the initial tube diameter and the material has a good ductility, the maximum expansion ratio can reach 100% under the optimal loading path. Figure 1.7a shows some variable diameters tubular parts with different expansion ratios produced by the SPS company in Germany. For aluminum alloys and low alloy high strength steels, the maximum expansion ratios are usually less than 50%. Moreover, due to the non-uniform deformation, the difficulty of hydroforming asymmetrical variable diameters tubular part is greater than that of hydroforming symmetrical ones with same materials and forming length. The expansion ratio of low carbon steel asymmetric tubular part is generally less than 60%. Figure 1.7b shows the variable diameters tubular part with the up-down asymmetry feature, of which the expansion ratio can reach 75% after a single operation of preforming. Therefore, to obtain a large expansion ratio for the asymmetrical variable diameters tubular part with hard-to form materials, a preform with a specific shape is usually required. In this case, the expansion ratio of the low carbon steel tube with left–right asymmetry can reach 75% if the "useful wrinkle" is used as the preform to accumulate materials in the forming zone [39]. The elongation of 5A02 aluminum alloy tube is only 12%, but its expansion ratio of the symmetrical variable diameters tubular part can reach 35% if the "useful wrinkle" preform is used [34].

The curved hollow parts with irregular cross-sections have two or three-dimensional axis curves. For these parts, the typical shapes of cross-section include rectangle, trapezoid, ellipse and transition shape among these shapes, as shown in Fig. 1.8. The part was manufactured by Schuler company in Germany, and has achieved a high volume production [65]. This part is a typical hollow structure with a 3D axis and irregular cross-sections. Compared to stamped parts, the number of hydroformed parts is reduced from 6 to 1, the weight is reduced by 30%. The production and die cost are reduced by 20% and 60%, respectively. Up to now, the longest hydroformed part is the low carbon steel truck side-member manufactured by General motors, which has a 12 m in length. The longest aluminum hydroformed part is the side-member used in the Volvo Jeep [66]. The length and diameter reach 5 m and 100 mm, respectively, as shown in Fig. 1.9. The largest hydroformed part is the rear

(a)

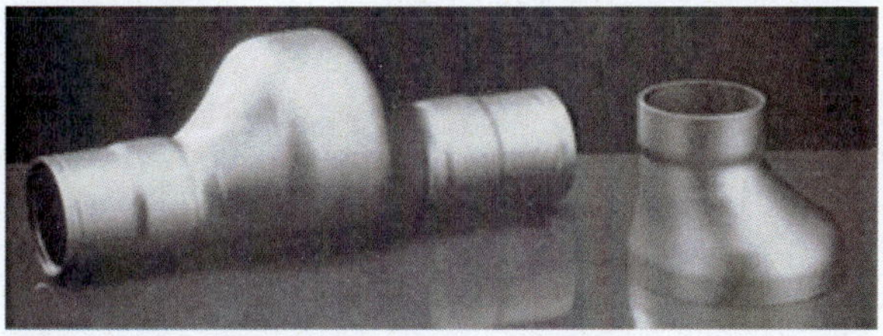

(b)

Fig. 1.7 Typical variable diameters tubular parts (source: SPS company in Germany). **a** Symmetrical variable diameters tubular parts; **b** asymmetrical variable diameters tubular parts

axle of 1.8 m heavy truck manufactured by APT, Sweden. The mass of this part reaches 60 kg, and the diameter and thickness are 200 mm and 10 mm, respectively.

The structures of multi-branch tubular parts mainly include T-shaped three-branch, Y-shaped three-branch, X-shaped (cruciform) four-branch and six-branch tube. Among these types, the forming of Y-shaped tubular part is the most difficult due to its asymmetrical structure. The main index of multi-branch tube hydroforming

(a)

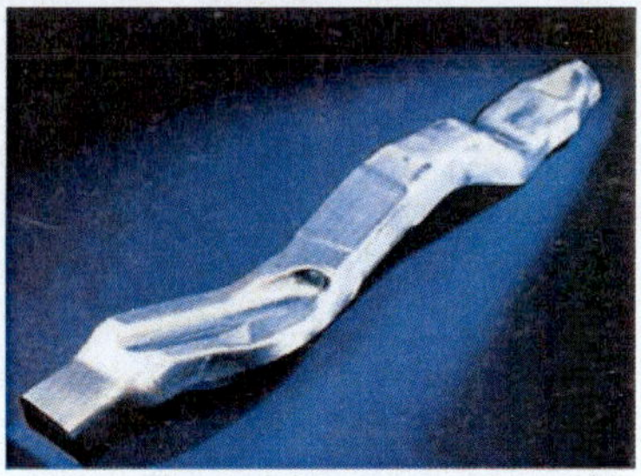

(b)

Fig. 1.8 Hollow curved parts with irregular cross-sections (*source* Schuler company in Germany). **a** Main part of sub-frame; **b** frame beam

Fig. 1.9 Hydroforming part of aluminum alloy [66]

is the height of branch, which can reach 1 and 0.75 times of the original tube diameter, for the T-shaped and the Y-shaped tubular part, respectively. For stainless steel or aluminum multi-branch tubular parts, due to the thickness becomes thinner, wrinkling is easy to occur at the beginning of hydroforming, which increases the difficulty of hydroforming requiring the higher control precision of coordinating pressure and displacement [24,67]. Currently, the wall thickness of thinnest part can reach about 1 mm for the tubes with an outside diameter ranging from 30 to 50 mm.

After tube hydroforming, the hydropiercing can be conducted on the parts directly under the support of internal pressure [68, 69]. The advantage is that, the high pressure liquid inside the part act as an inner soft die, which can support the wall of tube to avoid collapse, especially for the locations where the rigid die cannot be mounted. As shown in Fig. 1.8a, the holes on the engine cradle are processed by the hydropiercing technology.

1.2.2 Overview of Sheet Hydroforming Technology

To solve the above problems and meet the requirements of complicated sheet metal parts in various industries, in 1960s, a Japanese researcher firstly proposed the deep drawing method in which the liquid pressure was exerted directly on the blank. This method is the original prototype of so-called modern sheet hydroforming. In addition, it was found that the vertical drawing force of the sheet at the corner of female die was an important factor of determining the overflow pressure. Mathematical analytical solution was used to explain that friction retention and overflow lubrication effect were the fundamental reasons of increasing the forming limit during sheet hydroforming. In 1961, E. Buerke proposed and patented a new method. The sealing ring was designed on the die surface to prevent the liquid from flowing out of the die. In China, the development of sheet hydroforming technology was relatively late. Until 1970s, research on using sheet hydroforming for applications in pane parts was carried out. Impressive achievements in the process parameters, forming limit, and forming mechanism etc. have been made. At present, research institutes those have systematically conducted investigations of sheet hydroforming include the University of Stuttgart, Germany, Alburg University, Denmark Chiba University of Technology, Japan [12, 13, 70, 71], Harbin Institute of Technology, China [72–78] around the world. Moreover, some famous companies, such as Schuler (Germany), SPS (Germany), AP&T (Sweden), AMINO (Japan), can provide industrial production equipment and development of forming parts.

At present, sheet hydroforming technology has been used to manufacture parts with cylinder, cone, paraboloid, box and other complex shapes. The materials involved are carbon steel, high strength steels, stainless steel and aluminum alloy. The thickness of raw sheet ranges from 0.2 mm to 3.2 mm. Compared to the conventional deep drawing, a higher forming limit and drawing ratio can be obtained using sheet hydroforming. For example, the maximum drawing ratio is 2.6 for low carbon steel cylindrical part, and the maximum drawing ratios of stainless steel and aluminum alloy cylindrical parts are 2.7 and 2.5, respectively. Moreover, the deep drawing ratio can be further increased if some special processing innovations were used. For example, the double-sided pressures with a pressure difference can be given on the upper and lower surface of sheet material. In this process, the back liquid pressure on the lower surface is greater than that on the upper surface.

Toyota Motor corporation in Japan has established a production line of covering panel parts using the 40,000 kN large sheet hydroforming equipment as the primary press. A large panel part with a weight of 7 kg and plane size of 950 mm × 1300 mm can be formed. Figure 1.10 shows a car fender part formed using the sheet hydroforming. The material is 6061 aluminium alloys and thickness is 1.1 mm [79].

The sheet hydroforming with pre-bulging was proposed by the Schuler company in Germany, which has the advantages of obtaining uniform thickness and improving the collapse resistance due to the hardening of material of the top region of parts. The liquid flux of this hydroforming system could reach 12,000 L/min, and the liquid pressure was 16 MPa, which were the largest in the world at that time, In the 1990s, new technologies have been developed aiming to further improve the forming limit of sheet hydroforming, such as sheet hydroforming with controllable radial pressure

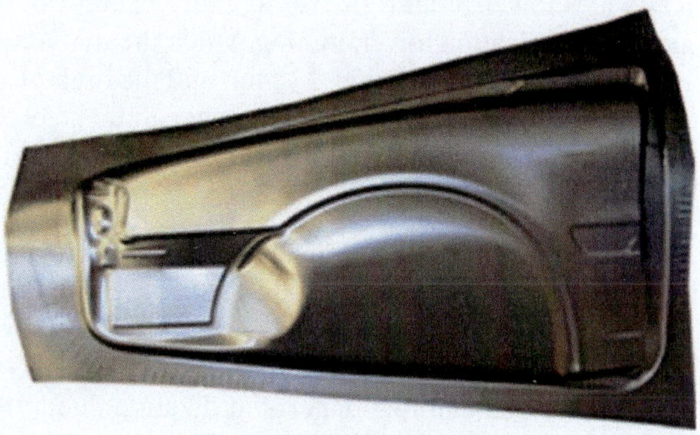

Fig. 1.10 Aluminum alloy panel part formed by hydro mechanical deep-drawing [79]

[80], reverse sheet hydroforming with peripheral pressure, sheet hydroforming with temperature difference fields, etc. Using these new technologies, the forming limit of can be improved effectively. For example, the drawing ratio of aluminum A1100 cylindrical parts can be increased from 2.6 to 3.3 using sheet hydroforming with radial pressure. For the rectangular box parts, their drawing ratio can be improved from 2.9 to 3.6. For the reverse sheet hydroforming with peripheral pressure, the ultimate drawing ratio can reach 4.9.

In recent years, the liquid punch deep drawing has developed with the significant progress of servo control technology of blank holding force and high-pressure sealing technology. In Germany, this technique is also known as high pressure sheet forming [81]. The main characteristic of this process is the instant variation of clamping force with the internal pressure. At the beginning of forming, the clamping force is small as the internal pressure is low, which facilitates the drawn-in of sheet material into the die cavity. While at the later stage of forming, once the sheet material is completely drawned into the die cavity, the clamping force should be increased to ensure the sealing, then the parts can be completely formed by increasing internal pressure. The highest calibration pressure can reach 150 MPa, by which the parts with large depth, complex shape and small transition corners can be formed. In the past, the instant control of the clamping force was very difficult. Therefore, in order to ensure the sealing, the clamping force exerted at the initial stage was relatively large, which constrains the draw-in of the sheet material into the die cavity. Essentially, the sheet material experiences a pure bulging deformation. Then, only the parts with small depth, nonuniform thickness and simple shape can be formed.

1.2.3 Overview of Shell Hydroforming Technology

The die-less hydroforming technology of spherical vessels was invented in 1985. Since then, significant progress has been achieved in shell structure, shell material, forming process, fundamental research and engineering applications. Shell hydroforming technology has been widely used in industrial structures, such as spherical water tower, liquid gas storage tank, stainless steel decoration ball structure, and communication tower, among which the maximum diameter reaches 9.4 m, the maximum thickness reaches 24 mm, and the highest working pressure is 1.77 MPa. The used materials mainly include low carbon steel, low alloy steel, stainless steel, aluminum alloy and copper alloy. For the development of shell hydroforming, there are three dominant stages [14, 82, 83].

(1) The shell structure has been extended from flat-type polyhedral shell to single-curvature polyhedral shell. In 1985, the soccer-shaped flat-type polyhedral shell with a spherical diameter of 600 mm was successfully formed. However, the main disadvantage of flat-type polyhedral shell is the easy occurrence of welding seam crack and low material utilization due to the large dihedral angle during hydro-bulging. In comparison, the soccer-shaped polyhedral shell with single

curvature has the main advantages of small dihedral angle, small distortion of weld seam and low thinning. The utilization of material can be improved by selecting a reasonable number of petals along the equatorial zone. Therefore, the polyhedral shells with single curvature have been widely used in industry, such as the 9.4 m diametered spherical water tower, liquefied petroleum gas storage tank and many stainless-steel decorative sphere structures.

(2) Development from low and atmospheric pressure spherical vessels to the high pressure vessels. In 1992, 200m^3 liquefied petroleum gas (LPG) storage tank was successfully developed using the shell hydroforming technology [84–87], as shown in Fig. 1.11. The diameter was 7.1 m and the wall thickness was 24 mm. This spherical tank has manufactured with material of 16MnR low alloy steel, which can work under a maximum internal pressure of 1.77 MPa. This successful application represents three technological breakthroughs of shell hydroforming. Firstly, this technology successfully achieves the progress from thin sheet to the medium thick plate from the perspective of thickness. Previously, the maximum thickness of the shells used in water tank and decorative sphere was 8 mm, while the thickness of the LPG spherical tank was 24 mm. With increasing thickness, the angular deformation increases near the weld seam, which easily induces the crack of inside weld seam. Secondly, the selection of material was extended from low carbon steels only to low alloy vessel steel. However, the low alloy steel exhibits a more significant hardening compared to low carbon steel, it is critical to strictly control the quality of the welded joint to ensure that enough plastic forming ability of joint zone. Thirdly, the storage medium is not limited to water, inflammable and explosive media such as like liquefied petroleum gas can be stored in the vessels. Consequently, the grades of vessels have been developed from atmospheric and category I and II pressure vessels to category III, which requires the effects of bulging deformation, such as residual stress, mechanical properties and thickness distribution, on the safety of spherical vessels to be addressed essentially.

Fig. 1.11 200m^3 liquefied petroleum gas (LPG) storage tank

(3) Extension from spherical shell to non-spherical shell. The non-spherical shells mainly include the ellipsoid, torus and other irregular shapes. The ellipsoidal shells have the advantages of small wind load coefficient and attractive appearance, which are used as the major structure of the 1000 m^3 water tower. However, the forming of ellipsoidal shell is more complicated compared to that of a spherical shell, as several sets of dies are needed for forming an ellipsoidal shell with variable curvatures. Extensive theoretical and experimental researches have shown that only the ellipsoidal shell with an axial length ratio (the ratio of the long axis to short axis) less than $\sqrt{2}$ can be manufactured by hydroforming. While for an ellipsoidal shell with an axial length ratio greater than $\sqrt{2}$, instability and wrinkling occurs near the equatorial zone [88–90].

1.3 Development and Perspective of Hydroforming Technology

1.3.1 Development and Perspective of Tube Hydroforming

Tube hydroforming technology has been widely used in automobile industry in the recent decade. In the meantime, the urgent demand of reducing weight and saving cost also promotes the continuous improvement of tube hydroforming technology, which brings a boost development of tube hydroforming. The future development directions can be summarized as follows:

(1) Ultra-high pressure forming. Currently, the maximum pressure of the intensifier of tube hydroforming equipment can usually supply a pressure of 400 MPa. In order to form parts with more complex shapes, higher precision, greater wall thickness and higher strength materials (e.g. ultra-high strength steels, titanium alloys and Ni-based superalloys), higher internal pressure is needed which can be up to 600 MPa, or even 1000 MPa. However, the challenges those are brought by the ultra-high pressure forming need to be solved, such as the sealing of the movable tube ends subjected to ultra-high pressure, reduction of friction under ultra-high pressures, die materials and control precision of ultra-high pressure liquid.

(2) Novel forming processes. Tube hydroforming of tailor-welded tubes [91]. In this process, tubes with different thickness or materials are firstly welded into a whole tube blank, then to be hydroformed. The process is able to further reduce the weight of structure. The conical tube with different diameters at the ends is adopted to manufacture special structures, such as the crash box; Double-layered tube hydroforming is used to manufacture the double-layered car exhaust pipe, which can improve the catalytic and purification performance of exhaust gas; Profiled tubes with non-circular cross-section can be used as a preform to manufacture the designed parts; Tube hydroforming with external pressure and the double-sided tube hydroforming [72].

(3) Hydroforming of ultra high-strength steel. With the increasing demand of structure lightweight, steels with higher strength are used in the car body. However, the ductility of high strength steels are poor. For example, the elongation of the steel is decreased from 45 to 12% with the strength is increased from 250 to 1000 MPa. The decrease of material ductility arises a higher tendency of splitting which increases the forming difficulty. Therefore, an in-depth and thorough investigation of bending, preforming, tube hydroforming, thickness distribution and lubrication need to be performed systematically in the future.
(4) Warm hydroforming. In order to solve the problems of poor ductility at room temperature of high-performance light alloys, hydroforming using internally heated warm medium to form parts with irregular cross-sections is an important development direction for tube hydroforming [92–99]. Currently, the temperature of heat-resistant oil can reach 300 °C and the pressure can reach 100 MPa, which can fully satisfy the requirements of hydroforming aluminum and magnesium alloy tubes. However, the main issue of warm hydroforming is the long forming cycle time and low efficiency. In addition, for titanium alloys, a forming temperature higher than 600 °C is required, which exceeds the thermal margin of heat-resistant oil, and using gas medium as the forming medium becomes the ideal solution.

1.3.2 Development Direction and Perspective of Sheet Hydroforming

(1) New forming processes with improved forming limit and part quality. Currently, the sheet hydroforming technology is developing new processes, such as sheet hydroforming with active radial pressure, double-sided sheet hydroforming, sheet hydroforming with pre-bulging and warm sheet hydroforming.

For sheet hydroforming with active radial pressured, in addition to the liquid pressure in the liquid cavity, additional liquid pressure is given independently in the radial direction on the flange area of sheet to aid the flow of material into the die cavity. The forming limit can be further improved to form parts a deeper cavity and more complex surface profiles. For the double-sided pressure sheet hydroforming, the liquid pressure is simultaneously exerted on the upper surface of the sheet to cooperatively function with deep drawing, which can partially or even completely compensate the reverse-bulging caused by liquid cavity pressure. This process is especially dedicated for the forming of conical parts with large unsupported area, which allows a higher liquid pressure to be used to compress thinning and increase the forming limit. For the sheet hydroforming with pre-bulging, obvious strain hardening can be obtained by the pre-bulging and subsequent deep drawing to improve the overall stiffness of large parts. The original reinforcing ribs can be avoided. The process is suitable for forming the roofs of SUV and commercial vehicle. For the warm sheet hydroforming, the combination of material thermal properties and technical advantages of sheet hydroforming, which increases the formability t of light materials with poor ductility, such as aluminum alloys and magnesium alloys. The applications in the automotive, aircraft and aerospace industries are promoted subsequently.

(2) Deep drawing of materials with poor ductility. High performance aluminum alloys, magnesium alloys and ultra-high strength steels usually have high strength, but poor ductility. For example, compared with steel, aluminum and magnesium alloy sheets tend to experience splitting and wrinkling is easier to occur due to their smaller anisotropy coefficients and hardening exponents. Therefore, multiple processes are required in the conventional stamping process resulting in the overall complex operation. Using sheet hydroforming, the insufficient formability of light materials can be compensated to simplify the processing and improve the production efficiency.

(3) Forming of large part with complex surface profiles. The conventional stamping of large complex parts normally requires large-sized punch and female die with profiles consistent with the part. The die cost is high, and a long time of adjusting dies is needed. While using sheet hydroforming, only the punch is needed. The pressure provided by the liquid cavity enables the deformed sheet to attach to the punch firmly, which plays the role of a soft female die. Sheet hydroforming of large part can reduce the die cost significantly, and adjust the die assembly easily, which has already started to be used in the large covering panel parts in the automotive industry.

(4) Combination with conventional deep drawing process to improve efficiency. Conventional deep drawing is used to form the majority of the part profile, and local features can be subsequently formed by the sheet hydroforming. For example, the knob of aluminium alloy door panel can be formed by this routine. Alternatively, sheet hydroforming can be conducted first followed by the conventional forming process. For example, regarding the flanging of a blank with holes, a higher straight edge can be obtained, if sheet hydroforming is performed first followed by the flanging during the unloading of liquid pressure.

1.4 Recent Developments of Hydroforming

1.4.1 Recent Developments of Tube Hydroforming

Since 2007, the application of tube hydroforming technology in the automotive industry of European exhibits a continuously increasing trend. However, the application in automobiles began to decrease after 2014, and the number of tube hydroforming production lines for automobile parts is remained at 110 approximately, as shown in Fig. 1.12 [100]. The main reason for this decrease is caused by the rapid development of hot stamping of ultra-high strength steel. Some hydroformed tubular parts are replaced by the hot stamped parts. According to the data statistics from Society of Automotive Engineers (SAE), USA, in 2015, the mass of ultra-high strength steel parts occupies about 30% of the new car, while that of low-carbon steel parts decreases from more than 50% to about 30%. Compared to the forming of low carbon steels, there still exists a series of challenges exist in the hydroforming of ultra-high strength steels, including the high forming pressure, severe die wear, large springback, low dimensional accuracy, and great difficulty in forming complex cross-sections. To address these challenges, Schuler in Germany and Vari-Form in the United States have developed low pressure sequence hydroforming technology (SHS) [101, 102]. Using such technology, the forming pressure can be reduced by about 30% compared to the conventional tube hydroforming, and in the meantime, the die wear can be reduced effectively. The SHS technology has achieved the tube hydroforming production of ultra high strength parts.

Although hydroforming started relatively late in China, great progress has been made in recent years due to the promotion of rapid development of the automotive industry. Figure 1.13 summarizes the evolution of tube hydroforming production lines in China. In 2005, BAOSTEEL imported the first 5000 tonnage tube hydroforming equipment from Schuler Germany, and started the development of hydroforming automobile chassis parts and small-volume production. In 2000, Institute of Fluid Pressure Forming Technology, Harbin Institute of Technology, has developed the first generation of tube hydroforming equipment and delivered the first tube hydroforming equipment for automotive Original Equipment Manufacturers (OEMs) since 2006. After more than 10 years continuous research and development, the 3rd generation of CNC tube hydroforming equipment series products have been established. The technical specifications and functions have reached the same level of similar equipment over the world. Particularly, the pressure control precision based on the compensation of liquid has an one order of magnitude increase compared to the highest index globally. According to incomplete statistics, there are about 43 sets of tube hydroforming equipment used in the automobile industry of China currently, among which, the equipment developed by Harbin Institute of Technology occupies about 60%, Schuler Germany and AP&T Sweden occupies about 30%, and the rest 10% for Japan companies.

Prof. Kuwabala et al. developed biaxial loading equipment to systematically investigate the yielding behavior of tubes under the tension-tension stress states, and proved the suitable yield criteria for anisotropic tubes [103, 104]. The intelligent control hydroforming of the three-branch tubular part was invented by Prof. Manabe et al., and the instant inspection and intelligent control of the corner radius was realized by mounting displacement sensors at the corner area [105, 106]. Micro-hydroforming is the new research direction that was developed in recent years, Prof. Hartl has made important progress in the fundamental theory, process and equipment of this research area [107, 108]. Moreover, extensive research has been conducted on tube forming limit, high strength steel tube forming technology and tooling die by University of Stuttgart [109] and Chemnitz University, Germany [110].

In recent years, the tube hydroforming technology has been developed rapidly in China. The application area has been extended from the automotive industry to aerospace and aircraft industries. Particularly, Harbin Institute of Technology has made important progress, which are summarized in details as follows:

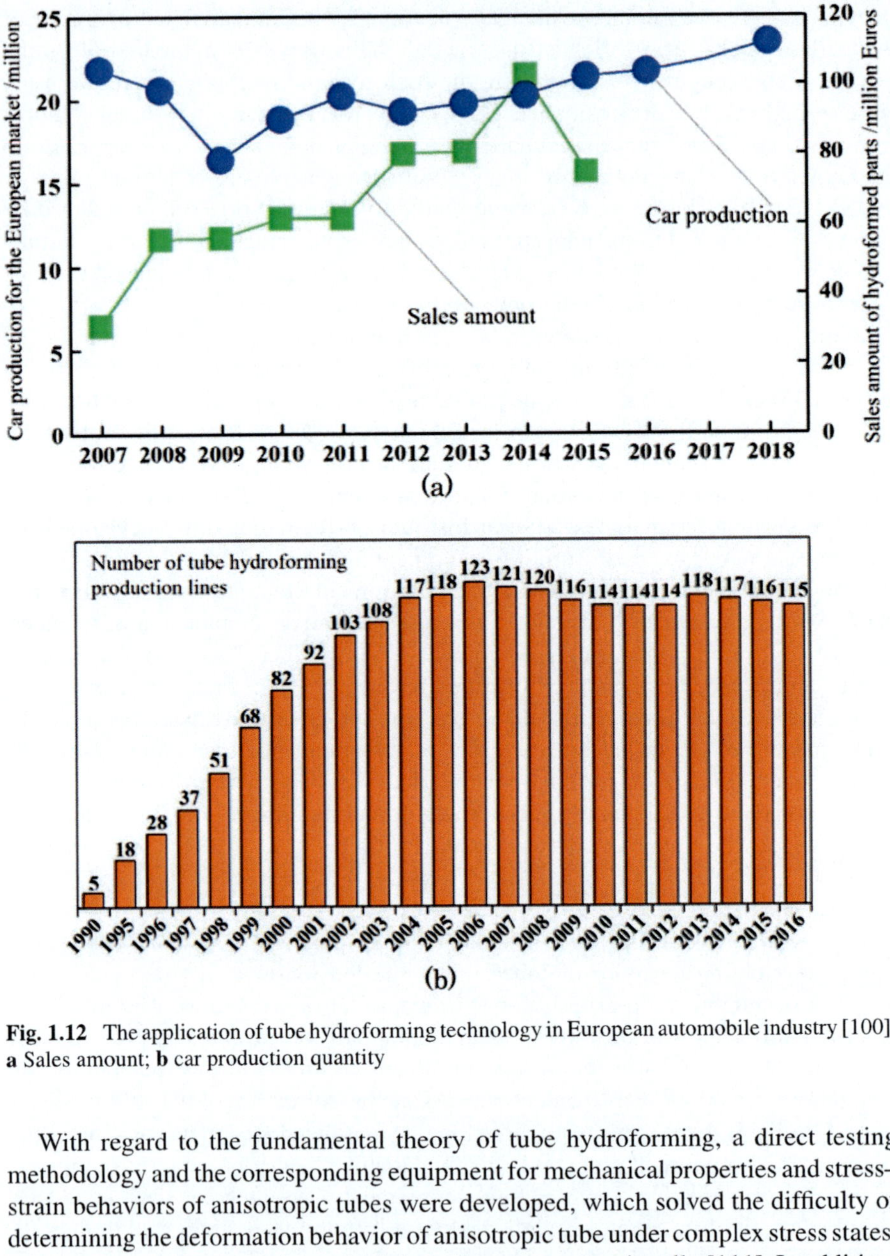

Fig. 1.12 The application of tube hydroforming technology in European automobile industry [100]. **a** Sales amount; **b** car production quantity

With regard to the fundamental theory of tube hydroforming, a direct testing methodology and the corresponding equipment for mechanical properties and stress–strain behaviors of anisotropic tubes were developed, which solved the difficulty of determining the deformation behavior of anisotropic tube under complex stress states, that cannot be achieved using uniaxial tensile tests conventionally [111]. In addition, the instability behavior of tubes under three-dimensional stress states was investigated, which reveals the influence of normal stress through-thickness on wrinkling and splitting [112–114].

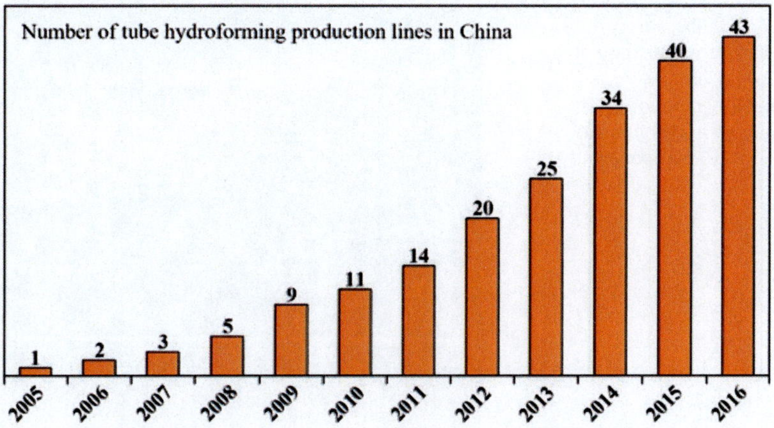

Fig. 1.13 The situation of tube hydroforming production lines in China

With regard to tube hydroforming process, the process of large-scale ultra thin-walled part was developed. The ratio of diameter to thickness up to 200 of the part was achieved. The optimization method of preforming and the hydro-compressing forming technology was investigated [115], which overcomes the difficulty of obtaining reasonable cross-sectional shapes during compressing due to the instability of large-scale thin-walled tubular. Moreover, the pressure forming technology using various media (e.g. water, oil, hot oil, and semi-liquid medium) has been developed to realize the forming from room temperature to high temperature, which provides a new forming technology for the complex structures of hard-to-form materials such as magnesium alloys, aluminum lithium alloys and titanium alloys [116, 117].

With regard to the tube hydroforming equipment, a fast and accurate variable clamping force technique and a pressure control method based on liquid volume compression were developed. A series of large CNC tube hydroforming equipment was developed. The overall performance of equipment has been in the leading group around the world, while the accuracy and feedback rhythm of pressure control are believed to be the forefront technology globally. Up to now, more than 30 tube hydroforming equipment have been provided for various companies.

With regard to the industrial applications, besides the extensive application of automobile parts, successfully applications of forming critical components in the aerospace and aircraft industries have been achieved, including the launching rocket, satellite, space vehicles and airplanes, as shown in Fig. 1.14. The raw materials used in tube hydroforming are stainless steels, aluminum alloys, titanium alloys and Ni-based super alloys. Better shape accuracy, performance and structural integrity of tube hydroformed parts are achieved compared to traditional manufacturing technology, which greatly improves the reliability of aerospace and aircraft vehicles.

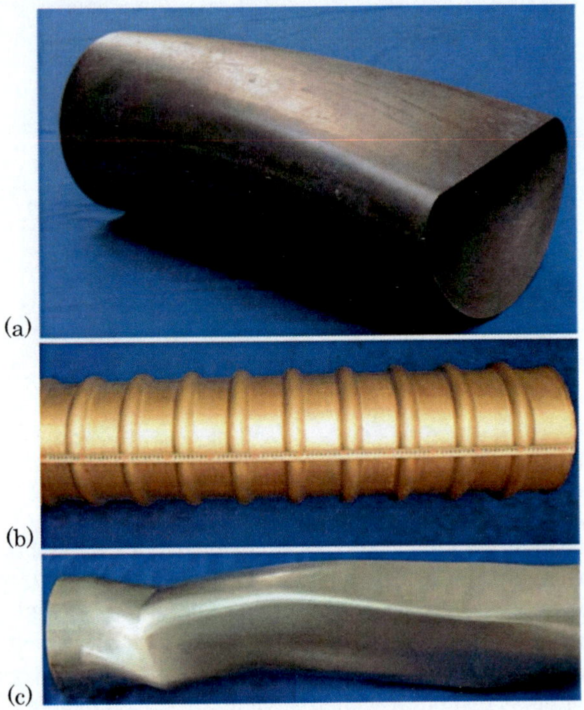

Fig. 1.14 Tubular parts formed by Harbin Institute of Technology, China. **a** Titanium alloy component; **b** Aluminum alloy component; **c** stainless steel components

1.4.2 Recent Developments of Sheet Hydroforming

The hydroforming processes and forming dies for aluminum alloy and high strength steel sheets have been systematically investigated by University of Stuttgart, Germany [118]. The north American division of Amino has developed a sheet hydroforming production line for automotive covering panel parts, using which, the mass productions of aluminum alloy (5182-O, 6111-T4, 6022-T4) engine hood, roof, outer and inner panels of door were started in 2011. These parts have already been used in pick-up trucks and luxury sports cars of General Motors and Ford [119]. The thickness of the 6xxx aluminium alloy outer panel ranges between 0.8 and 1.0 mm, while that of the 5xxx aluminium alloy inner panel is 1.5 mm.

Recently, Harbin Institute of Technology has been systematically investigated the sheet hydroforming, including wrinkling behavior of large-size curved part, the hydroforming of high-performance aluminum alloy and the evolution of microstructure and properties [120, 121]. The multi-axial loading hydroforming technology was proposed, in which the deformation zone is adjusted into a reasonable stress state by controlling the ratio of forward pressure to reverse pressure. Aided by the double-sided pressure, the simultaneous occurrence of wrinkling and cracking in aluminum alloy curved part with deep cavity was solved cooperatively. Figure 1.15 shows the successfully formed complex-shaped curved part with a deep cavity. Both

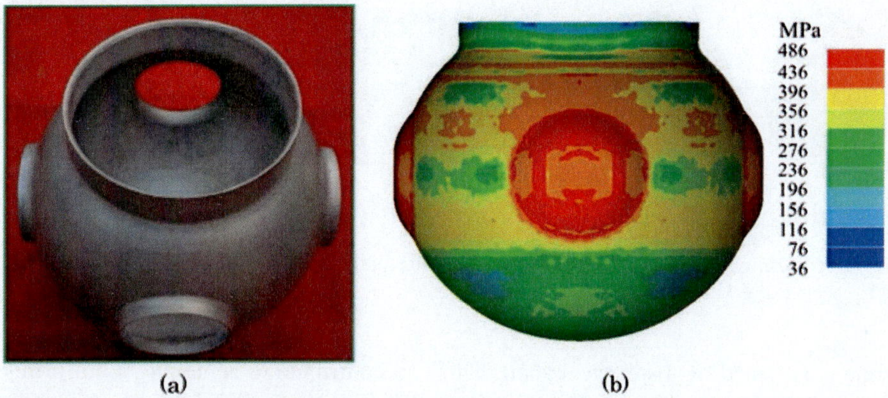

Fig. 1.15 Curved part with complex shape formed by Harbin Institute of Technology. **a** Photograph; **b** simulation (equivalent stress)

the dimensional accuracy and mechanical properties of this part satisfy the service requirements and guarantee the successful application of practical products. The investigation of such part has achieved the breakthrough of the pressurization of large-volume high-pressure liquid medium and closed-loop control technology. The largest sheet hydroforming system with a 5m^3 liquid capacity has been developed, which can achieve the integrated forming of curved parts with a diameters greater than 3 m.

1.4.3 Recent Developments of Shell Hydroforming

Reasoning from the previous theoretical analysis and a great quantity of large experimental investigations, it has been approved that only the ellipsoidal shell with the axial length ratio (the diameter ratio of the long axis to the short axis) is $1 \leq \lambda \leq \sqrt{2}$ can be formed soundly using the shell hydroforming. While for an ellipsoidal shell with an axial length ratio $\lambda > \sqrt{2}$, wrinkling will occur during hydroforming due to the influence of zonal compressive stress near the equatorial zone. For the long ellipsoidal shell with an axial length ratio $\lambda < 1$, the sound shell product also cannot be formed because the stiffness of pole plate is too high to be deformed, as shown in Fig. 1.16.

To address this issue, a hydroforming method of an ellipsoidal shell with double axial length ratios was proposed [122–124]. The basic idea of this method is to assemble an ellipsoidal shell with double generating lines using a segment shell with axial ratio $1 < \lambda < \sqrt{2}$, using such a structure, the original compressive stress at the equatorial zone was avoided, and the whole shell experiences a biaxial tension stress states which results in the avoidance of wrinkling. Moreover, this method can be also used to form long ellipsoid shells using hydroforming, in which the plate in polar

Fig. 1.16 Shell shape with different axial length ratios for the shell hydroforming. **a** $\lambda < 1$; **b** $\lambda = 1$; **c** $1 < \lambda \leq \sqrt{2}$; **d** $\lambda > \sqrt{2}$

zone is replaced by a spherical shell. This can contribute to achieving a uniformly distributed yield pressure within the shell, and solving the difficulty of deforming plates during the hydroforming of long ellipsoid shell hydroforming [125].

Chapter 2
Hydroforming of Variable Diameter Tubular Parts

2.1 Process Stages and Application Fields

2.1.1 Process Stages

A variable diameter tubular part refers to one that the diameter at one or several positions in the middle is larger than that of the two ends. Its main geometric feature is the varied diameter or circumference of the tubular part along the axis, and the axis is either a straight line or a two-dimensional curve.

Hydroforming of variable diameter tubular parts is usually divided by three stages, which are filling, forming, and calibration, as schematically shown in Fig. 2.1. In the filling stage (Fig. 2.1a), a tube blank is first placed on the lower die, and the upper die moves down to clamp firmly. Then liquid is filled into the tube to discharge the internal air, and two ends of the tube blank are securely sealed by horizontal punches. In the forming stage (Fig. 2.1b), along with the bulging of tube under increased internal pressure, the punches at both ends push the tube to achieve axial feeding under a given loading path. Under the cooperative function of internal pressure and axial feeding, the tube is gradually formed to attach to a die, most of the area has been formed except for the corners. In the calibration stage (Fig. 2.1c), the internal pressure is increased. Material at the corner zone is deformed to zone fully attach to the die surface, and required shape is formed finally. Basically, there is no axial feeding in the calibration stage.

For the cross-section shapes, the circular section of the tube can be changed into a rectangle, trapezoid, oval or other irregular sections by hydroforming, as shown in Fig. 2.2. According to the characteristics of stress and deformation, the parts are divided into forming and feeding zone, respectively. The forming zone represents the positions where the tube diameter changes with plastic deformation. While for the feeding zone, the outer diameter of the tube is constant due to the die constraint. The main function of this zone is to feed the material to the forming zone.

S. Yuan, *Modern Hydroforming Technology*,
https://doi.org/10.1007/978-981-19-5775-8_2

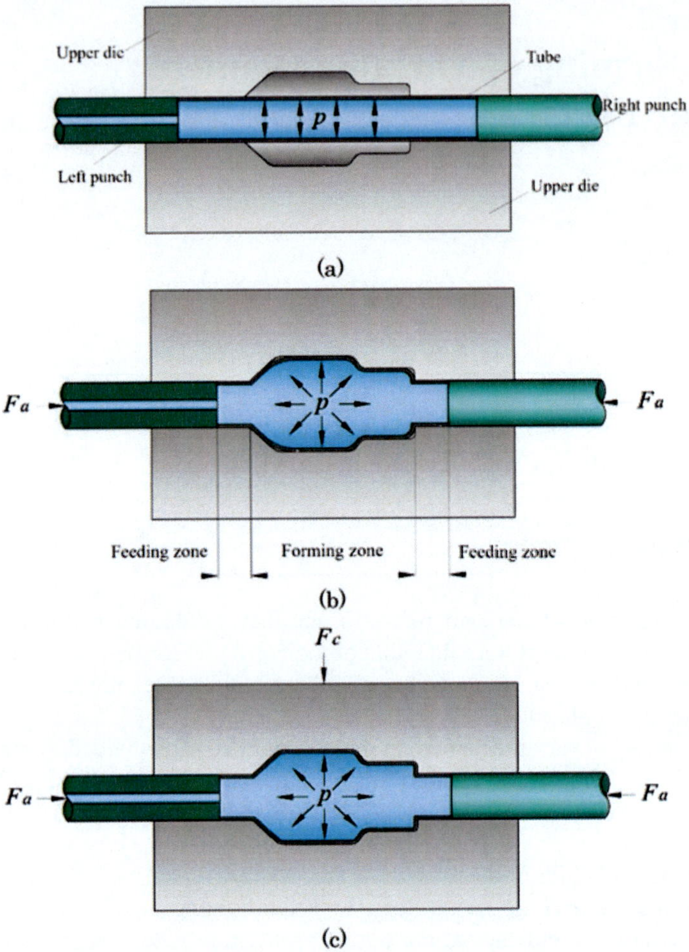

Fig. 2.1 The stages of hydroforming variable diameter tubular parts. **a** Filling; **b** forming; **c** calibration

Fig. 2.2 Variation of cross-sections in hydroforming of tubular parts

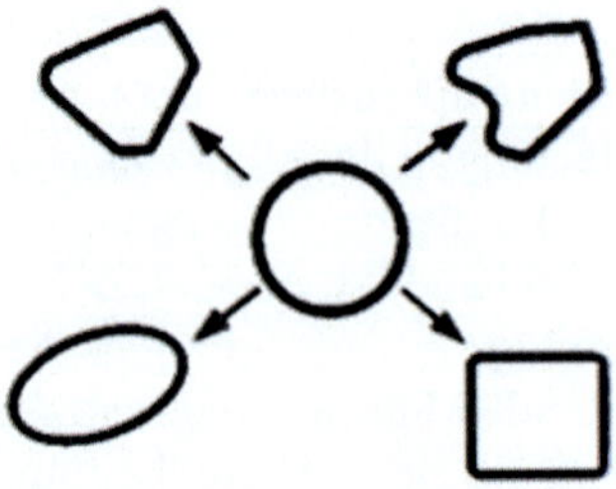

2.1.2 Application Fields

The hydroforming technology of variable diameter tubular parts is suitable for manufacturing automotive intake and exhaust piping systems, aircraft piping systems, rocket power piping systems, and tubular parts with complex cross-sections used in bicycles and air conditioning. It is mainly used for functional components in piping system, or tubular parts connecting with different diameters. In addition, the hollow crankshafts of small aircraft engines and the hollow stepped shafts in the drive system can also be manufactured. From the view of structure, the variable diameter tubular parts can be divided into symmetrical and asymmetrical forms, it may be up-down, left–right or complete asymmetry for the asymmetrical variable diameter tubular parts. Compared to symmetrical variable diameter tubular parts, the forming of asymmetrical variable diameter tubular parts is more difficult, especially the up-down asymmetric structure.

The suitable tube for hydroforming usually has a diameter of 25–200 mm, and a thickness of 1–8 mm. The ratio of tube diameter to thickness is usually 10–50 correspondingly.

2.2 Determination of Main Process Parameters

The main process parameters of hydroforming include initial yielding pressure, splitting pressure, calibration pressure, axial feeding force, die clamping force and axial feeding length. In this section, to facilitate the preliminary selection of process parameters in practical applications. The calculations of these parameters are given based on certain assumptions.

2.2.1 Initial Yielding Pressure

The initial yielding pressure refers to the internal pressure required to deform the tube plastically. The tube is a cylindrical shell subjected to uniform internal pressure. The ratio of axial stress σ_z to circumferential stress σ_θ is $\sigma_z/\sigma_\theta = \eta$, and the sequence of principal stresses is $\sigma_1=\sigma_\theta$, $\sigma_3 = -\sigma_z$. In this case, the calculation formula of initial yielding stress according to Tresca yield criterion is given as follows:

$$p_s = \frac{1}{1-\eta}\frac{2t}{d}\sigma_s \tag{2.1}$$

where

σ_s Yield strength of materials (MPa);
t Tube thickness (mm);

d Outside diameter of tube (mm);
η Ratio of axial stress σ_z to circumferential stress σ_θ.

Because the axial force is given during hydroforming, the tube experiences compressive stress, the range of η is $-1 \leq \eta \leq 0$. When $\eta = -1$, the initial yielding pressure is:

$$p_s = \frac{t}{d}\sigma_s \tag{2.2}$$

When there is no axial force, $\eta = 0$, the initial yielding pressure is given in Eq. (2.3), which just corresponds to the free-bulging process.

$$p_s = \frac{2t}{d}\sigma_s \tag{2.3}$$

In practical application, Eq. (2.3) is often used to estimate the initial yielding pressure due to the simplicity and reliability.

2.2.2 *Splitting Pressure*

The splitting pressure p_b for pure bulging can be estimated by the following formula:

$$p_b = \frac{2t}{d}\sigma_b \tag{2.4}$$

where

σ_b Tensile strength of materials (MPa).

Figure 2.3 shows the comparison between experimental and calculated values from Eq. (2.4) of the splitting pressure for different materials and thickness. It can be seen from Fig. 2.3 that the calculated splitting pressure is in good agreement with the experimental values. Moreover, the Eq. (2.4) is suitable not only to steel tubes, but also to aluminum and copper alloy tubes.

2.2.3 *Calibration Pressure*

At the later stage of tube hydroforming, most area of the part has been formed. Subsequently, to form section transition corners and ensure a satisfying dimensional accuracy, a higher internal pressure is needed. This stage is usually called calibration, as shown in Fig. 2.4. Since calibration is the last stage of tube hydroforming, the calibration pressure is also called forming pressure that can be estimated by the following formula:

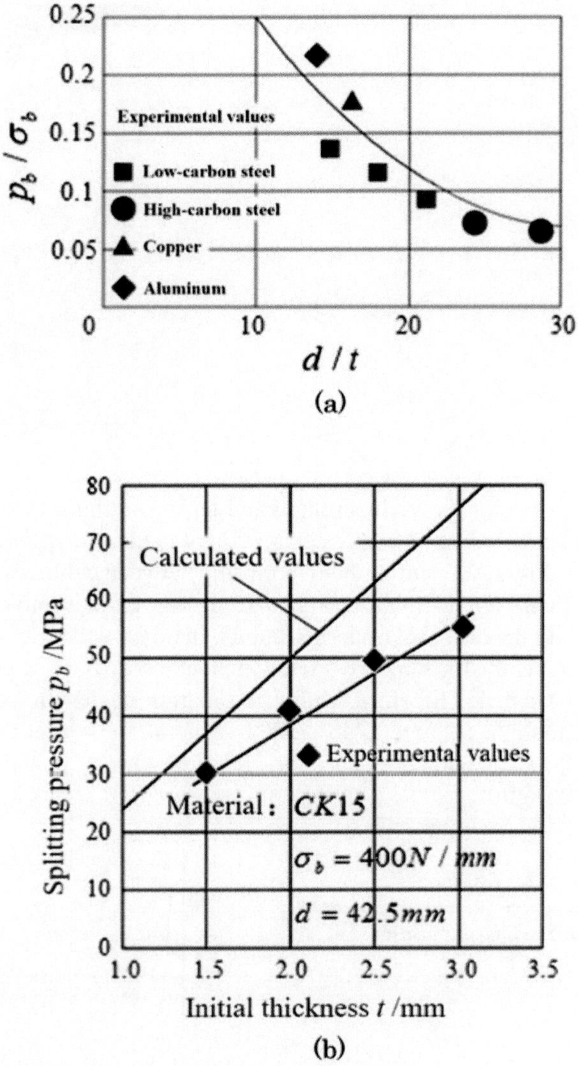

Fig. 2.3 Comparison between the experimental and calculated values of splitting pressure. **a** Different materials; **b** different thicknesses

$$p_c = \frac{t}{r_c}\sigma_s \tag{2.5}$$

where

r_c The minimum radius of cross-section of transition corner (mm);
t Average thickness at transition corner (mm);
σ_s Flow stress of materials (MPa).

For work-hardened materials, the calibration pressure is needed to consider the strain hardening. As an engineering estimation, the flow stress is assumed to be the

Fig. 2.4 Schematic diagram of calibration pressure calculation at the transition corner

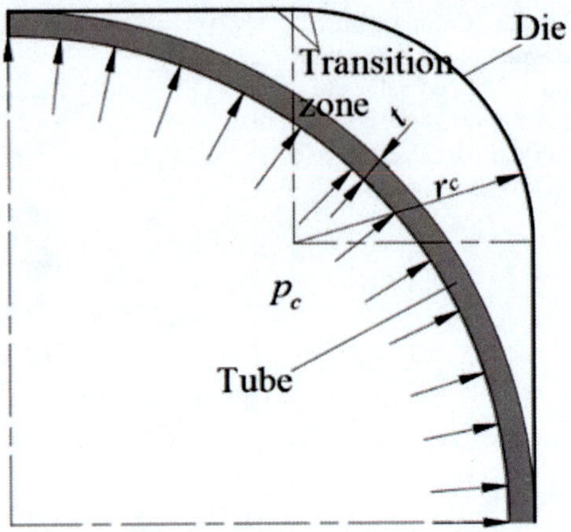

average of the yield strength and tensile strength of the material is regarded as the flow stress for a simple calculation of calibration pressure.

Figure 2.5 shows the relationship between calibration pressure and radius of transition corner. It can be found from Fig. 2.5 that the calibration pressure increases with the decrease of corner radius. In other words, the smaller the corner radius, the higher the forming pressure, which indicates that a larger clamping force of the press is needed. Therefore, within the margin of design, the radius of transition corner should be as large as possible. In general, the corner radius is usually designed as r_c = 4t–10t, and the calibration pressure is approximately 1/4–1/10 of the initial yield strength.

Fig. 2.5 Relationship between calibration pressure and radius of transition corner

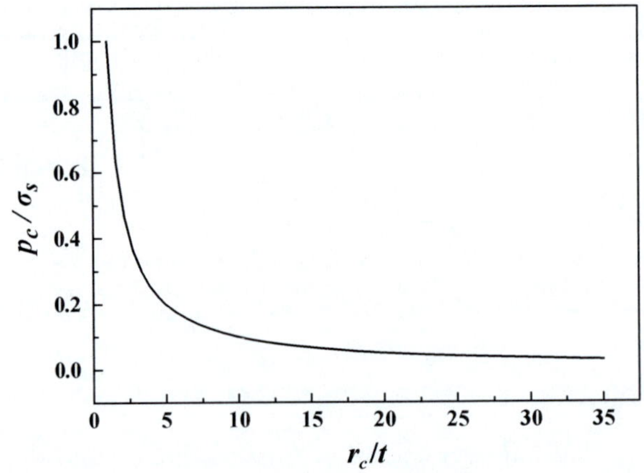

2.2.4 Axial Feeding Force

The axial feeding force F_a consists of three portions: the counter-force F_p acting on the punch by internal pressure, the frictional force F_μ and the force required to maintain plastic deformation of the tube F_t, as shown in Fig. 2.6. Axial feeding force is the main process parameter for the selection of the horizontal cylinder. Assuming that the normal pressure between the tube and the die is equal to the internal pressure, F_a can be calculated by the following formula.

$$F_a = (F_p + F_\mu + F_t) \times 10^{-3} \tag{2.6}$$

$$F_p = \pi \frac{d_i^2}{4} p_i$$

$$F_\mu = \pi d l_\mu p_i \mu$$

$$F_t = \pi d t \sigma_s$$

where

- d_i Inner diameter of tube (mm);
- l_μ Contact length between tube and die (mm);
- μ Friction coefficient.

For the above three force compositions, the counter-force acting on the punch by internal pressure is the greatest, followed by the friction force between the tube and die. The minimum force is the force required to deform the tube. The following formula can be used for estimating the axial feeding force for practical application.

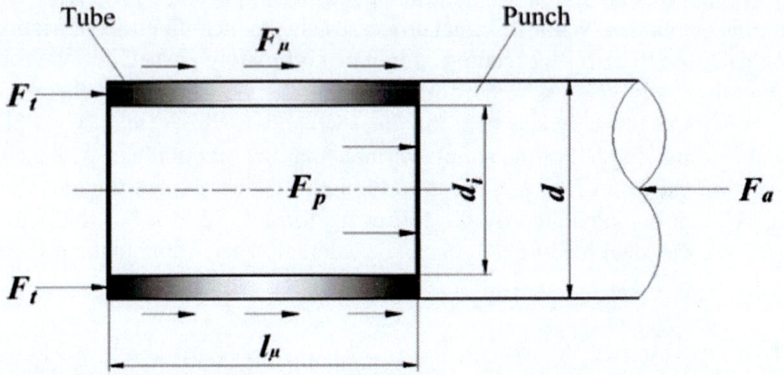

Fig. 2.6 The illustration composition of axial feeding force

$$F_a = (1.2 - 1.5)F_p \tag{2.7}$$

2.2.5 Clamping Force

Clamping force F_c is one required to clamp the upper die and lower die firmly and eliminate the initial gap. The main purpose of calculating the clamping force is to determine the capacity of hydroforming press. The calculation formula is:

$$F_c = A_p P_c \times 10^{-3} \tag{2.8}$$

where

p_c Calibration pressure (MPa);
A_p The horizonal projected area of tube (mm^2). For the parts with curved axis, the projected area A_p is the product of the width and the projected axis length on the horizontal plane.

2.2.6 Axial Feeding Length

One of the main characteristics of hydroforming of variable diameter tubular parts is that the thinning in the forming area can be alleviated and the expansion ratio increases with the axial feeding. Therefore, the axial feeding length is an important parameter to determine the horizontal cylinder stroke.

An ideal feeding length is defined as a feeding value to keep the tube thickness constant during hydroforming. According to the volume consistency, the surface area of formed part is equal to that of the tube blank. In this case, the ideal axial feeding length can be obtained. While in actual processes, due to the influences of friction and loading path, the material for feeding cannot be completely pushed into the forming zone, and the wall thickness of the forming area must be decreased. Therefore, the actual value is necessarily less than the ideal value, which is generally 60–80% of the ideal feeding length. For the simple variable diameter tubular part with a circular cross-section shown in Fig. 2.7, the initial length of tube, l_0, and the ideal feeding length, Δl, can be calculated by the following formula (2.9b). For the parts with other shapes, the ideal feeding length can be calculated according to the principle of equaled surface areas.

$$l_0 = \frac{Dl'}{d} + \frac{D^2 - d^2}{2d \sin \alpha} + (l_1 - l) \tag{2.9a}$$

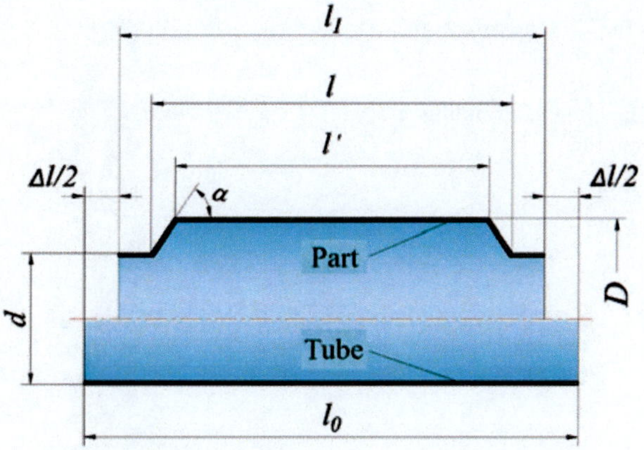

Fig. 2.7 Schematic diagram of calculation of ideal feeding length

$$\Delta l = l_0 - l_1 = \frac{Dl'}{d} + \frac{D^2 - d^2}{2d \sin \alpha} - l \quad (2.9b)$$

where

Δl Ideal feeding length (mm);
l_0 Initial length of tube (mm);
l_1 Length of the part (mm);
l Length of the forming area (mm);
α Semi-angle of cone-shaped transition area (°);
l' Length of forming area (mm), $l' = l - (D - d)/\tan \alpha$;
d Outer diameter of tube (mm);
D Outer diameter of the part (mm).

2.3 Defects and Loading Path

2.3.1 Type of Defects

The hydroforming of variable diameter tubular part is a complex process due to the combined actions of internal pressure and axial feeding. On one hand, if the internal pressure is too high, the excessive thinning and even splitting will occur on the tube. On the other hand, buckling or wrinkling may occur at a too large axial feed. The main defects in hydroforming of variable diameter tubular part are shown in Fig. 2.8. It can be seen that the qualified parts can be obtained only by using a reasonable combination of internal pressure and axial feeding.

Buckling is a defect that caused by an excessive axial force applied when the length of the tube in forming area is too long and internal pressure is insufficient at the initial forming stage. This defect can be avoided by choosing a reasonable length of tube, appropriate preforming process and controlled process parameters.

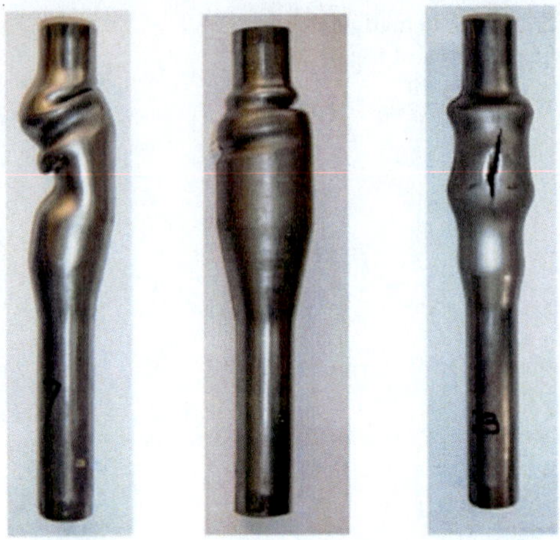

Fig. 2.8 The defects in hydroforming of variable diameter tubular parts. **a** Buckling; **b** wrinkling; **c** splitting

Wrinkling is also resulted from the excessive axial force. At the beginning of forming, the number, location and shape of wrinkles are related to the geometric dimensions of the tube and loading conditions. Wrinkles can be divided into two types. One is the wrinkle that cannot be flattened under higher internal pressure in the last calibration stage, which is usually called as "dead wrinkle". It is a defect can be prevented by adjusting the complex loading path. The other type of wrinkle can be flattened under higher internal pressure in the last calibration stage, which is called as "useful wrinkle". Such wrinkle is not a defect, which is regarded as a mean of preforming, by which the tube can be pushed inward to form the wrinkles to accumulate materials in the forming area at the beginning of forming.

For low carbon steel materials, when the expansion ratio is bigger than 40%, the tubular part will be prone to splitting due to excessive internal pressure. The splitting pressure can be estimated by formula (2.4). It is known that the splitting is caused by local thinning of the tube wall. The beginning of thinning depends on the tube thickness, mechanical properties of materials and loading conditions. In order to avoid splitting, it is necessary to ensure that the tube wall is attached to the die cavity before the splitting occurs. Moreover, for the parts with large expansion ratios, using preformed blank or intermediate annealing are the main methods to avoid splitting.

2.3.2 Process Window and Loading Path

The process window refers to a reasonable range of combinations between the axial stress and internal pressure, in which the tube experiences neither wrinkling nor splitting, as shown in Fig. 2.9. Using this process window, the axial critical stress for wrinkling and splitting pressure can be determined. The process window is an area enclosed by three lines a, b and c. Line a represents the relationship between axial stress and internal pressure at the beginning of deformation tube. The equation for this curve is determined by the Mises yield criterion. On this line, the point a_1 represents the initial yielding pressure, which can be calculated using the formula (2.3). Line b represents the splitting pressure, and point b_1 on this line represents the splitting pressure without axial stress, which can be estimated by formula (2.4). Line c represents the axial critical stress for wrinkling, where point c_1 is the axial critical stress for wrinkling without internal pressure. However, the critical stress for wrinkling under internal pressure, or the specific form of line c, needs to be determined by experimental or finite element analysis or complex estimation using analytical formula. There are two possible forms of line c. One is that the axial stress increases with increasing internal pressure. The other is that the axial stress firstly declines within a certain pressure range and then increases again. For such a case, only early stages can be finished for some experiments indicated by the critical axial stress of wrinkling decreases with increasing of internal pressure. For practical process, the loading path is determined by the geometrical dimension of tube, material mechanical properties and loading condition (ratio of axial stress to internal pressure).

The process window is divided into four regions: A, B, C and D. Among them, region A is the elastic region, where the tube is still in the elastic range. Region B is the splitting region. When the internal pressure locates in this region, the tube will split. Region C is the wrinkling region, and the wrinkling will occur when the axial stress is within this region. Region D is the forming region. Only when the matching relationship between internal pressure and axial stress is within this range, the tube can be guaranteed to undergo plastic deformation without wrinkling and splitting.

However, in actual processes, due to the influence of friction and other factors, it is difficult to accurately control the axial force. Therefore, the relationship between internal pressure and axial feeding length shown in Fig. 2.10 is usually adopted in

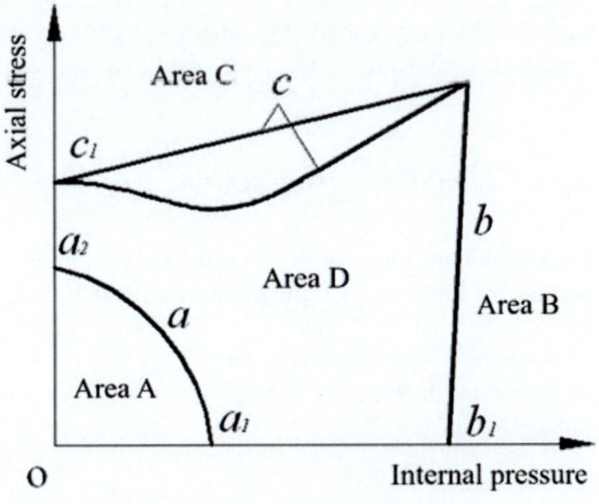

Fig. 2.9 Schematic diagram of the relationship between axial stress and internal pressure

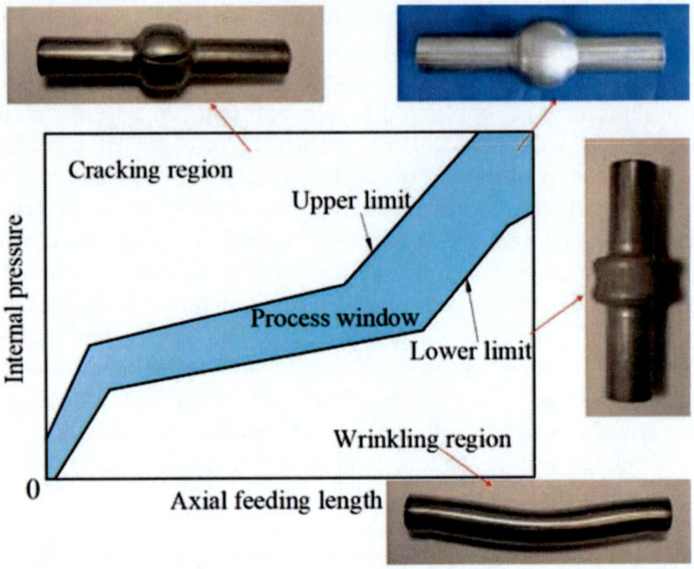

Fig. 2.10 Process window and loading path

production. This relationship is also known as the loading curve or the loading path. The key of determining the loading path is to determine the upper and lower limits of the internal pressure. The common method is to obtain the initial loading path by numerical simulation, and then determine the formal loading path by experiments. In the process window, qualified parts can be obtained by any loading path within this window. However, if the loading paths at different positions are used, the thickness thinning rates of parts are different. When the loading path approaches the upper limit, the thickness thinning is serious. While approaching to the lower limit, the thickness thinning is small. For a part, a larger internal pressure range in the process window is beneficial to form parts due to the easy implemented process control.

2.3.3 Limit Expansion Ratio

Expansion ratio refers to the ratio between perimeter of tube section and the initial perimeter. While for the limit expansion ratio, it is the maximum expansion ratio of the tube that can be obtained from the initial circular cross-section at one operation without any preforming. The limit expansion ratio is shown in Fig. 2.11, which is an important index to indicate the formability of hydroforming variable diameter tubular parts.

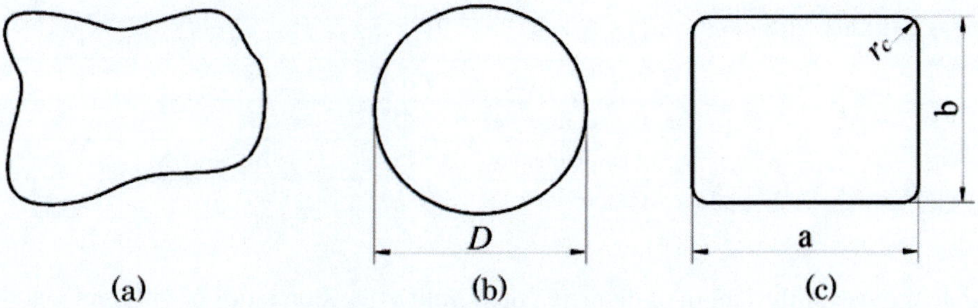

Fig. 2.11 Schematic diagram of definition for limit expansion ratio. **a** Cross-section with arbitrary shape; **b** circular cross-section; **c** rectangular cross-section

The expansion ratio can be calculated by the following formula:

$$\eta = \frac{C - \pi d}{\pi d} \times 100\% \tag{2.10a}$$

For the variable diameter tubular part with a maximum circular cross-section, the formula for calculating the expansion ratio can be simplified as follows:

$$\eta = \frac{D - d}{d} \times 100\% \tag{2.10b}$$

For the variable diameter tubular part with a maximum rectangular cross-section, the formula for calculating the expansion ratio can be simplified as follows:

$$\eta = \frac{(2a + 2b - 4r_c) - \pi d}{\pi d} \times 100\% \tag{2.10c}$$

where

η Expansion ratio;
C Sectional perimeter of the part (mm);
d Tube diameter (mm);
D Sectional diameter of the part (mm);
a, b Length and width of rectangular cross-section (mm);
r_c Transition corner radius of rectangular cross-section (mm).

Moreover, in order to calculate the limit expansion ratio $\eta_{\max}$, the sectional perimeter in formula (2.10a) should be replaced by the maximum sectional perimeter or diameter of cross-section. The main factors influencing the limit expansion ratio are mechanical properties of tube (including elongation, hardening exponent n, coefficient of normal anisotropy r), shape of the part, length of forming zone, and loading path. In addition, the thickness and the location of maximum section can also affect. For the variable diameter tubular parts with a circular cross-section and the forming

Table 2.1 Limit expansion ratio of tubular parts with circular cross-section

Materials	η_{max} (%)	Pure bulging η_{max} (%)
Stainless steel	100	35
Low-carbon steel	80	25
Aluminum alloy	40	20
Copper alloy	120	60

zone is twice of the length of diameter, their limit expansion ratios of different materials can be obtained through experiments, as summarized in Table 2.1. In this table, the pure bulging represents no axial feeding is used.

It can be found that the higher the elongation, the greater the allowable deformation before splitting, which indicates that a higher limit expansion ratio can be obtained. For the material with a greater hardening exponent, its strong strain hardening ability enables a uniform strain distribution of the deformation zone and improved local deformation capacity of the material. Therefore, it has a higher expansion ratio. For example, the limit expansion ratio of stainless steel is obviously higher than that of aluminum alloy. Moreover, the location of the maximum cross-section also has a great influence on the limit expansion ratio. When the maximum section locates at both ends of the part, the material can be easily put into this location. Therefore, a higher limit expansion ratio can be obtained. However, when the maximum cross-section is in the middle, the feeding of material is difficult resulting in relatively small limit expansion ratio. Generally, the limit expansion ratio increases with the increase of tube thickness, but the increase rate is small. Moreover, the limit expansion ratio of the asymmetric structure is smaller than that of the symmetrical structure due to the inhomogeneous deformation.

2.3.4 *Control and Use of Wrinkles*

As mentioned above, wrinkles can be divided into two types. One is "dead wrinkle" that is a defect; and the other is "useful wrinkle" that can be flattened under higher internal pressure in the subsequent calibration stage. In this section, the control and use of wrinkles will be discussed from the view of geometry.

The principle of tube hydroforming using useful wrinkles is shown in Fig. 2.12. The main procedure is that: firstly, the internal pressure is imposed on the tube as a supporting medium, as shown in Fig. 2.12a; Then, the tube is horizontally pushed inwards a length Δl, which is the axial feeding length, through which, the tube is formed into a preformed tubular blank with wrinkles, as shown in Fig. 2.21b); thirdly, these wrinkles are flattened under higher internal pressure to attach the die cavity so that part is formed in the calibration stage, as shown in Fig. 2.12(c).

In Fig. 2.12, l_0 and d are the length and diameter of initial tube respectively. After hydroforming, the length and diameter of the part are l_1 and D, respectively. The forming area has a length of l, and the semi-angle of cone-shaped transition area in

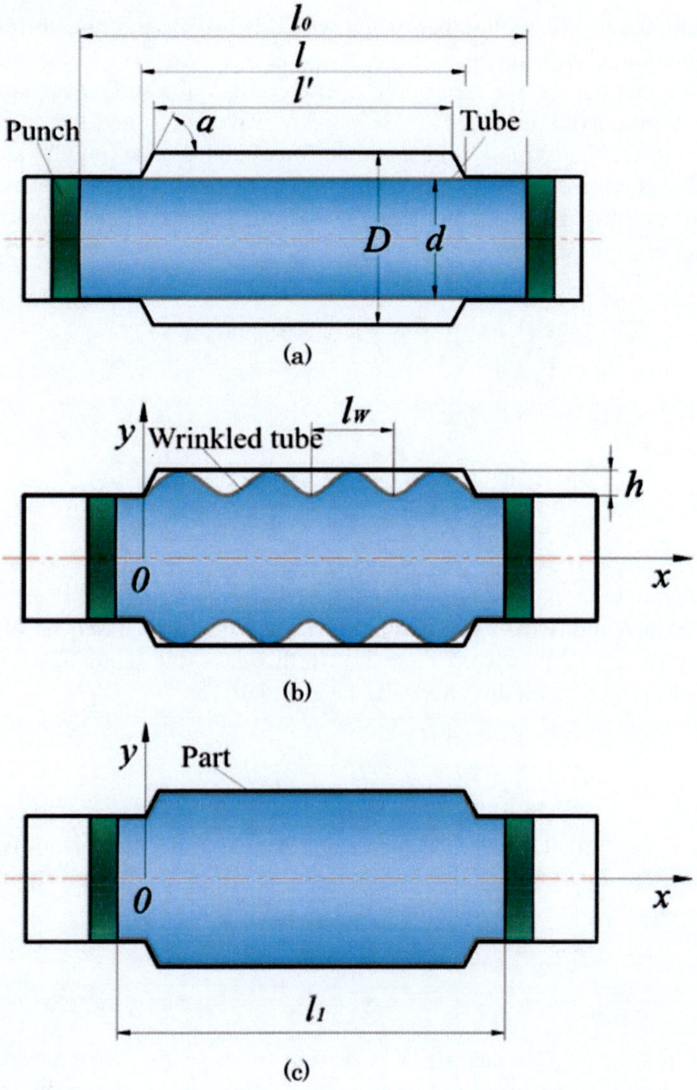

Fig. 2.12 Controlling and utilization of wrinkling. **a** Initial stage; **b** wrinkling stage; **c** calibration stage

die cavity is α. Therefore, the length of forming area with the maximum diameter can be expressed as $l' = l - (D - d)/\tan\alpha$.

It is assumed that wrinkles with a number of n are evenly distributed in the die cavity, and the wave profile of wrinkles is described by a sine curve. The top of the wrinkle wave is closed to the die and its bottom is on the outside surface of the original tube. In this case, the required feeding length for such wrinkles are named axial feeding length. If the number of the wrinkles is different, the corresponding axial feeding length is different.

It is given that the surface area of the workpiece is S, the surface area of a single wrinkle is S', the axial length of single wrinkle wave is l_W, and the amplitude of the sine curve for the wrinkle is h. It can be found that the maximum value of h is $(D - d)/2$. When the semi-angle of cone-shaped transition area is relatively small, the value of h in this area is smaller than $(D - d)/2$. Then, the sine equation for one wrinkle can be expressed as

$$y = h\sin\left(\frac{\pi}{l_W}x\right) + \frac{d}{2} \quad (0 \le x \le l_W) \tag{2.11}$$

The surface area of single wrinkle:

$$S' = 2\pi\int_0^{l_W}\left[h\sin\left(\frac{\pi}{l_W}x\right) + \frac{d}{2}\right]\sqrt{1 + \frac{h^2\pi^2}{4l_W^2}\cos^2\left(\frac{\pi}{l_W}x\right)}dx \tag{2.12}$$

If the thickness of the tube remains unchanged during wrinkling, the surface area of the wrinkled tube is the same as that of the original tube based on the volume consistency principle, that is $nS' = \pi d l_0$. In this case, the required axial feeding length for n wrinkles with this shape can be calculated as

$$\Delta l = l_0 - l_1 = \frac{nS'}{\pi d} - l_1 \tag{2.13}$$

However, the surface area of the tube after wrinkling, nS', may be different with the surface area of part S. When $nS' < S$, thinning will occur in the forming area of the tube, and the average thinning ratio can be expressed as

$$\bar{\delta}_t = \frac{S - nS'}{S} \times 100\% \tag{2.14}$$

For a specific part, $D = 88$ mm, $d = 65$ mm and $\alpha = 20°$, three different length of forming area $l = 65$ mm, 97.5 mm and 130 mm, correspondingly, the length-diameter ratio $l/d = 1$, 1.5 and 2, the wrinkles number, axial feeding length, and average thinning ratio (a negative value represents an increase in wall thickness) are calculated and given in Table 2.2.

It can be seen from Table 2.2 that the thinning ratio of thickness becomes smaller with increasing the number of wrinkles and feeding length, and thickening even occurs in some particular condition. This is true for all parts with three different length-diameter ratios. The materials can be accumulated in advance in the forming area through formation of wrinkles. The key factor is the number of wrinkles. As long as the number of wrinkles is appropriate, the thickness after forming remained basically unchanged, or the minimum value meets the requirements.

Table 2.2 The relationship between the number of wrinkles and the average thinning ratio of thickness

Length-diameter ratio	Number of wrinkles n	Axial feeding length Δl /mm	Average thinning ratio $\overline{\delta}_t$/%
1	2	23.2	9.4
	3	28.7	6.1
	4	41.0	− 1.2
1.5	3	35.3	7.2
	4	44.7	2.7
	5	58.1	− 3.6
2	3	43.4	7.3
	4	49.9	4.7
	5	61.2	0.3
	6	75.2	− 5.2

For experimental verification of the control and use of wrinkles, the 5A02 aluminum alloy tube with a diameter of 63 mm and thickness of 2.0 mm was used. The real axial feeding length used in the experiment is 80% of the ideal feeding length, that is, 30 mm. The wrinkles with different number or shape were obtained by changing the internal pressure. Figure 2.13 shows the wrinkle shapes of 5A02 aluminum alloy tube under different internal pressures.

It can be found from Fig. 2.13 that the wrinkles only occur on both ends of the deformation zone of the tube when the internal pressure was lower (1.0 p_s). The

(a) (b) (c) (d)

Fig. 2.13 The wrinkle shape of 5A02 aluminum alloy tube under different internal pressures p_i. **a** $p_i = 1.0\ p_s$; **b** $p_i = 1.2\ p_s$; **c** $p_i = 1.4\ p_s$; **d** $p_i = 1.8\ p_s$

wrinkles at one end of the deformation zone show an axisymmetric form, while the other end is folded, but it shows the tendency of changing into an axisymmetric form. Meanwhile, another wrinkle, called middle wrinkle, starts to develop between the two previous wrinkles, but the location is not at the centroid of the tube. Moreover, the middle wrinkle is incompletely developed in the circumferential direction. For the cases with higher internal pressure (1.2 p_s, 1.4 p_s and 1.8 p_s), three axisymmetric wrinkles can be formed. The difference is that, with increasing internal pressure, the diameters of peak and valley of middle wrinkles becomes larger, and the width of three peaks increase.

It can be seen from the above that the wrinkling behavior of 5A02 aluminum alloy tube can be controlled by changing the combinations between internal pressure and axial feeding length. In addition, different number and shapes of wrinkles can also be obtained by changing the length of the forming zone. For example, four or more wrinkles can be formed by increasing the length of forming area. In order to use wrinkles as preform in tube hydroforming, it is necessary to investigate the shape change of wrinkles obtained under different conditions during flattening. Therefore, the wrinkled 5A02 aluminum alloy tubular samples shown in Fig. 2.13 were flattened under higher internal pressure in calibration stage. The maximum calibration pressure is 80 MPa, and the tubular parts obtained after flattening is shown in Fig. 2.14. It can be seen from Fig. 2.14 clearly that the wrinkled tubes obtained under lower internal pressure (1.0 p_s), tend to splitting fracture before contacting with the die cavity in the calibration stage, which is due to the lack of sufficient materials in the middle of the tube. With the increase of internal pressure in wrinkling stage, the three axisymmetric wrinkles formed in advance are free from fracture during the subsequent flattening process, i.e., calibration stage. However, when the internal pressures in wrinkling stage are 1.2 p_s and 1.4 p_s, the wrinkles cannot be flattened completely; as a result, the so-called dead wrinkles appear at the position of original wave bottom for the hydroformed tubes after calibration stage. Furthermore, it can be found that higher wrinkling internal pressure will alleviate the occurrence of dead wrinkles. Therefore, a sound hydroformed tubular specimen can be obtained at an internal pressure of about 1.8 p_s, at which the three axisymmetric wrinkles can be flattened completely without any dead wrinkle formation and fully contact with the die cavity in the end. The corresponding wrinkles obtained under internal pressure of about 1.8 p_s can be referred as the "useful wrinkles" for the 5A02 aluminum alloy tubes.

2.4 Thickness Distribution and Influencing Factors

2.4.1 Thickness Distribution

For a variable diameter tubular part formed by hydroforming, the thickness distribution is that the value in the forming area decreases while that in the feeding area

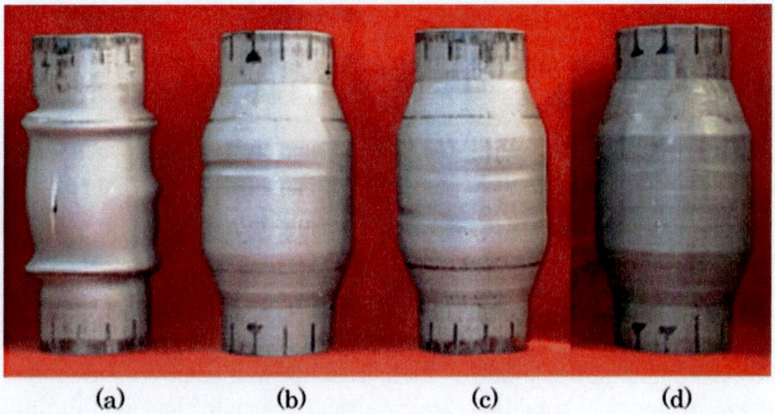

Fig. 2.14 Effect of internal pressure (p_i) on the flattening of wrinkles. **a** $p_i = 1.0$ p_s; **b** $p_i = 1.2$ p_s; **c** $p_i = 1.4$ p_s; **d** $p_i = 1.8$ p_s

increases, as shown in Fig. 2.15. Thinning ratio refers to the ratio between the thickness reduction at a certain point after deformation and the initial wall thickness. A positive value means thinning and a negative value means thickening. The maximum thinning ratio is defined as follows:

$$\delta_{tm} = \frac{t - t_{\min}}{t} \times 100\% = (1 - \frac{t_{\min}}{t}) \times 100\% \tag{2.15}$$

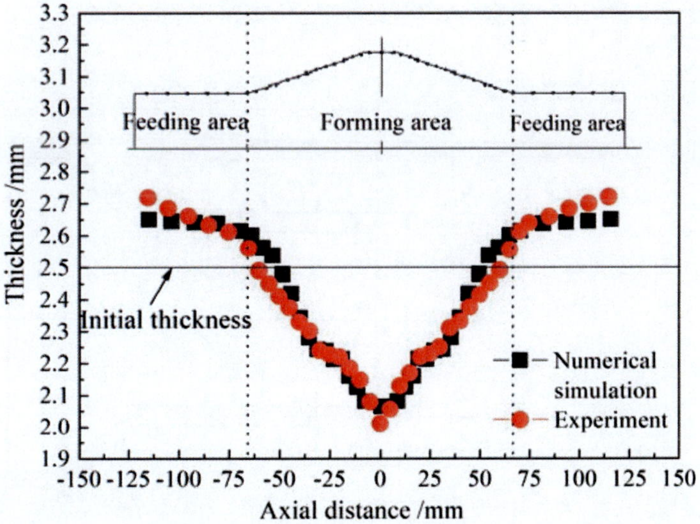

Fig. 2.15 Thickness distribution of a variable diameter tubular part

where

t_{min} Minimum thickness in forming area (mm);
t Initial tube thickness (mm).

For the biconical tubular part shown in Fig. 2.15, the initial thickness is 2.5 mm and the minimum thickness in the forming area is 2 mm, then the maximum thinning ratio is 20%. The maximum thickness in the feeding area is 2.72 mm, corresponding to a thickening ratio of 8.8%.

The main factors affecting the maximum thinning ratio of the forming area are the part shape, expansion ratio, length of the forming area, loading curve and mechanical properties of material (n value and r value). Generally, the larger n and r could induce a smaller maximum thinning ratio and a more uniform thickness distribution in the forming area when other conditions are consistent. When the length of forming area is the same, the greater the expansion ratio, the greater the thinning ratio. Under the condition that the shape and material of parts are determined, the most effective approach to reduce the thickness thinning is to control the loading curve, reduce the friction in the feeding area, and feed more material to the forming area.

2.4.2 Thickness Dividing Circle

For the variable diameter tubular part, the thickness distribution trend is that thinning in the forming area and thickening in the feeding area. Therefore, there exists a cross-section with unchanged thickness between the feeding area and forming area. Such a cross-section with unchanged thickness is called the thickness dividing circle, as shown in Fig. 2.16.

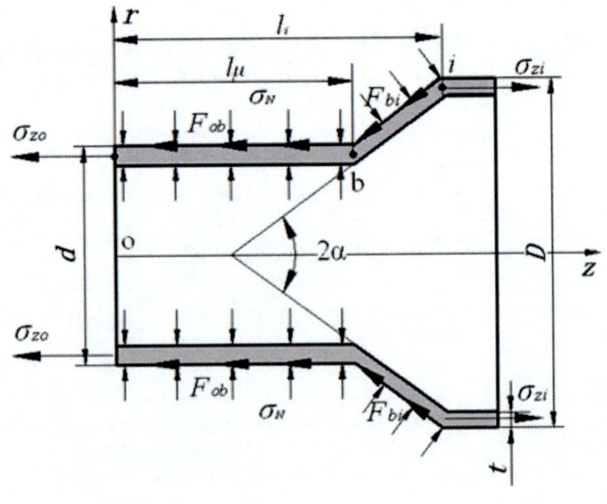

Fig. 2.16 Thickness dividing circle and the schematic diagram for force analysis

According to the force analysis and the incremental plastic constitutive equation, the formula of the distance from the thickness dividing circle to the tube end can be deduced and given as follows:

$$\frac{l_i}{d} = \frac{l_\mu}{d} + \frac{1}{\mu \cos \alpha}\left(-\mu \frac{l_\mu}{d} - \frac{D}{2d} - \frac{t}{d}\lambda\right) \tag{2.16}$$

where

- l_1 The distance from the thickness dividing circle to the tube end (mm);
- α Half cone angle of transition zone;
- t The initial thickness or thickness at the dividing circle (mm);
- λ Ratio of axial stress to internal pressure.

2.5 Tubes for Hydroforming

2.5.1 Material

There are a variety of material candidates for hydroforming, such as mild steel, low alloy high strength steel, stainless steel, aluminum alloy, copper alloy and Ni-based alloy, etc.

At present, for the automobile applications, the tensile strength of mild steel and low alloy high strength steel commonly used ranges between 300 and 450 MPa. With a further demand for weight reduction, the tensile strength of the hydroformed parts will reach 500–600 MPa or even 1000 MPa.

The commonly used stainless steels are mainly the austenitic stainless steel 304 and 1Cr18Ni9Ti. For the engine part, heat and oxidation resistant ferritic stainless steel such as the 429 and 309, etc., are normally required.

The commonly used aluminum alloys mainly include the 5000 series, 6000 series and 7000 series. The 5000 series aluminum alloy tubes are mainly used for the tubular parts of aircrafts and rockets; the 6000 series aluminum alloy tubes are mainly used in automobiles and bicycles. In comparison, the application of hydroformed tubular parts using 7000 series aluminum alloy has just started.

For the titanium material, the pure titanium tubes and TC4 are mainly used. However, due to the high deformation resistance, low ductility and severe spring-back of titanium alloy at room temperature, only parts with simple shapes can be manufactured using the hydroforming.

2.5.2 Requirement of Tube Materials

For the tubes used in hydroforming, the tubes required to own not only enough mechanical properties for the structure requirements, but also formability, precision of diameter and wall thickness.

In the aspect of formability, better ductility and larger n and r values are required for the tubes. The fabrication of tubes has great influence on mechanical properties and formability. Therefore, work hardening should be minimized during the fabrication of tubes, so as to preserve the ductility of material and improve the r value to be best, and obtain tubes suitable for subsequent hydroforming.

The precision of outer diameter and wall thickness of tubes for hydroforming are usually higher than those of common structural steel tubes. These tubes are specifically fabricated according to the sealing method adopted in hydroforming and the accuracy requirements of thickness for the parts. In general, the precision tolerances of outer diameter and thickness of hot and cold rolled steel tubes can meet the requirements of hydroforming, and the tolerances of resistance welded tubes are also easy to meet the requirements. The difference in the tolerance of each batch, it can also be solved by adjusting the dimensions of forming dies. The tube ends must be cut perpendicularly to the center line of tube, and the verticality difference between the end section and the center line should be within 1.5°.

In addition, it has a cleanliness requirement on both the external and internal surface of the tubes. For the steel tubes, acid pickling is used to ensure the rust-free condition on the surface. This is mainly because the oxide and rust on the surface of the tube would arise the die wear and contaminate the pressurized medium in hydroforming. Moreover, of the various kinds of debris attached on the tube surfaces could fall into the die cavity which would affect the quality of the parts.

2.5.3 Types and Specification of Tubes

At present, the main types of tubes used for hydroforming are seamless tubes, resistance welded tubes (ERW), drawn tubes (DOM) and laser welded tubes. For steel tubes, the ERW tube is usually preferred because of its lower cost and better formability compared to the seamless tube and drawn tube. Laser welded tube has the best formability, but the cost is high. Therefore, they are mainly used in the forming of complex parts with great deformation requirement.

Normally, the fabrication of seamless tubes can guarantee a fine grain microstructure and uniform streamline. Moreover, the mechanical properties and surface quality of the tube can be guaranteed through subsequent heat treatment, e.g. and annealing, tempering and acid pickling process. Due to the absence of welding seams, the seamless tubes have a good consistency in mechanical properties. Therefore, they are suitable for the hydroformed parts required a certain forming pressure range. However, a certain eccentricity is inevitable in the manufacture of seamless tube, which could result in the thickness variation along the circumferential direction of the tube. Further, it is easy to induce the splitting and nonuniform thickness distribution during forming, and deteriorate the part performance.

For the ERW tubes, they are fabricated using the hot- or cold-rolled coils. To meet the requirements of surface quality of the hydroformed parts, the surface of coils should be free of oxidized skin and treated with acid pickling and oil coating. The manufacture process of ERW tube has a certain influence on the work hardening, and different process parameters lead to different yield stresses at different circumferential positions of the tube. Therefore, this difference needs to be reduced by strictly controlling the manufacturing process. Both the inner and outer wall of the ERW tube has certain weld protrusion. Hence, the weld protrusion is usually scraped by specific tool directly after welding. Sometimes, the weld protrusion on the inner wall of the ERW tube can be kept with the height of the weld being limited within a certain range. However, the biggest problem of ERW tube used in hydroforming is the splitting at welding seam and heat-affected zone. Therefore, many manufacturers have developed ERW tubes dedicated for hydroforming, and the elongation of tubes can be significantly improved by changing the coiling, welding and post-welding heat treatment processes. Compared to the original process, the elongation of the SUS409L stainless steel tube manufactured by CBR is basically above 50%, which is a significant improvement, as shown in Fig. 2.17. Moreover, the elongation and r value of the ERW steel tube can be increased by 30–100% by rolling at a high temperature of 900 °C. For example, for steel tubes with a tensile strength of 450 MPa, the r value is increased from 0.8–0.9 to 1.7–2.1, which greatly improves the formability of tubes.

DOM tubes are produced by a cold drawing process using fully annealed thick-walled ERW tubes. During tube drawing, the use of a die with mandrel is beneficial for obtaining high dimension accuracy and uniform material properties. After cold drawing, the tube can be normalized or annealed to obtain the mechanical properties required by tube hydroforming.

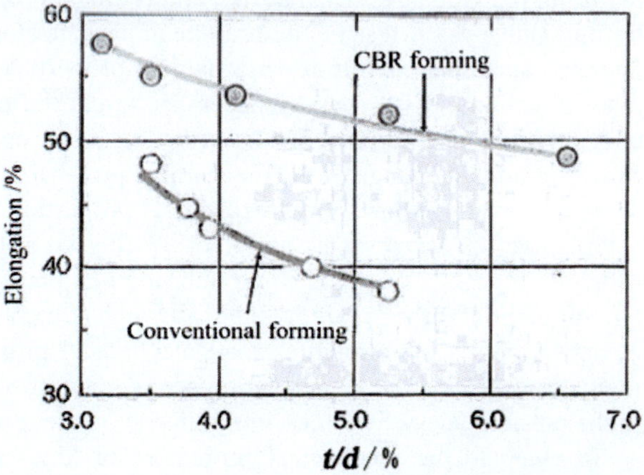

Fig. 2.17 Effect of tube coiling process on the elongation of ERW tube

Laser welded tubes are used for hydroforming of complex parts. Generally, the elongation of laser welded tubes and the cracking resistance of welded joints are better than those of ERW tubes, which is caused by the smaller heat affected zone compared to that of ERW tube. Therefore, the laser welded tube has obvious advantages in the manufacturing of hydroformed parts with large expansion ratio. However, the high cost is the main problem.

2.5.4 Measurement of Mechanical Properties of Tubes

Due to the influence of tube fabrication, the mechanical properties of the tube in axial and circumferential directions are usually different, especially the elongation, n value and r value. For the testing of mechanical properties, the difficulty is the accurate test of properties in the circumferential direction. At present, the main test methods include uniaxial tension test, ring hoop tension test and hydraulic bulging test.

The uniaxial tension test is the most commonly used method. For welded tube, the axial and circumferential tensile properties can be approximately represented by the tensile properties of the initial sheet in different directions. However, for the tubes with smaller diameters and higher strength, the performance of the tube, particularly that in the circumferential direction, will change obviously due to the serious work hardening during coiling. In the meantime, the welding process also has a great impact on the tube performance. Therefore, large errors can be inevitably induced by using initial sheet performance to represent that of welded tubes. In this case, arc-shaped specimens can be cut directly from the welded tubes and then tensioned along the axial direction to avoid work hardening effect. However, this method can only be used for the axial tensile test, while is not valid for the testing of circumferential properties. For seamless tubes, such as extruded tubes or cold-drawn tubes, the mechanical properties of tubes along the axial and circumferential directions vary significantly requiring individual testing. The axial mechanical property of tubes can be obtained by uniaxial tension test of the arc specimen cut along the axial direction or the flattened specimen. While for mechanical properties along the circumferential or other directions, it is generally necessary to cut the tube and then flatten it before uniaxial tension. Similarly, work hardening resulting from flattening process will cause significant change of the mechanical properties, thus making it impossible to accurately evaluate the overall mechanical properties of tubes.

To solve the above issue, a method of ring hoop tension test was proposed, as shown in Fig. 2.18. In this method, the ring hoop tension specimen is cut along the circumferential direction of the tube, and then stretched using the ring hoop tension device to obtain the circumferential mechanical properties directly. However, the main disadvantage is that the friction between the specimen and the D-shaped block in the tension device has a certain influence on the test results.

To evaluate the mechanical properties of tubes directly, the method of tube hydraulic bulging test has gained increased attention. During the test, the pressured

Fig. 2.18 Method for ring hoop tension test of tubes. **a** Schematic diagram of machining specimen; **b** specimen for ring hoop tension test; **c** schematic diagram of ring hoop tension device

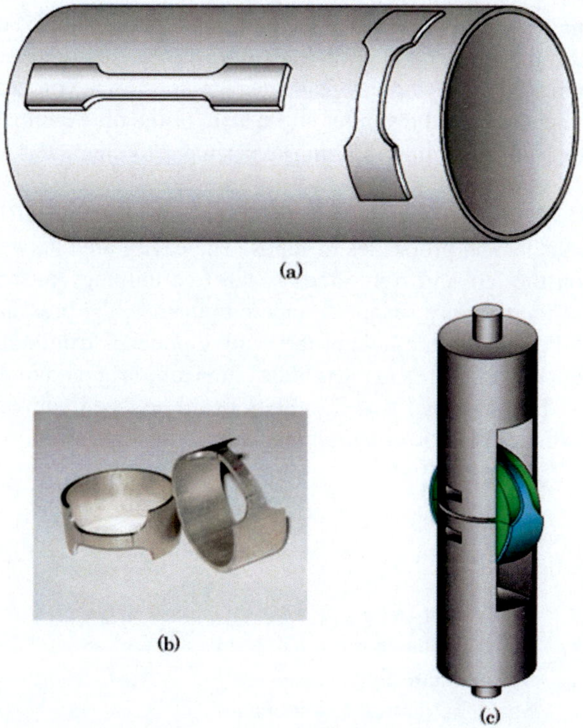

liquid was injected into the tube to cause its deformation. By measuring the bulging height of the tube under different pressures, the axial and circumferential tensile curves and the equivalent stress- strain relationships could be calculated. Using the hydraulic bulging test, further processing of tube is not needed and work hardening can be neglected. At the same time, the stress–strain state during bulging is more similar to that of hydroforming. The dominant advantage of this method is that the mechanical properties and stress–strain curves of both circumferential and axial directions can be obtained simultaneously. However, the main disadvantage is there exists an error in the mechanical property parameters calculated according to the analytical formula. In addition, the specific hydraulic bulging test device, die and instrument must be manufactured resulting in high cost.

2.5.5 Testing Methods and Apparatus for Mechanical Properties of Anisotropic Tubes

Accurate measurement and characterization of mechanical properties and flow stress curve of tubes is the foundation of accurate numerical simulation analysis of tube hydroforming. Because of the large difference between the axial and circumferential mechanical properties of the tube and its strong anisotropy, it is difficult to accurately test the mechanical properties of the tube under complex stress states by using either uniaxial or ring hoop tension test methods. In the process of hydraulic bulging experiment, due to the similar stress state of the tube compared to hydroforming, method of using hydraulic bulging experiment to characterize tube mechanical properties was proposed.

Figure 2.19 shows the experimental principle of hydraulic bulging for testing mechanical properties of tubes. The cavity area between the ring-shaped die block on the left and right sides is the free bulging zone of the tube. When both ends of the tube are sealed by tapered punches, the pressured liquid is pumped into the tube through the hole at the sealing punches from a high-pressure source, then the unconstraint area in the middle of the tube is bulged freely under the internal pressure.

It is assumed that the relationship between equivalent stress and strain of tube satisfies the power hardening law:

$$\sigma_i = K\varepsilon_i^n \tag{2.17}$$

where

σ_i Equivalent stress;
ε_i Equivalent strain;
K Strength coefficient;
n Strain hardening exponent.

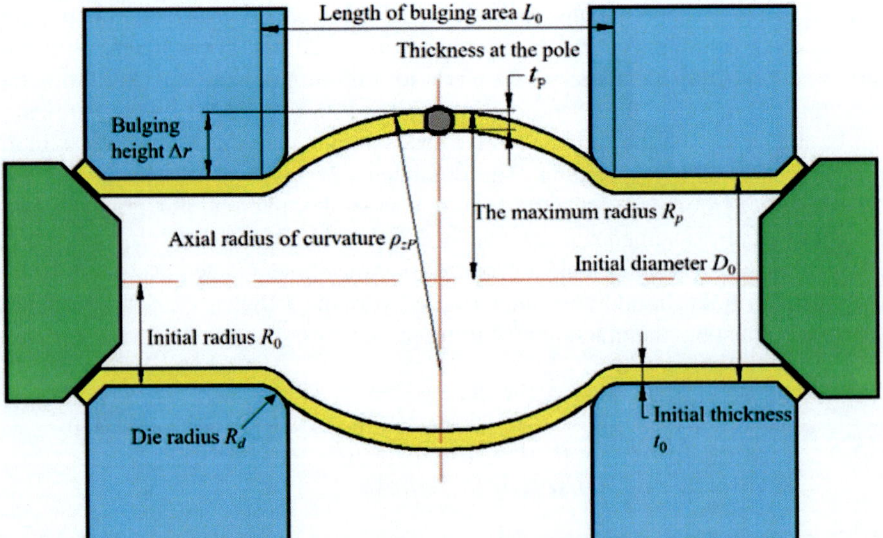

Fig. 2.19 Experimental principle of tube hydraulic bulging and geometric parameters

In order to obtain this relationship, the equivalent stress and equivalent strain should be obtained respectively.

Because the tubes used in hydroforming are very thin and the diameter-thickness ratio are relatively large, the normal stress in the thickness direction of tube can be ignored, and the plane stress state is approximately considered. During the hydraulic bulging process, the circumferential stress is greater than the axial stress, $\sigma_\theta > \sigma_z > 0$. Therefore, the circumferential stress is the first principal stress and the axial stress is the second. According to the plane stress state anisotropic yield criterion of Balart89, the equivalent stress can be expressed as:

$$\sigma_i = \left\{ \frac{1}{r_{90}(1+r_0)} \left(r_{90}|\sigma_\theta|^M + r_0|\sigma_z|^M + r_0 r_{90}|\sigma_\theta - \sigma_z|^M \right) \right\}^{1/M} \tag{2.18}$$

where

σ_θ Circumferential stress at the pole of tube (MPa), which corresponds to the first principal stress;

σ_z Axial stress at the pole of tube (MPa), which corresponds to the second principal stress;

r_0, r_{90} Anisotropy coefficient of tube along the axial and circumferential direction, respectively;

M Material parameter, for the BCC lattice $M = 6$ while for the FCC lattice $M = 8$.

According to Drucker's associated flow rule, the circumferential and axial strain increment can be expressed as:

$$\begin{cases} d\varepsilon_\theta = \dfrac{\partial g}{\partial \sigma_\theta} d\lambda = \dfrac{M}{r_{90}(1+r_0)} \left[r_{90}(\sigma_\theta)^{M-1} + r_0 r_{90}(\sigma_\theta - \sigma_z)^{M-1} \right] d\lambda \\ d\varepsilon_z = \dfrac{\partial g}{\partial \sigma_z} d\lambda = \dfrac{M}{r_{90}(1+r_0)} \left[r_0(\sigma_z)^{M-1} - r_0 r_{90}(\sigma_\theta - \sigma_z)^{M-1} \right] d\lambda \end{cases} \tag{2.19}$$

In the case that the first-order increment of plastic work per unit volume is the same, the equivalent plastic strain increment can be expressed as:

$$d\varepsilon_i = \frac{\sigma_\theta d\varepsilon_\theta + \sigma_z d\varepsilon_z}{\sigma_i} \tag{2.20}$$

By combining the Eqs. (2.19) and (2.20), according to the incompressibility, the equivalent plastic strain increment can be expressed as:

$$d\varepsilon_i = \frac{[r_{90}(1+r_0)]^{1/M}(\sigma_\theta d\varepsilon_\theta + \sigma_z d\varepsilon_z)}{\left[r_{90}\sigma_\theta^M + r_0\sigma_z^M + r_0 r_{90}(\sigma_\theta - \sigma_z)^M \right]^{1/M}} \tag{2.21}$$

When the hydraulic bulging is conducted on the tube with fixed ends, the axial and circumferential stress components are expressed as:

$$\begin{cases} \sigma_z = \frac{p(\rho_{\theta P}-t_P)^2}{2t_P(\rho_{\theta P}-t_P/2)} & \text{(a)} \\ \sigma_\theta = \frac{p(\rho_{\theta P}-t_P)}{2t_P(\rho_{zP}-t_P/2)}(2\rho_{zP} - \rho_{\theta P} - t_P) & \text{(b)} \end{cases} \tag{2.22}$$

where

t_P Thickness at the pole (mm);
p Internal pressure for bulging (MPa);
ρ_{zP} Axial radius of curvature at the pole (mm);
$\rho_{\theta P}$ Circumferential radius of curvature at the pole (mm),

$$\rho_{\theta P} = R_P = R_0 + \Delta r.$$

By analyzing the strain of the tube during hydraulic bulging, the circumferential strain and normal strain are obtained respectively:

$$\begin{cases} \varepsilon_\theta = \ln\left(\frac{R_P-t_P/2}{R_0-t_0/2}\right) = \ln\left(\frac{R_0+\Delta r-t_P/2}{R_0-t_0/2}\right) & \text{(a)} \\ \varepsilon_t = \ln\left(\frac{t_P}{t_0}\right) & \text{(b)} \end{cases} \tag{2.23}$$

where

t_0 Initial thickness of tube (mm);
R_0 Initial outer radius of tube (mm), $R_0 = D_0/2$;
Δr Bulging height (mm);
t_P Thickness at the pole (mm).

Therefore, in order to obtain the stress and strain components, not only the internal pressure p during bulging experiment process should be measured, but also the bulging height (radial displacement) Δr, thickness at the pole t_P and the axial radius of curvature at the pole ρ_{zP}. During testing, the values of Δr, ρ_{zP}, t_P and p should be measured, as shown in Eq. (2.24):

$$\begin{cases} \Delta r = \Delta r(p) \\ t_P = t_P(p) \\ \rho_{zP} = \rho_{zP}(p) \end{cases} \tag{2.24}$$

During the hydraulic bulging experiment, the internal pressure p and the bulging height Δr, namely $\Delta r = \Delta r(p)$, are easy to achieve real-time measurements through sensors. However, it is difficult to measure the axial radius of curvature at the pole ρ_{zP} and the thickness at the pole t_P sinstantly. In order to avoid the direct measurement of axial curvature radius, the axial profile of the tube is assumed to be an arc tangent to the profile of the die, as shown in Fig. 2.20. The coordinate axis r-z is taken as the reference coordinate system. In addition, the initial thickness of tube t_0, initial outer radius of tube R_0, and bulging height Δr are known. In this case, the outer radius at the pole of bulging area is known, $R_P = R_0 + \Delta r$.

Since the axial profile of the tube is assumed to be an arc and tangent to the contour of the die, the axial profile of the bulging area in the middle can be expressed as:

$$z^2 + (r - a)^2 = r_z^2 \tag{2.25}$$

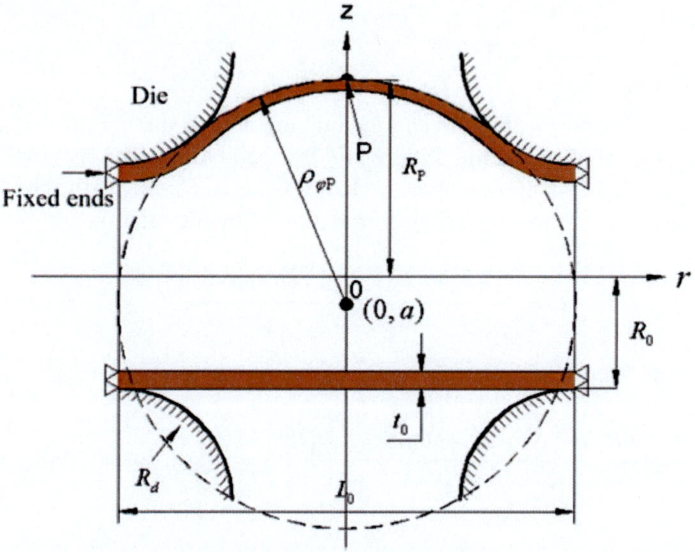

Fig. 2.20 Schematic diagram of the geometry relationship

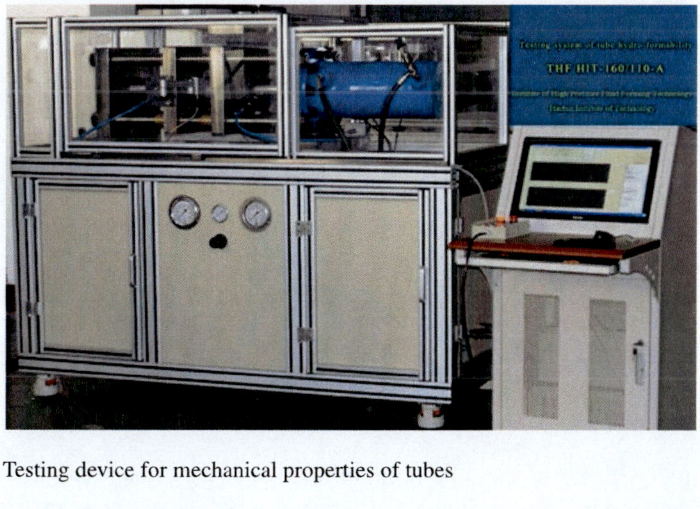

Fig. 2.21 Testing device for mechanical properties of tubes

where

a Vertical coordinates for the center of the profile arc;
r_z Arc radius of axial profile of the tube, $r_z = R_P - a$.

The profile arc of the die can be expressed as:

$$(z - L_0/2)^2 + (r - R_0 - R_d)^2 = R_d^2 \tag{2.26}$$

where

R_d Die radius.

Because the profiles of the die and the bulging area in the middle are tangent to each other, a quadratic equation with regard to r can be obtained by combining the Eqs. (2.25) and (2.26) to eliminate the z. On this basis, according to the condition that two circles are tangent to each other $\Delta = 0$, the following relationship is obtained

$$a = \frac{\Delta r R_d + (\Delta r)^2/2 + R_0 \Delta r - L_0^2/8}{\Delta r} \tag{2.27}$$

Then the arc of the middle bulging area can be expressed as:

$$z^2 + \left(r - \frac{\Delta r R_d + (\Delta r)^2/2 + R_0 \Delta r - L_0^2/8}{\Delta r} \right)^2 = \left(\Delta r/2 - R_d + L_0^2/(8\Delta r) \right)^2 \tag{2.28}$$

According to the deformation theory of plasticity, the relationship between the plastic strain component and the stress deviator is given as follows:

$$\frac{\varepsilon_\theta^p}{\sigma_\theta'} = \frac{\varepsilon_\varphi^p}{\sigma_z'} = \frac{\varepsilon_t^p}{\sigma_t'} \tag{2.29}$$

Based on the incompressibility condition, it can be further calculated that:

$$\frac{\varepsilon_\theta}{\varepsilon_z} = -\frac{\varepsilon_\theta}{\varepsilon_\theta + \varepsilon_t} = \frac{\sigma_\theta'}{\sigma_z'} \tag{2.30}$$

By substituting Eq. (2.23) into the above equation, the ratio of circumferential to axial strain in the left half of Eq. (2.30) can be obtained:

$$\frac{\varepsilon_\theta}{\varepsilon_z} = -\frac{\varepsilon_\theta}{\varepsilon_\theta + \varepsilon_t} = -\frac{\ln \frac{R_P}{R_0}}{\ln \frac{R_P}{R_0} + \ln \frac{t_P}{t_0}} = -\frac{\ln \frac{R_P}{R_0}}{\ln \frac{R_P t_P}{R_0 t_0}} \tag{2.31}$$

Since the thickness of the tube is small relative to the circumferential and axial curvature radius, which can be ignored. Therefore, when the two ends are fixed during the bulging process, the axial and circumferential stress equations in (2.22) can be simplified as:

$$\begin{cases} \sigma_z = \frac{p\rho_{\theta P}^2}{2t_P\rho_{\theta P}} = \frac{p\rho_{\theta P}}{2t_P} = \frac{p(R_0+\Delta r)}{2t_P} & \text{(a)} \\ \sigma_\theta = \frac{p\rho_{\theta P}}{2t_P\rho_{zP}}(2\rho_{zP} - \rho_{\theta P}) & \text{(b)} \end{cases} \tag{2.32}$$

If the ratio of the axial curvature radius to the circumferential curvature radius at the pole is given as α, then

$$\alpha = \frac{\rho_{zP}}{\rho_{\theta P}} = \frac{(\Delta r)^2/2 - \Delta r R_d + L_0^2/8}{\Delta r(R_0 + \Delta r)} \tag{2.33}$$

Then the axial curvature radius at the pole ρ_{zP} can be expressed as

$$\rho_{zP} = \alpha\rho_{\theta P} \tag{2.34}$$

In order to express the circumferential stress as the function of the initial conditions and α of the tube, the Eq. (2.34) was substituted into (2.32b), and

$$\sigma_\theta = \frac{p(R_0 + \Delta r)}{2t_P\alpha}(2\alpha - 1) \tag{2.35}$$

According to Eqs. (2.32) and (2.35), it can be obtained that the ratio of circumferential to axial stress deviators is:

$$\frac{\sigma_\theta'}{\sigma_z'} = \frac{2\sigma_\theta - \sigma_z}{2\sigma_z - \sigma_\theta} = 3\alpha - 2 \tag{2.36}$$

By substituting Eqs. (2.31) and (2.36) into Eqs. (2.30), the expression of the thickness at the pole of the bulging area can be obtained:

$$t_P = t_0\left(1 + \frac{\Delta r}{R_0}\right)^{-1-\frac{1}{3\alpha-2}} \tag{2.37}$$

where

α The ratio of axial to circumferential curvature radius at the pole with the value given in Eq. (2.33).

It can be seen from Eq. (2.37) that the thickness at the pole t_P is only related to the bulging height (radial displacement at the pole) Δr. In this way, it can be calculated just by obtaining Δr the during the hydraulic bulging process.

Based on the above theoretical analysis, a testing device for mechanical properties of tubes was developed, as shown in Fig. 2.21, by the Institute of Fluid pressure Forming Technology of Harbin Institute of Technology. It consists of five parts: main

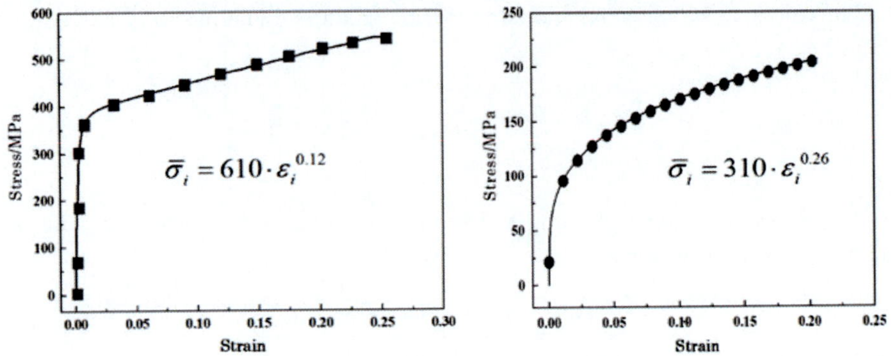

Fig. 2.22 Typical stress–strain curves of tubes. **a** Low carbon steel tube; **b** 6063 aluminum alloy tube

body, hydraulic system, high pressure system, computer control system and calculation system. The main technical parameters of this testing equipment are given as follows: the maximum bulging pressure: 160 MPa; the maximum thrust of horizontal cylinder: 50 ton; measuring range of tube diameter: 20–110 mm; measurable tube length: 120–300 mm.

The testing device for determining mechanical properties of tubes has the following characteristics. Firstly, without cutting the tube, the mechanical properties and formability of the tube can be obtained directly. Secondly, the stress–strain curves and formability of anisotropic materials under complex stress state can be tested. Thirdly, it only needs to test the initial thickness of the tube and the thickness at the pole after bulging, and the results are calculated by the software according to the linear formula. In this case, the error caused by repetitive measurement of thickness and multi-sample method by interrupting the tests is avoided.

The following mechanical property parameters of the tube can be obtained through the hydraulic bulging test: Ultimate bursting pressure, the maximum expansion ratio, true stress–strain curve, engineering stress–strain curve, yield strength, tensile strength and working hardening exponent, *n*. Figure 2.22 shows the typical stress–strain curve of tubes obtained by this bulging test.

2.6 Friction and Lubrication During Hydroforming

Friction plays an important role in the thickness distribution, limit expansion ratio and defects in tube hydroforming. Therefore, the proper lubrication for reducing friction is a key problem in tube hydroforming.

Generally, for low carbon and stainless steel tubes, the friction coefficient ranges from 0.02 to 0.07, while the range is 0.0–0.15 for aluminum. For a specific material, the friction coefficient depends on lubricant, internal pressure, feeding speed, feeding length, and hardness and surface treatment condition of die.

There are three kinds of lubricants commonly used in tube hydroforming:

(1) solid lubricants, MoS_2 and graphite;
(2) lubricating oil and paraffin;
(3) emulsifier and polymer-based lubricant.

The current status of these three lubricants usage in the production is that: solid lubricants and lubricating oil account for about 40% and 30%, respectively, and the rest are emulsifiers and polymer-based lubricants. Most lubricants are sprayed onto the tube surface and soaked for a while expect for lubricating oil. The lubricant coating on the tube surface should be dried and hardened, and a uniform coating thickness of lubricant is required.

Figure 2.23 shows the effect of internal pressure on friction coefficient. The material used was low carbon steel St37-2 tube with a diameter of 70 mm. The feeding speed of material was 50 mm/s. It can be found that, using soap as the lubricant, the internal pressure has no effect on the friction coefficient, which remains constant within 200 MPa. While using MoS_2 as the lubricant, the friction coefficient decreases with the increase of internal pressure. The reason is that, the shear stress required for the sliding of MoS_2 solid lubricant is constant, and the friction coefficient is equal to the shear stress divided by the normal stress. The greater the internal pressure, the greater the normal stress, subsequently the friction coefficient becomes smaller.

Moreover, some specific measures can be also used to reduce the friction coefficient. For example, using a pressurized hydraulic oil film on the tube outside is an effective measure to reduce friction (see Fig. 2.24). Besides, the higher the pressure, the more significant reduction of the friction coefficient. The friction coefficient under the pressure of 50 MPa is about 15% of that without using oil film. In addition, it was also found that the friction coefficient increased significantly with the increase of feeding length, which indicates that the friction coefficient at the later stage of tube hydroforming is significantly higher than that at the early stage.

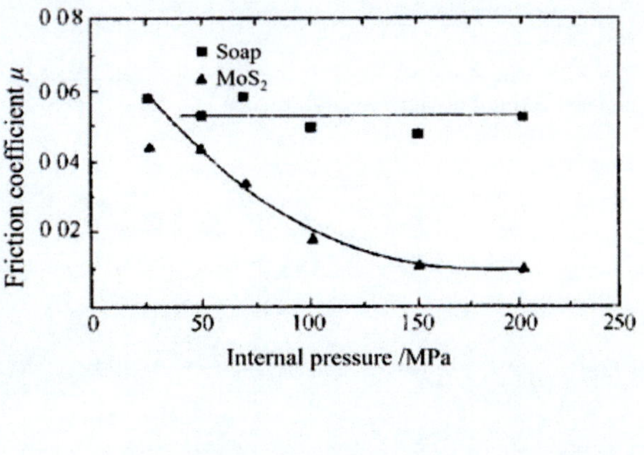

Fig. 2.23 Effect of internal pressure on friction coefficient

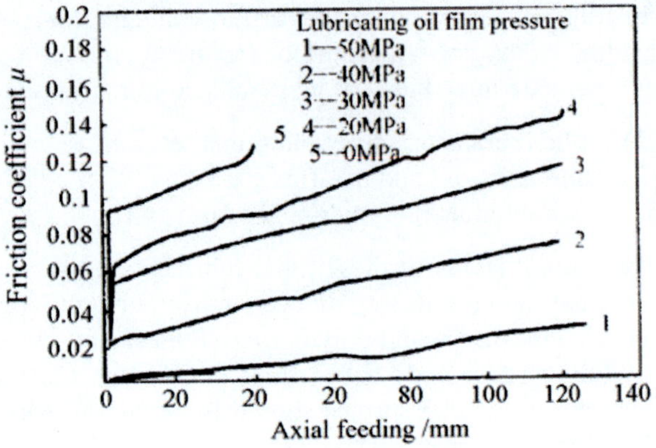

Fig. 2.24 Effect of oil film pressure on friction

2.7 Hydroforming Process of Typical Variable Diameter Tubular Parts

2.7.1 Hydroforming of an Aluminum Alloy Variable Tubular Part

Figure 2.25 shows a variable diameter tubular part of aluminum alloy. The diameter of the forming area (D) is 88 mm. The length of the forming area with maximum diameter (l') is 65 mm, and the length of forming area (l) is 128 mm. The part has a total length (l_1) of 200 mm. the half cone angle in the transition area is $\alpha = 20°$. The initial tube has an outer diameter ($_d$) of 65 mm and a thickness (t) of 1.5 mm. Therefore, the expansion ratio is 35%. This material is 5A02 aluminum alloy with a yield strength of 81.5 MPa and tensile strength of 183.6 MPa, respectively. The elongation is $\delta = 22.4\%$.

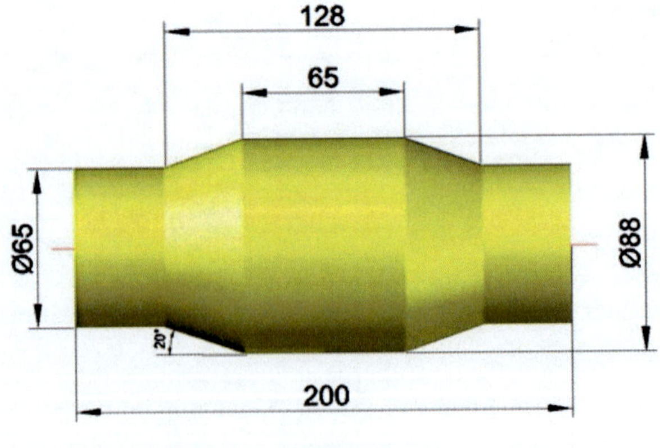

Fig. 2.25 Shape and dimension (mm) of variable diameter tubular part with aluminum alloy

The main process parameters are calculated as follows:

(1) Initial length of tube and the axial feeding length.

According to Eq. (2.10a), the initial length of the tube can be calculated as

$$l_0 = Dl'/d + \frac{D^2 - d^2}{2d \sin\alpha} + (l_1 - l) = 239.1 \text{ mm}$$

According to Eq. (2.10b), the axial feeding length can be calculated as

$$\Delta l = l_0 - l_1 = 239.1 - 200 = 39.1 \text{ mm}$$

(2) Initial yielding pressure.
According to Eq. (2.3), the initial yielding pressure is calculated as

$$p_s = \frac{2t}{d}\sigma_s = \frac{2 \times 1.5}{65} \times 81.5 = 3.8 \text{ MPa}$$

(3) Calibration pressure.
Considering the work hardening of material during hydroforming, the calibration pressure can be estimated according to Eq. (2.5).

$$p_c = \frac{t}{r_c}\sigma_s = \frac{1.5}{5} \times \frac{\sigma_{0.2} + \sigma_b}{2} = 39.8 \text{ MPa}$$

(4) Die clamping force.
According to Eq. (2.8), the die clamping force is calculated, in which the projected area of the part is 12817 mm^2.

$$F_c = A_p p_c \times 10^{-3} = 12817 \times 39.8 \times 10^{-3} = 510 \text{ kN}$$

(5) Assuming the friction coefficient is 0.1, the axial force is calculated according to Eq. (2.6). The length of the feeding area is

$$l_\mu = \frac{l_1 - l}{2} + \frac{\Delta l}{2} = \frac{200 - 128}{2} + \frac{39.1}{2} = 55.55 \text{ mm}$$

Then

$$F_p = \pi \frac{d_i^2}{4} p_i = 3.14 \times \frac{62^2}{4} \times 39.8 = 120 \text{ kN}$$

$$F_\mu = \pi d l_\mu p_i \mu = 3.14 \times 65 \times 55.55 \times 39.8 \times 0.1 = 45.1 \text{ kN}$$

$$F_t = \pi t d \sigma_s = 3.14 \times 1.5 \times 65 \times 81.5 = 12.6 \text{ kN}$$

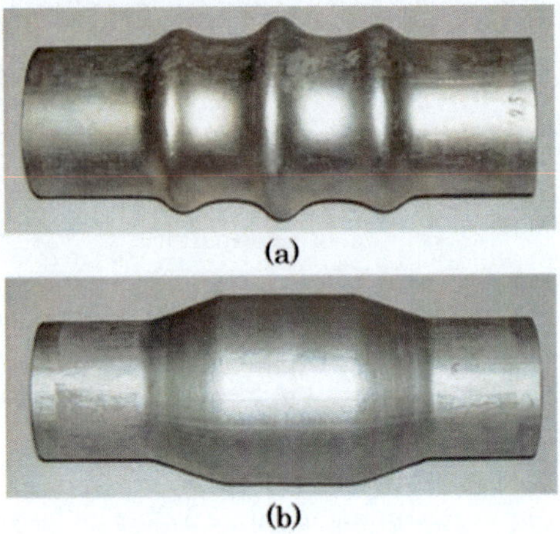

Fig. 2.26 Tube hydroforming process by the useful wrinkles. **a** Shape of the useful wrinkles; **b** variable diameter tubular part

Therefore, the axial force can be obtained as follows

$$F_a = (120 + 45.1 + 12.6) = 190.3 \text{ kN}$$

If the equation of $F_a = (1.2 - 1.5)F_p$ was used to estimate the axial force, it is obtained that $F_a = (144 - 180)$kN. It can be seen that Eq. (2.7) is relatively simple for an estimation.

During hydroforming, the "useful wrinkles" are firstly obtained by feeding materials under a certain internal pressure (see Fig. 2.26a) to complete the target feeding length. Then the final part can be obtained by increasing the internal pressure for calibration (see Fig. 2.26b). Due to the sufficient feeding of materials, thickness reduction is effectively controlled, and the maximum thinning is less than 10%. In addition, Fig. 2.27 shows the forming defects during hydroforming process.

2.7.2 *Hydroforming of a Low Carbon Steel Tubular Part*

Figure 2.28 shows a bottle-shaped part. The outer diameter and thickness of the initial tube is 40 mm and 2.5 mm, respectively. The part is made of 20 steel, and the diameter of forming area is 183 mm. The difficulty for the hydroforming of this part is listed as bellow:

(1) The left and right sides of this part are asymmetrical, which require different feeding length at the ends tube;
(2) The length of forming area is too long, which is 4.6 times of the tube diameter, and the feeding length needed is up to 80 mm. The length of the initial tube is 403 mm, which is nearly 10 times of the outer diameter of the tube resulting in the easy occurrence of bulking;

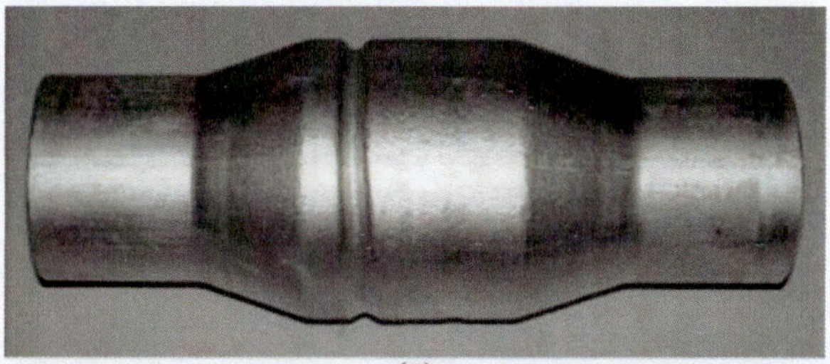

(a)

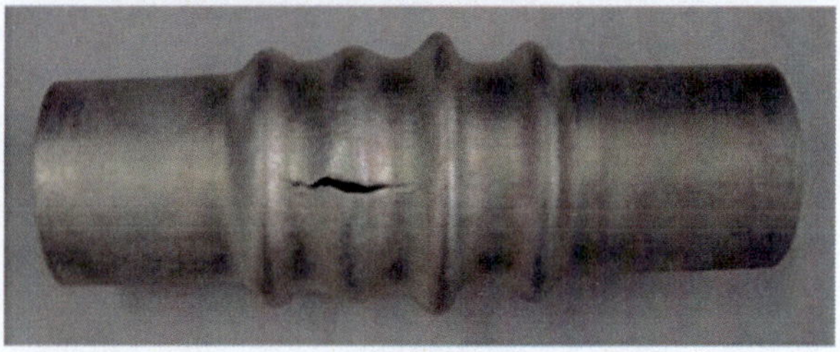

(b)

Fig. 2.27 Forming defects. **a** Dead wrinkle; b splitting of wrinkles

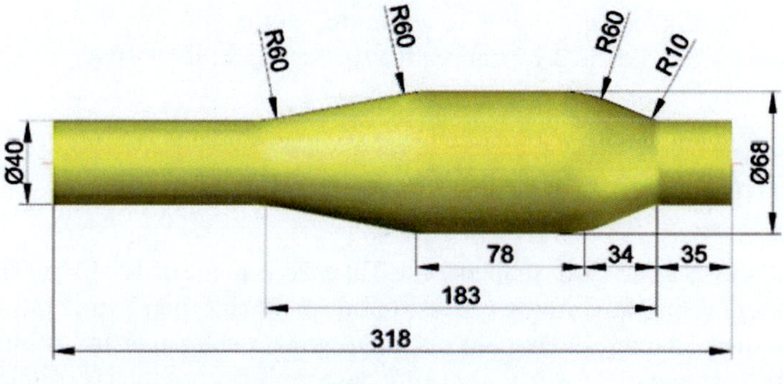

Fig. 2.28 Shape and dimension (mm) of the bottle-shaped part

(3) Large expansion ratio. The expansion ratio is 70%, while the elongation of material is only 28%.

Because the material of this bottle-shaped part is low carbon steel, and the length of forming area is long, a suitable preform must be used to obtain a large expansion ratio. Therefore, the "useful wrinkles" are also used as a preform to accumulate materials in the forming area, which is shown in Fig. 2.29a. Then the bottle-shaped part can be obtained by flattening the wrinkles under higher internal pressure in the calibration stage, as shown in Fig. 2.29b. Regarding the thickness distribution, thickening happens in the feeding area at both ends while thinning occurs in the middle forming area. The longer the length of feeding area, the severer the thickening.

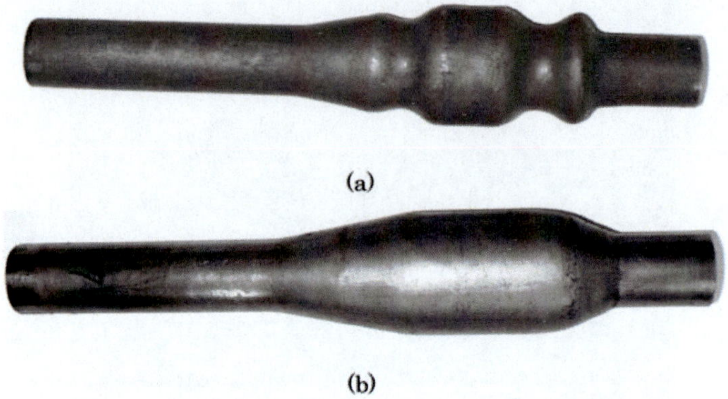

Fig. 2.29 Bottle-shaped part. **a** Shape of the useful wrinkles; **b** bottle-shaped part

For this part, the thickening of the left feeding area is greater than that of the right feeding area. In the forming area, the thickness distribution fluctuates along the axial direction, which corresponds to the location of wave top and bottom in the forming process. The thickness is slightly thinner at the wave top and slightly thicker at the wave bottom, which is also a typical feature of the thickness distribution of parts formed by "useful wrinkles". For this bottle-shaped part, the thickness at the thinnest position is 1.92 mm, corresponding to a thinning ratio of 25.2%, and the maximum thickness on the left side is 2.75 mm with a thickening ratio of 10%.

2.7.3 Hydroforming of a Ω-shaped Joint Part

Figure 2.30 shows a part of Ω-shaped joint. This part is made of 1Cr18Ni9Ti stainless steel. In the conventional forming process of this joint, half rings were firstly stamped using rigid die with edges being cut off, then the two half rings are assembled and welded into a complete circle. The conventional forming process has certain problems summarized as follows:

(1) The width of convex rib on the half ring formed by die press bending is not consistent, so the accuracy is difficult to be guaranteed;
(2) Welding induce serious distortion of the half rings, and the butt surface are warped, which results in the out of dimension specification. The two ends could not be further machined;
(3) The two half rings are welded by argon tungsten-arc welding manually, which has a large amount of welding leakage and easy to damage the part resulting in waste products. If the cleaning is not well, the roll welding process of steel wire sheath is insufficient, and the reliability is reduced;
(4) The overlapped part structure makes it impossible to carry out X-ray inspection of the whole part, which is easy to miss of welding defects.

Fig. 2.30 Ω-shaped joint part

Due to the large diameter and thin thickness, seamless tube is not available. Therefore, laser welding is used to fabricate tubes. As the laser welding technology has advantages of high weld elongation, small heat affected zone, narrow weld pass and good surface quality, small cleaning work after welding, and good repeatability.

The problems in the hydroforming of Ω-shaped joint are listed as bellow: (1) Ultra-thin parts with a large diameter. The diameter-thickness ratio reaches 107, which make the sealing of tube ends become more difficult; (2) The inner corner is small, which is only 3.3 times of the sheet thickness, a higher forming pressure is needed; (3) Large axial reaction force. When the forming pressure is 150 MPa, the axial reaction force is close to 3000 kN, which exceeds the capacity of the existing hydraulic servo cylinder; (4) There are no transition area, or the transition area are just the vertical edges. Moreover, it is required that the minimum thickness at the position of Φ190mm convex is larger than 1.3 mm. These difficulties are combined to make the forming of parts. In order to form small corners, high forming pressure is needed. Subsequently, the axial reaction force is large, and the sealing of tube ends is difficult. Furthermore, to meet the requirement of minimum thickness and forming corners under high pressure, certain axial feeding is needed. However, excessive feeding or improper process parameter control of internal pressure will produce inward concave defect, as shown in Fig. 2.31.

For this successfully formed Ω-shaped part, the thickness thinning is larger in the middle expansion area, and the thickness in the feeding area at both ends is basically unchanged. The minimum thickness is 1.37 mm.

2.7.4 Hydroforming of an Irregular Tubular Part with Double Cones

The irregular tubular part with double cones is manufactured using the material of 1Cr18Ni9Ti with a uniform elongation of 45%. The geometrical shape is shown

Fig. 2.31 Defect of inward concave

in Fig. 2.32. The diameter and thickness of the initial tube are 56 and 2.5 mm, respectively. There are two difficulties in the hydroforming of this part:

(1) The maximum expansion ratio is 111%, about three times of the uniform elongation of the material;
(2) Shapes are asymmetry in up-down and left–right directions. This asymmetry feature makes the local plastic deformation large and easy to cause splitting. Therefore, a reasonable preform is required.

Table 2.3 shows the expansion ratio of each cross-section of this tubular part with double cones. The schematic of each cross-section position is shown in Fig. 2.33, in which the cross-sections are selected every 10 mm from the center. It can be seen from this table that, at cross-sections from A5 to A7, the expansion ratio in the lower half is much higher than that in the upper half, which indicates that the deformation is not uniform. At cross-sections from A1 to A4, the expansion ratio in the lower half is more than 2 times of the uniform elongation of the material, which is prone to experience splitting in the lower half.

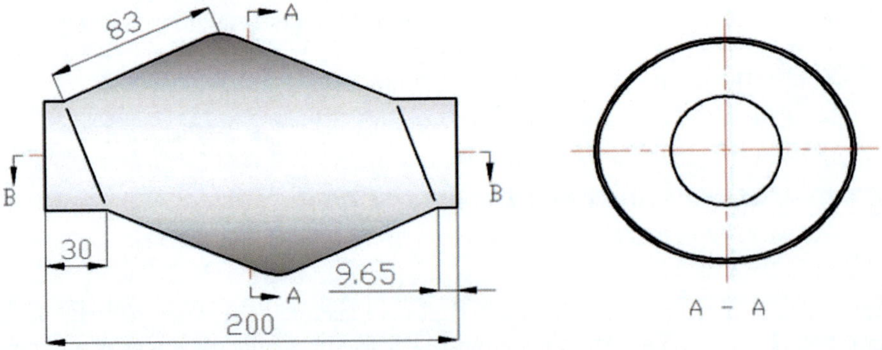

Fig. 2.32 Shape and dimension (mm) of an irregular tubular part with double cones

Table 2.3 The expansion ratio of each cross-section of the tubular part with double cones

Cross-sections	A1	A2	A3	A4	A5	A6	A7
Expansion ratio /%	111.47	111.56	101.57	84.75	67.92	51.11	34.32
Expansion ratio of the upper half /%	111.47	102.99	89.13	72.80	56.45	40.12	23.79
Expansion ratio of the lower half /%	111.47	120.13	114.01	96.70	79.39	62.10	44.84
Ratio of the upper half to the lower half	1:1	1:1.17	1:1.28	1:1.33	1:1.41	1:1.55	1:1.89

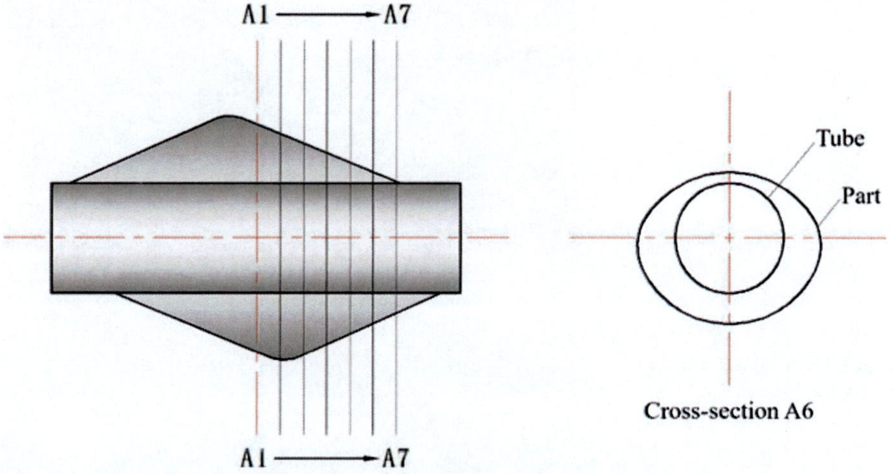

Fig. 2.33 Schematic diagram of cross-sections of the irregular tubular part with double cones

Figure 2.34a shows a symmetrical biconical preform tube blank obtained by the "useful wrinkles" method. The maximum expansion ratio of this preform is 78%. If this preformed tube is used to form the final part, the non-uniform deformation at cross-sections A5 to A7 will lead to splitting even though the average circumferential deformation has been reduced and annealing treatment has been carried out, as shown in Fig. 2.34b. In order to obtain qualified parts, it is necessary to preform the tube twice on the basis of such a obtaining double cones to overcome the difficulty of non-uniform plastic deformation. After using a reasonable secondary preformed tube blank, qualified parts can be obtained, as shown in Fig. 2.35.

2.7.5 *Hydroforming of a Long Waved Pipe*

The long waved pipe is a typical large variable diameter tubular part. The material is 5000 aluminum alloy. Because of the long distance between each corrugation,

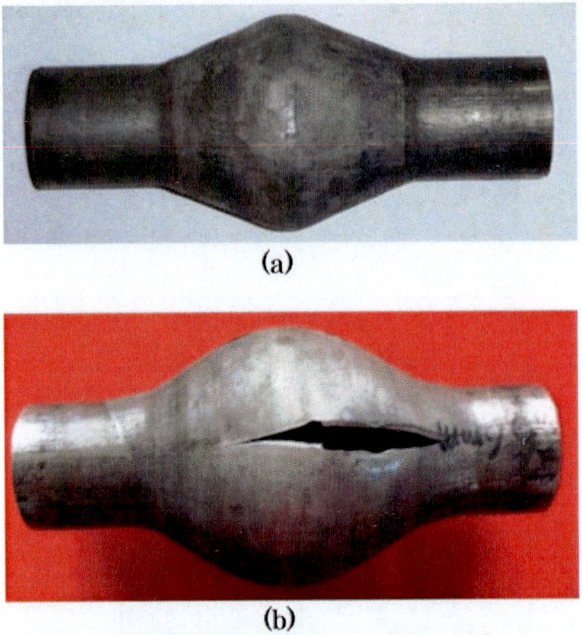

Fig. 2.34 Primary preforming tube blank and splitting defect. **a** Primary preforming tube blank; **b** splitting

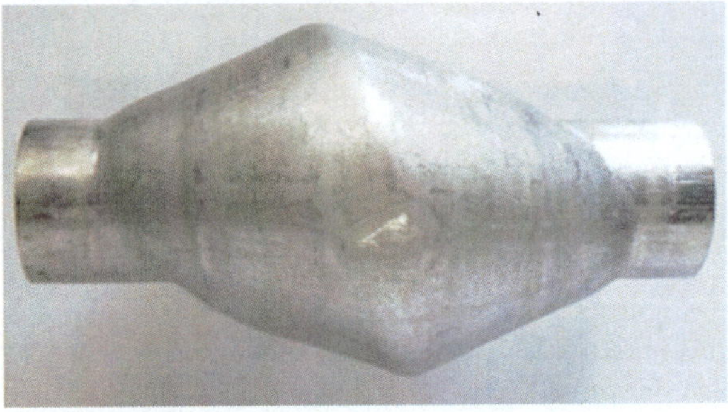

Fig. 2.35 Irregular tubular part with double cones

generally greater than 100 mm, the existing technology for making bellows is not capable of manufacturing such long continuous waved part.

The difficulties of forming this part are summarized as follows. Firstly, the diameter is large, which results in axial force up to 8000 kN; Secondly, the displacement precision during axial feeding is difficult to control, which is easy to cause wrinkling or splitting; Thirdly, when the long part with more than 10 waves is hydroformed integrally, the middle zone cannot obtain sufficient material feed, which makes the thinning ratio too big to meet the design requirements. For this part, if all the waves are formed at the same time, the maximum thinning ratio is as high as 24% because the central wave cannot be fed with sufficient material, which does not meet the design requirement of 15% thinning ratio. Moreover, due to the poor ductility of

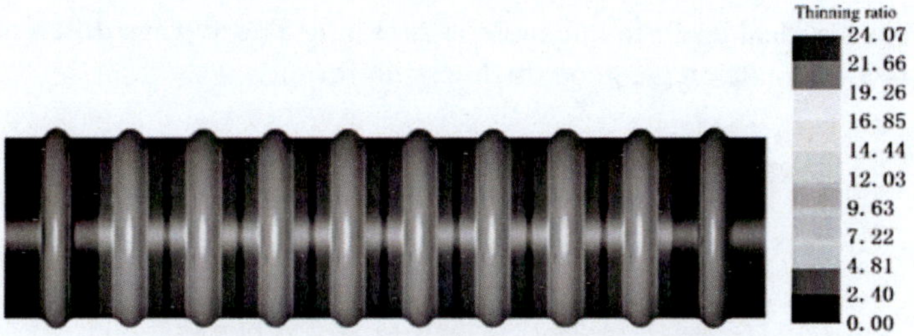

(a)

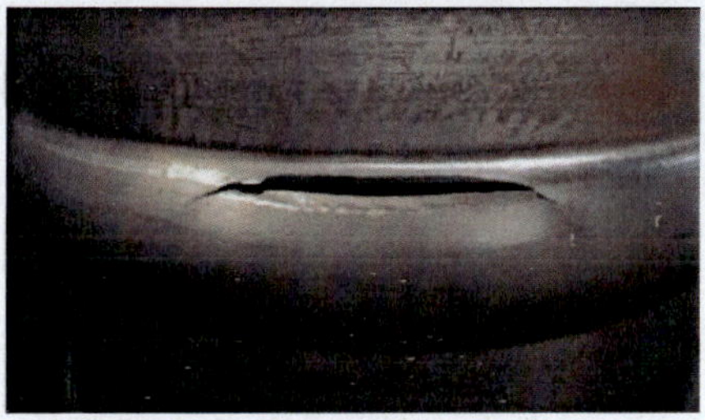

(b)

Fig. 2.36 Thickness variation and splitting defect in the integrate forming of a long waved tubular part. **a** Thinning ratio distribution; **b** splitting defect

5A03 aluminum alloy, serious thinning is also prone to cause splitting, as shown in Fig. 2.36.

To solve the problem of integrate forming, a local forming method with self-feeding material was invented. In this method, by the axial tension generated by the internal pressure in the corrugated area, the tube material is able to be pulled into the corrugated area to realize the self-feeding and improve the thickness uniformity. Because the internal pressure is only applied to the local bulging area for each corrugated area, the required closing force is only 300 tons, which is 1/10 of that of the integrate forming. In addition, since the circumferential hydraulic counter force only acts on the narrow annular region, the axial thrust is only 30 tons, which is reduced to 1/20 of that of integrate forming.

Figure 2.37 shows the thickness comparisons between sound part formed by the local forming method with self-feeding and the integrate forming tubular part. It can be seen that the thinning of the self-feeding locally formed wave in corrugated area is much smaller than that of the integral formed part. After hydroforming, the total length of the tubular part is 107 mm shorter than the length of the initial tube, and the average axial feeding length of each ripple is 10.7 mm. The feeding length of each ripple is slightly different, which results in slightly different thinning ratio. However, all the thinning ratios are less than 10% satisfying the design requirements. Moreover, the minimum thickness appears at the top of the ripple after forming. The minimum thickness of the whole part is 2.83 mm, which corresponds to a maximum thinning ratio of 9.9%. Under the axial feeding length of 10.7 mm, the wave deformation is uniform, and the ideal uniform thickness is 2.93 mm. The average difference of the thickness is 3.4% indicating a good thickness uniformity.

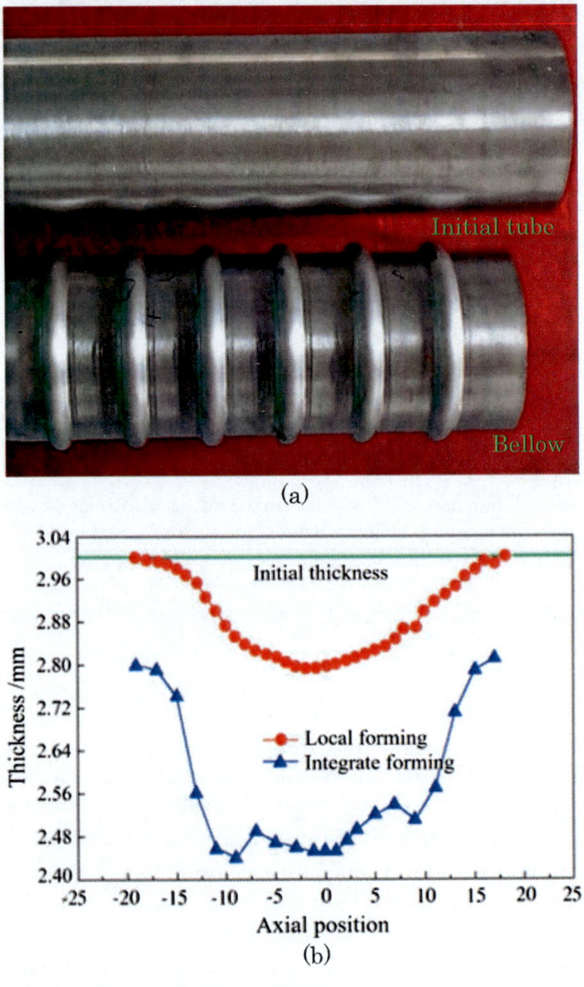

Fig. 2.37 The long bellows pipe and the thickness changes in corrugated area. **a** The long bellows pipe; **b** thickness distribution in corrugated area

Chapter 3
Hydroforming of Curved Parts with Irregular Cross-Sections

Check for updates

3.1 Process and Typical Cross-Sections

3.1.1 Process Description

The main process of hydroforming of curved parts with irregular cross-sections includes bending, preforming and hydroforming, as shown in Fig. 3.1. Hydropiercing is combined in some cases, during which, the holes can be punched directly under the supporting pressure after hydroforming.

Since the axis of the part is a two or three-dimensional curve, the bending process needs to be firstly conducted to bend the tube into the same or similar shape compared to the axis of the part. Compared to parts manufactured using conventional bending processes, besides the requirements of shape and dimensions of bending axis, the control of thickness reduction during bending is more important, which is the premise of a successful hydroforming process.

After bending, if the cross-section of the part is simple or the tube diameter, d, is smaller than the minimum width of die cavity, w, the bent tube can be directly subjected to hydroforming, as shown in Fig. 3.2. However, if the cross-section of the part is complex or the tube diameter is greater than the minimum width of the die cavity, the tube cannot be mounted into the die cavity, the preforming of tube is needed to obtain appropriate cross-section shapes, as shown in Fig. 3.3. The preforming has three main functions: first, regarding the process with a dimension condition of $d > w$, the preforming flattens the tube so that the tube can be placed into the hydroforming die smoothly, and the flash defect during the die clamping is avoided; The second is to obtain a reasonable material distribution, which enables a uniform deformation of the parts to be achieved, as well as the avoidance wrinkles and splitting; The third is to reduce the calibration pressure for the transition corner and control the thickness through reasonable preforming shape. Subsequently, the die clamping force of the press is reduced, which contributes to saving the die cost and improving the production efficiency.

S. Yuan, *Modern Hydroforming Technology*,
https://doi.org/10.1007/978-981-19-5775-8_3

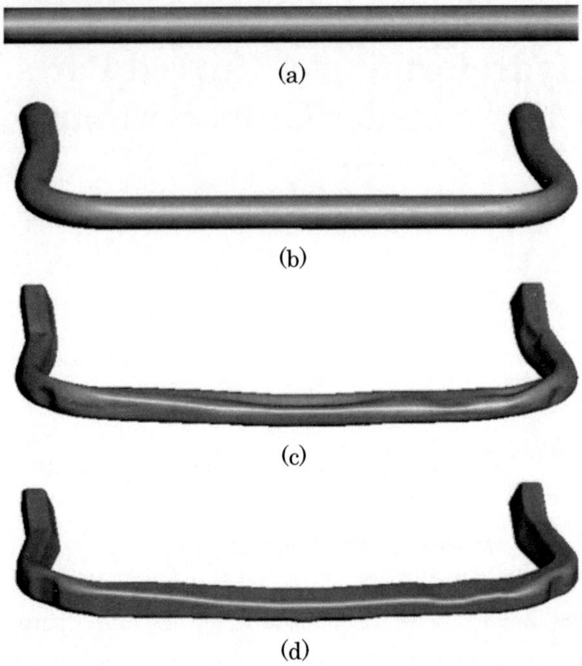

Fig. 3.1 The main process of hydroforming of curved parts. **a** Initial tube; **b** bending; **c** preforming; **d** hydroforming

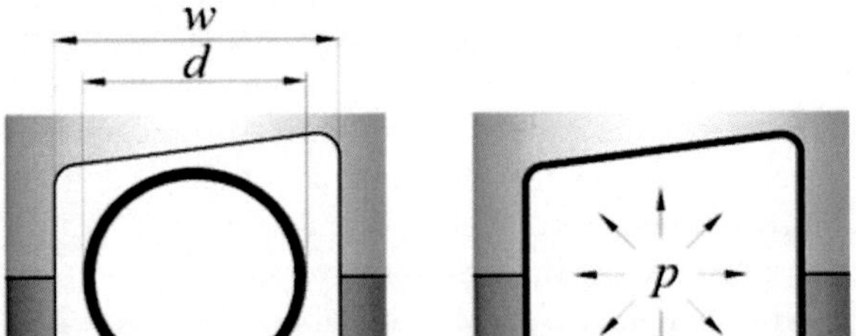

Fig. 3.2 Direct tube hydroforming. **a** Initial stage; **b** hydroforming

During hydroforming, the preformed tube is placed into hydroforming die cavity. First, the tube is filled with emulsion using a quick filling system, and then a high-pressure liquid is given through a punch at one end of tube. The inside pressure is increased following a specific loading path, by which, the tube or the preformed tube is deformed to attach the die cavity to form the parts with desired shapes.

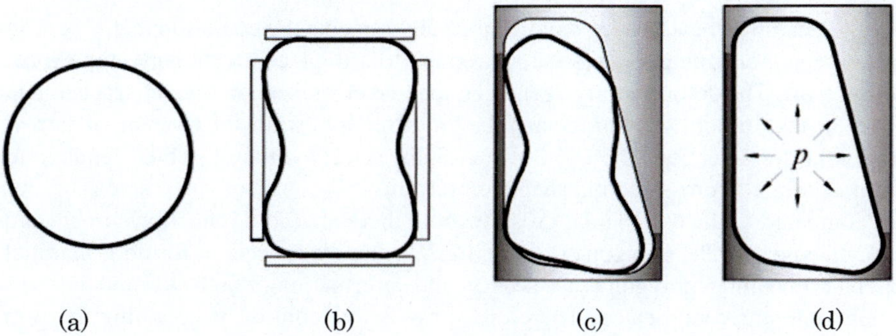

Fig. 3.3 Shape changing of cross-section. **a** Tube; **b** preforming; **c** die closing; **d** hydroforming

3.1.2 Typical Cross-Sections

The typical cross-section shapes of hydroformed parts are mainly quadrilateral, polygon, ellipse and irregular geometries, as shown in Fig. 3.4. More specifically, quadrilateral cross-sections include square, rectangle and trapezoid shapes. Polygon cross-sections include regular pentagon, regular hexagon and other shapes of the pentagon and hexagon. Ellipse cross-sections are usually different at the ratios of the short and long axis length, or long ellipses. The irregular cross-sections refer to complex shapes those are not included in above cases, and mainly used as the transition between two regular cross-sections.

Among these cross-sections, the most common and representative cross-sections are rectangular and square. To define a rectangular cross-section, the dimensions include the length a, the width b and the transition corner radius r. When the length a is located either horizontally or vertically, the corresponding requirements of preforming are different. For the rectangular cross-section with the length a in the horizontal direction, normally it is greater than the tube diameter, $a > d$ (outer diameter of tube). Then, preforming is not required. For the rectangular cross-section

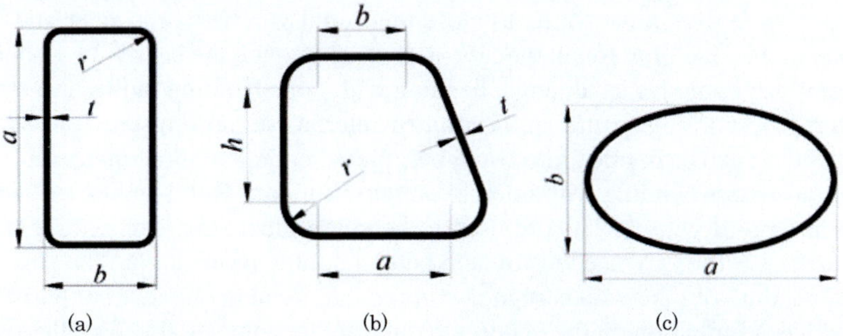

Fig. 3.4 Typical shapes of cross-sections. **a** Quadrilateral cross-sections; **b** trapezoid cross-sections; **c** elliptical cross-sections

with the width b located horizontally, normally it meets the condition that $b \leq d$. In this case, a preforming process is required in order to place the tube into the hydroforming die. The deformation of general trapezoid cross-section (the left trapezoid in Fig. 3.4b) is similar to that of rectangular cross-section, while the forming of narrow trapezoid cross-section (the right in Fig. 3.4b) is very difficult, which requires to form a particular cross-section shape of preform.

Compared to the rectangular cross-section, there are more transition corners and straight edges for the polygon cross-section. During the process of forming an initial circular tube into a polygon cross-section, the deformation of both side and the transition zone are complicated. Therefore, an effective control of obtaining uniform and cooperative deformation of tube edges is the key to achieve the successful hydroforming of polygon cross-sections.

Regarding the irregular cross-sections, the cross-section shape is normally very complex and not regular geometrical shape. During forming, the non-uniform deformation often induces splitting due to excessive localized thinning, or wrinkling due to the local folding of material. Therefore, how to control the deformation uniformity of cross-section is the key to realize the hydroforming of irregular cross-sections. For the forming of such kind of cross-section, preforming is normally needed to make reasonable transition shapes.

3.2 Tube Bending Processes

3.2.1 Common Bending Processes and Characteristics

The commonly used bending processes for tube hydroforming include press bending, roll bending and CNC rotary-draw bending. For the parts with large bending radius and simply shaped two-dimensional axis, the press bending and roll bending processes are usually used. While the CNC bending process is selected in terms of parts with complex three-dimensional axis.

Press bending is a process in which the tube is firstly placed on the lower die, then the upper die moves downwards to close the initial gap between upper and lower die. Due to the clamping force, the tube is pressed into the die cavity. Press bending is generally suitable for small angle bending with large bending radius and variable curvatures. According to the application of internal supporting pressure support, press bending can be divided into two types: press bending without internal pressure support and press bending with internal pressure support. Roll bending is a process in which three driving rollers are used to bend the tubes. Generally, three or four basic driving rollers are used to roll and bend the tube. By changing the spacing of rollers, bending of tube with certain curvature radius can be made. The roll bending method has a limitation on the bending radius, and is only suitable for thick-walled tubular parts with large curvature radius and simple axis shape. One of the main disadvantages of these two processes is that the cross-section shape after bending becomes ellipse, which is easy to cause splitting within the thinning zone, i.e. outside the bending, in the subsequent hydroforming process.

CNC bending is an advanced rotary-draw bending process. In CNC bending, the shape of the tube axis is firstly input into the numerical control system of the bending machine. Then, the bending machine is controlled by the numerical control program to achieve automatic bending of the tubes by rotating the tube around the die. Furthermore, CNC bending can be also used to bent tubes with complex three-dimensional axis. The tube can be bent continuously with different angles. In addition, such bending process have the advantages of good product quality and high production efficiency.

The tooling of CNC bending is generally composed of machine bed, clamping die, pressing die, wrinkle-resistance die and mandrel, as schematically shown in Fig. 3.5. The bending machine shown in Fig. 3.6 mainly includes machine tool bed, hydraulic oil tank, machine tool spindle, movable platform and control system. The bending die is connected with the bending machine spindle by a connecting mechanism. During tube bending, one end of the tube is clamped firmly on the bending die by clamping die. The supporting die is mounted at the outside of the tangential point between the tube and the bending die, and the wrinkle-resistance die is mounted inside. A mandrel is inserted inside the tube. The bending die rotates around the machine spindle, and the tube is gradually bent around the bending die. By clamping the back-end of the tube through the clamping device, the tube can move forward along the axial direction or rotate along circumferential direction, enabling to be loaded axially exert the axial force and change the spatial angle. Finally, the bending of tube in the three-dimensional space can be achieved.

CNC bending machine uses hydraulic servo control, which can realize space bending. In addition, continuous and multi-curvature radius bending can also be realized using multi-layered dies. The main parameters of CNC bending machine include: maximum bending radius, maximum thickness, bending radius, bending angle and bending speed. Among these parameters, the performance of the bending machine is determined by the maximum bending radius and maximum thickness, while the bending speed and axial feeding speed determine the production efficiency.

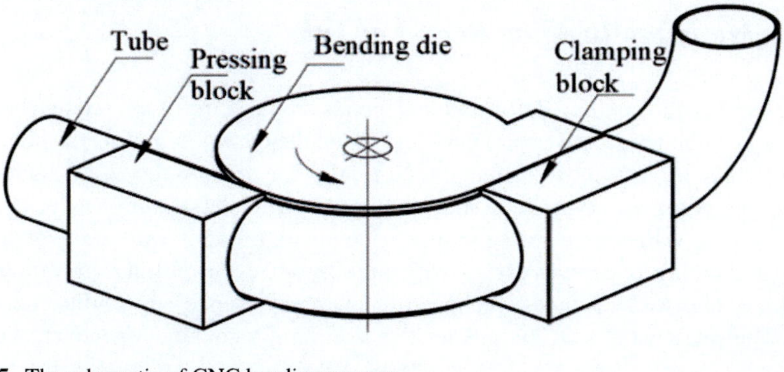

Fig. 3.5 The schematic of CNC bending process

Fig. 3.6 CNC tube bending machine

Table 3.1 Main technical parameters of tube bending machine

Technical parameters	Model 1	Model 2	Model 3	Model 4
Maximum diameter/mm	42	76	101	150
Minimum diameter/mm	10	20	25	50
Maximum thickness/mm	2	2	2.5	3
Maximum radius (center line)/mm	200	250	250	400
Minimum radius (center line)/mm	25	25	35	75
Maximum bending angle/(°)	193	193	193	193
Maximum length of tube/mm	3000	3000	3000	3000

Table 3.1 shows the main technical specifications of several types of tube bending machines used for automotive structural parts.

3.2.2 *Minimum Bending Radius of Tube*

The degree of tube bending deformation depends on the value of the relative bending radius R_b/d and the relative thickness t/d (R_b is the curvature radius of the neutral layer of the tube), as shown in Fig. 3.7. Normally, the smaller the R_b/d, the greater the bending deformation degree. The smaller the value of t/d, the smaller the relative thickness. When the bending radius is too small, the tube wall at the outside of curved neutral layer becomes excessively thinning and even splitting. In comparison, the inside tube wall becomes thickened or even experiences instability, i.e. wrinkling. The smaller the t/d, the greater the wrinkling tendency. Moreover, with the

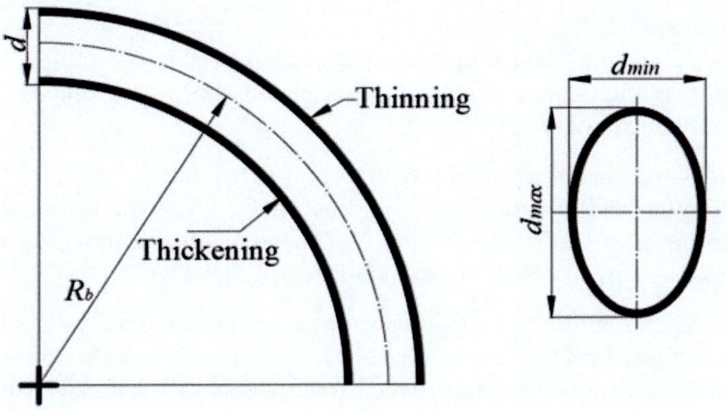

Fig. 3.7 Shape and thickness variation during tube bending process

Table 3.2 The minimum bending radius of the tube (d is tube diameter)

Bending method	Minimum bending radius (center line)
Press bending	$3d$–$5d$
Roll bending	$6d$
Rotary-draw bending	$1.25d$–$2d$

decrease of R_b/d, the deformation degree increases and the cross-section distortion (e.g. ellipticity) becomes severer.

To ensure the bending quality, the relative bending radius must be designed within a certain range. Table 3.2 gives the minimum relative bending radius for different bending processes. The actual minimum relative bending radius depends not only on the bending process, mandrel and equipment, but also on the mechanical properties of the material. Moreover, the actual thinning ratio and ellipticity of tubular part need to be determined through experimentation.

3.2.3 Cross-Section Distortion and Prevention Approaches

During tube bending process, cross-section distortion occurs, which makes the circular cross-section become elliptic as shown in Fig. 3.7. In severe conditions, the cross-section collapses. The distortion degree can be evaluated by the ellipticity, given as follows:

$$\eta = \frac{d_{\max} - d_{\min}}{d} \times 100\% \tag{3.1}$$

where

η Ellipticity (%);
d_{max} Long axis length of the elliptical cross-section of bent tube (mm);
d_{min} Short axis length of the elliptical cross-section of bent tube (mm);
d Outer diameter of tube (mm).

If the cross-section distortion of tubular part after bending is too large, the subsequent preforming and hydroforming could be affected, and the splitting is easily occurred on the outside of bent tube. For CNC bending, the common approaches to prevent cross-section distortion are summarized as follows:

1. A mandrel can be used to support the cross-section to prevent distortion. The commonly used mandrels are ball-headed mandrel, cone mandrel, spoon mandrel or multi-head mandrel. In mass production, utilization of mandrel with reasonable structure is the main approach to prevent cross-section distortion and inside wrinkling.
2. The tube inside can be filled with granular medium, fluid medium, elastic medium or low melting point alloy to prevent cross-section distortion. This method is easy and widely used in the small volume production.

3.2.4 Calculation of the Bending Moment

Bending moment is an important technical parameter to characterize equipment capability and select equipment. The bending moment of the tube not only depends on the mechanical properties of tube material, tube diameter and the bending radius, but also has a correlation with the bending method and tooling structure. In practical production, the following formula can be used to estimate the bending moment.

$$M = k_w w \sigma_b \sqrt[3]{\frac{d}{R_b}} \times 10^{-3} \quad (3.2)$$

where

M Bending moment (N · m);
d Tube diameter (mm);
σ_b Tensile strength of materials (MPa);
w Bending resistance modulus (mm^3), For the circular tube, $w = \frac{\pi(d^4 - d_i^4)}{32D}$ (d_i is the inner diameter of tube);
R_b Bending radius (mm);
k_w Empirical coefficient relating to mandrel and lubrication. When the rigid mandrel without lubrication is used, $k_w = 5$–8 is recommended; When a rigid hinged mandrel is used, $k_w = 3$ is recommended.

3.2.5 Thickness Variation

At the outside of the bending neutral layer, the thickness is reduced due to the tangential tensile stress in the tangential direction, while for the inside of the neutral layer, the thickness is increased due to the tangential compressive stress. The thickness reduction at the outside of bent tube has a great influence on the hydroforming process and the loading-bearing capacity of the parts in service. If the outside is excessively thinned, even if no splitting occurs during bending, it is prone to experience splitting in subsequent hydroforming, which increases the difficulty of hydroforming process. Regarding the bending parts used for hydroforming, in addition to ensuring the shape accuracy of the axis, the thinning ratio of thickness should be controlled within a certain range. Generally, the maximum thinning ratio should be lower than 20%. The thinning ratio is mainly related to the relative bending radius. Moreover, the mechanical properties of material and the bending process (e.g. mandrel type) also affect the thinning ratio. For the rotary-draw bending, the thinning ratio can be estimated by the following equation.

$$\delta_t = \frac{1}{1 + 2R'} \times 100\% \tag{3.3}$$

Regarding bending tubes with same relative bending radius, if the mechanical properties of materials are different, the thickness distributions also vary correspondingly. For example, the greater the strain hardening exponent n and the normal anisotropy coefficient r, the less the tendency of thinning. In addition, the thickness distribution after bending also varies with the type, diameter and position of the mandrel. The lubrication also has an effect on the thickness distribution. With the improvement of lubrication conditions, it is beneficial to the flow of materials. Subsequently, the deformation uniformity increases, and the thickness distribution tends to be more uniform.

3.2.6 Ultimate Diameter-Thickness Ratio of Tube Bending

In addition to the minimum bending radius, the diameter-thickness ratio of tube, is another important index to evaluate the difficulty of tube bending. As the diameter-thickness ratio of tube increases, the relative thickness of tube becomes thinner. Therefore, there is a serious wrinkling tendency during bending, which gradually increases the difficulty of tube bending. The ultimate diameter-thickness ratio refers to the maximum diameter-thickness ratio without wrinkling under given relative bending radius and bending process. At present, CNC bending process is commonly used, and the ultimate diameter-thickness ratios for different materials are summarized in Table 3.3. The main factors that determine the ultimate diameter-thickness ratio are the material and the bending radius. For example, for the same relative

Table 3.3 Ultimate diameter-thickness ratios of CNC tube bending

Materials	Relative bending radius	Specification $d \times t$ (mm)	Diameter-thickness ratio d/t	bending angle (°)
Stainless steel 321	1.7	63.5 × 0.5	127	90
Stainless steel AM350	1.0	38 × 0.71	53.5	180
Aluminum alloy 6061-O	1.7	127 × 1.65	77	90
Aluminum alloy 6061-T6	1.4	50.8 × 0.71	71.5	90
Nickel-based superalloy tube	1.0	38 × 0.457	83.2	90
Nickel-based superalloy tube	1.5	89 × 0.71	125.3	45
Pure titanium AMS 4902	1.5	101.6 × 0.89	114	90

bending radius (1.7), the ultimate diameter-thickness ratio of stainless steel tube is 127, while that of 6061-O aluminum alloy tube is only 77. The results show that under the same bending radius, the higher the elastic modulus and strength of the material, the higher the ultimate diameter-thickness ratio. For the same materials (Ni-based superalloy tubes), the smaller the relative bending radius, the smaller the ultimate diameter-thickness ratio.

3.3 Hydro-Bending Process of Tubes

3.3.1 Principle and Characteristics of Hydro-Bending

Tube hydro-bending is an integrated process of tube press bending and hydroforming. This process has its own particularity in addition to the general characteristics of bending. The forming principle is shown in Fig. 3.8. During the hydro-bending process, the liquid is firstly filled into the tube from one end, and the internal pressure is established within the tube as a flexible supporting. Then, the bending of tube is performed in the forming dies, and the internal pressure is kept at a certain magnitude along with the die pressing process, to avoid wrinkling and cross-section distortion. After the die gap is closed, the internal pressure can be increased for the calibration, so that the tube is able to be deformed to completely attach to the die cavity.

The tube hydro-bending can be divided into two types, that are tube hydro-bending without end constraint and tube hydro-bending with end constraint, respectively. For the forming process with constraints at the tube ends, pressing plates are used to fix

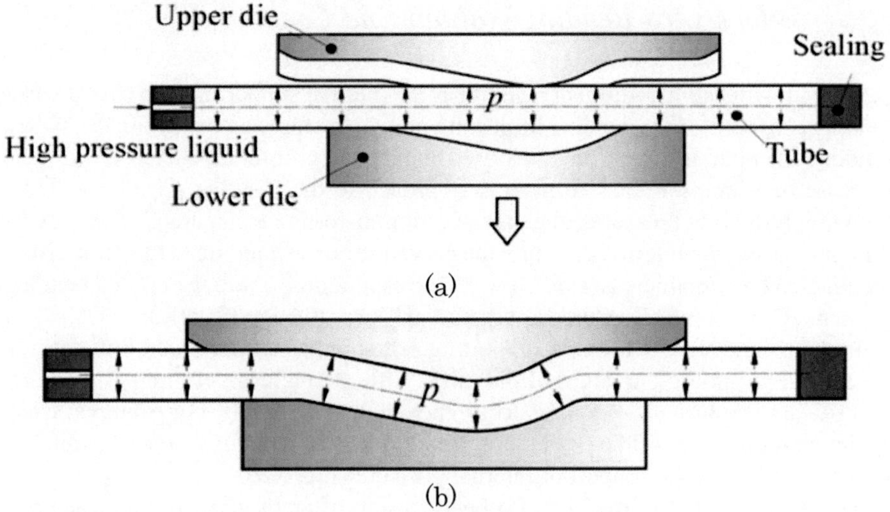

Fig. 3.8 Tube hydro-bending process. **a** Initial stage; **b** press bending

both ends of the tube, and the displacement in the vertical direction is restrained. Compared with tube hydro-bending without end constraint, the constraints limit the upwards deformation and axial movement of both ends, which achieves the forming of three arcs at the same time.

In the process of press bending, the inside of the bent tube sustains an axial compressive stress, while the outside sustains the axial tensile stress. The axial compressive stress results in the thickening of the tube wall inside the bending curvature. When the compressive stress reaches the critical value, wrinkling instability perpendicular to the axial direction of the tube will occur. For the outside of bent tube, the axial tensile stress decreases the thickness at the outside of the bent tube. Splitting failure occurs when this tensile stress reaches the critical value. Moreover, under the combined effect of the resultant force generated by the above opposite stresses, distortion occurs for the tube cross-section. To address such kind of defect, hydro-bending introduces liquid medium at a certain pressure into the tube, then, the internal pressure acts as the supporting force opposite to the resultant force, thus the cross-section distortion of the tube is suppressed. At the same time, due to the internal pressure, certain axial tensile stress can be generated through the sealing at tube ends to compensate part of the compressive stress inside, which effectively reduces the wrinkling tendency at the bending inside. For a certain bending radius, there is a critical value of internal pressure. The wrinkles are able to be completely eliminated once the pressure magnitude is greater than this critical value.

3.3.2 Tube Hydro-Bending Without End Constraint

Compared with using traditional mandrel as the internal supporting, the CNC hydro-bending can increase the bending limit and improve the forming quality. In the hydro-bending, the internal supporting pressure should be appropriately increased with the decrease of bending radius, to make it approach to the yielding pressure of tube. Using high internal pressures, the cross-section distortion exaggerates very slowly. With increasing the internal pressure, the curvature corresponding to the same cross-section deformation increases obviously. Correspondingly, both the critical bending moment and ultimate curvature for tube failure increase significantly.

In the meantime, the internal supporting pressure has an important influence on the critical wrinkling curvature of the tube. With the increase of internal pressure, the critical curvature of wrinkling occurrence increases greatly. For example, when the internal pressure is 0.76 p_s, the limit relative bending radius reaches 7, which is 1/6 of that without the support of internal pressure (Fig. 3.9).

The wrinkling behaviors in hydro-bending are different under different internal supporting pressures. Figure 3.10 shows the forming limit of $AlMg_{0.3}Mn$ aluminum alloy tube processed by hydro-bending. The tube diameter is 95 mm, the thickness is 1.5 mm, and the bending radius is 1500 mm. When the relative bending radius is 15, the maximum diameter-thickness ratio of the tube can reach 63 when the ultimate internal supporting pressure is 2 MPa. With increasing the internal pressure, the wrinkling occurrence during bending thin-walled tube is delayed, and the forming limit is gradually increased. Moreover, the bending limit of tube decreases rapidly with the increase of the diameter-thickness ratio. For example, when the relative bending radius is 15, the critical yield internal pressure is used as the supporting

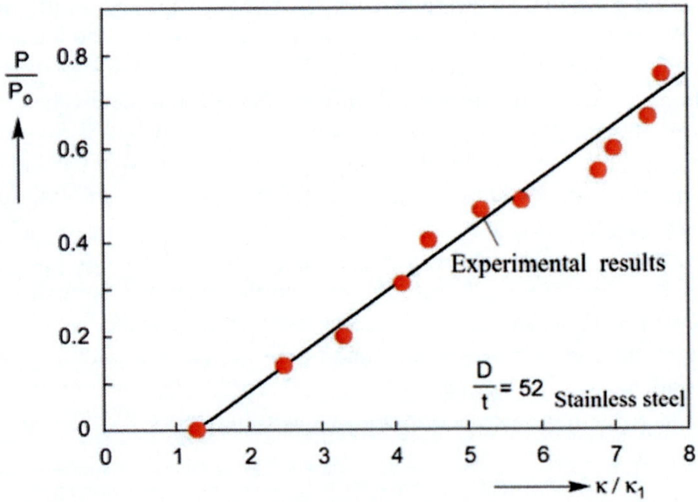

Fig. 3.9 Effect of internal pressure on critical wrinkling curvature of tube [126]

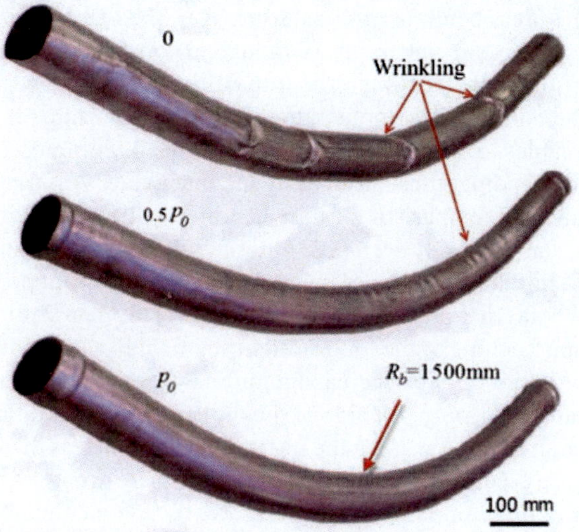

Fig. 3.10 Effect of internal pressure on wrinkling in thin-walled tube hydro-bending [126]

Table 3.4 The ultimate diameter-thickness ratio of tube hydro-bending

Relative bending radius	Low-carbon steel	5A02 aluminum alloy
6	60	50
5	50	45
4	45	40
3	40	36
2	36	30

pressure, the ratio of diameter to thickness ratio of the tube manufactured by hydro-bending is smaller than 100.

The ultimate ratio of diameter to thickness of tube hydro-bending is related to not only the material but also the bending radius. Table 3.4 shows the ultimate diameter-thickness ratio of low carbon steel and aluminum alloy tubes in hydro-bending under different bending radii. It can be found clearly that the ultimate diameter-thickness ratio decreases gradually with the decrease of bending radius. When the relative bending radius is 2–3, the ultimate ratio of diameter to thickness of low carbon steel and aluminum alloy tubular parts are smaller than 40.

3.3.3 Tube Hydro-Bending with End Constraint

The main defect of hydro-bending aluminum alloy tube is wrinkling on the inside of bending. The cross-section deformation of thin-walled elbow can be controlled by the press bending process. Moreover, the bending limit can be greatly increased by the tube end constraint. Without end constraints, severe wrinkling still occurs on the inside of the bending even if the ultimate internal supporting pressure is given. After the constraint is applied, the limit bending radius is decreased significantly, which enables to effectively reduce the degree of compressive deformation on the inside of the bending, thus eliminating the inner side wrinkling. Figure 3.11 shows the experimental comparison of tube bending without and with end constraint, respectively. It can be clearly seen that, without tube end constraint, wrinkles appear at both sides of middle of the inner bent arc when the internal pressure is less than 2.4 p_s. When the internal pressure is equal to or greater than 2.4 p_s, only one side of the bent arc approaching to the long straight segment is wrinkled. Thus, it is difficult to eliminate wrinkles only by increasing internal pressure. In comparison, using the constrain at tube ends, only one side of the inner arc of the middle bend is wrinkled near the long straight section when the internal pressure is less than 2.2 p_s. When the internal pressure is equal to or greater than 2.2 p_s, there are no wrinkles on both sides of the inner arc of the middle bend. Therefore, it can be concluded that tube end constraint can effectively suppress the formation of wrinkles, and thus sound parts can be obtained.

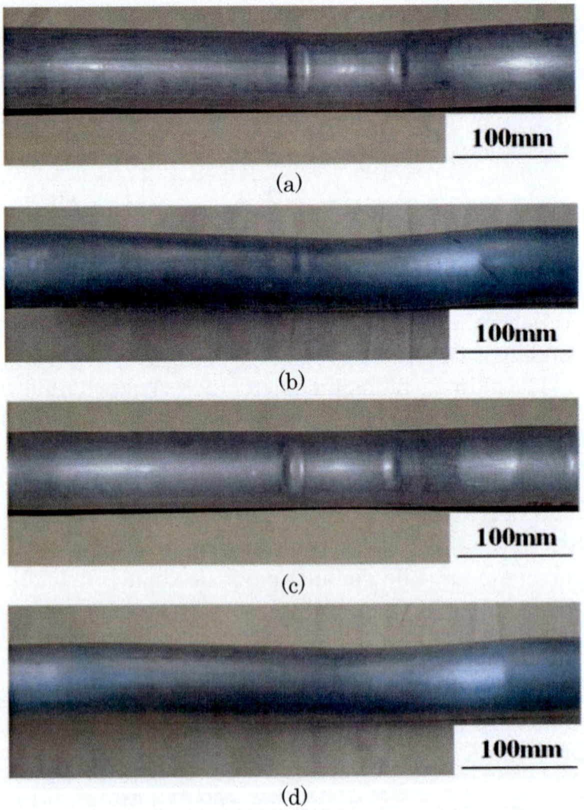

Fig. 3.11 Effect of end constraint on the wrinkling defect in tube hydro-bending. **a** Tube end free (pressure = 2.0 p_s); **b** with tube end constraint (pressure = 2.0 p_s); **c** tube end free(pressure = 2.2 p_s); **d** with tube end constraint (pressure = 2.2 p_s)

In conventional press bending, the part has some defects such as dead wrinkles and severe cross-section distortion. In tube hydro-bending, the wrinkles usually generate on the inner arc of the middle bend near the long straight section. When the internal pressure is less than 2.2 p_s, the inner arc of the middle bend near the long straight section is wrinkled. When the internal pressure is equal to or greater than 2.2 p_s, there are no wrinkles, and sound parts can be obtained. Therefore, the critical internal pressure for this part is 2.2 p_s. If the internal pressure is too high, for example, when the magnitude is equal to or greater than 2.6 p_s, plastic deformation will occur in the part during bending, which will result in an increase in diameter and the flash in parting surface, as shown in Fig. 3.12.

In tube hydro-bending, the wrinkle depth, *h*, is defined as the vertical distance between the top and bottom of a wave, while the wrinkle width, *b*, is expressed by the horizontally projected distance of the wrinkle along the circumferential direction of the specimen, as shown in Fig. 3.13. With the increase of internal pressure, both the

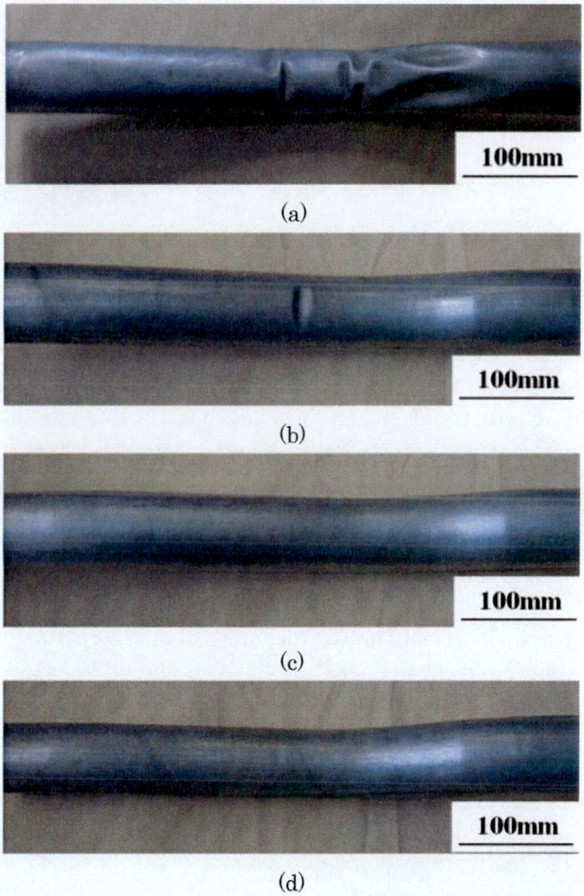

Fig. 3.12 Wrinkling on the inside of bending under different internal pressures. **a** Without internal pressure; **b** pressure = 1.8 p_s; **c** pressure = 2.2 p_s; **d** pressure = 2.6 p_s

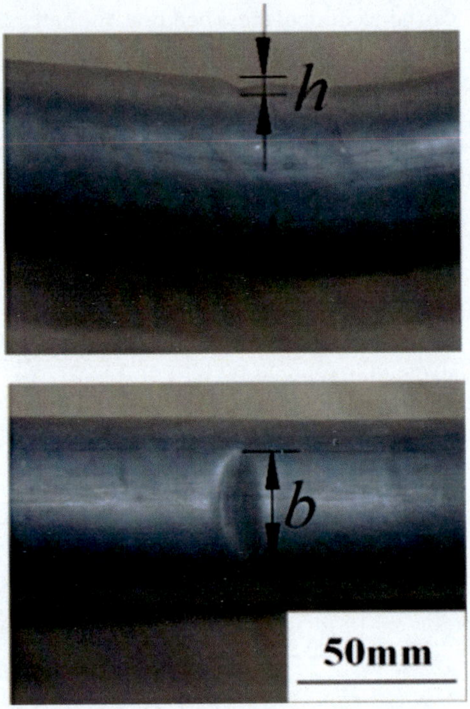

Fig. 3.13 Schematic for measuring wrinkle dimensions

depth and width of wrinkles gradually decrease. Furthermore, filling liquid in tube during bending process can effectively reduce cross-section distortion. The cross-section distortion decreases gradually with the increase of internal pressure. When the tube hydro-bending is conducted under the critical internal pressure of 2.2 p_s, the ellipticity is only 1.37%, which can be regarded as approximately a circle, as shown in Fig. 3.14.

3.4 Type of Defects

The main defects in hydroforming of curved parts with irregular cross-sections are splitting, dead wrinkles and flash, as shown in Fig. 3.15. For the splitting defect, it usually occurs on the outside of the curved area (see Fig. 3.15a), in transition area of polygon cross-section (see Fig. 3.15b) and in the heat-affected zone of welding seam (see Fig. 3.15c).

The reason for splitting is due to the excessive thinning of the wall thickness caused by bending and insufficient ductility of the material due to work hardening. The main prevention approach is to control excessive thinning during bending. The main reason for splitting of welding seam is that, when ERW welded tube is used for hydroforming, the poor welding quality induces cracking in the weld and the

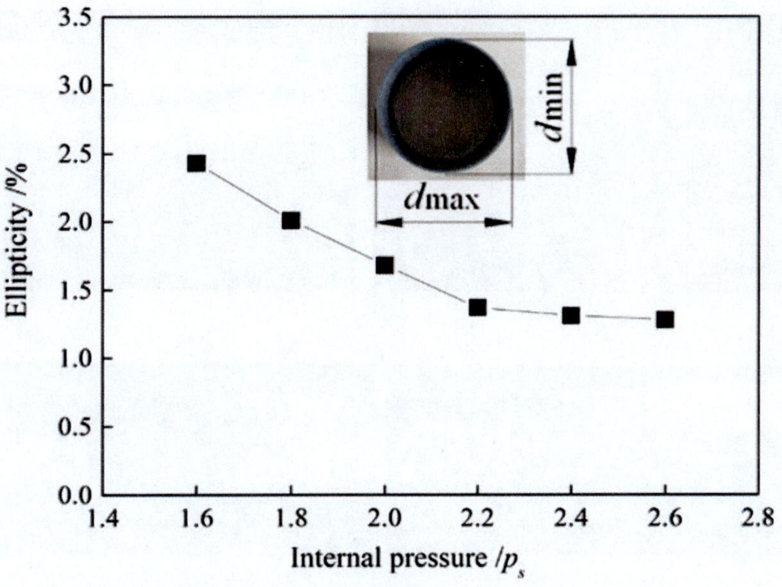

Fig. 3.14 Effect of internal pressure on ellipticity of tube cross-section

neighboring heat-affected zone. In practical production, the cracking of welding seam is the main cracking defect in tube hydroforming. The reason of cracking at the transition area is explained as follows by combining with the curvature and forces of the transition area shown in Fig. 3.16. From the equilibrium equation of the shell, it can be known that the circumferential stress is

$$\sigma_\theta = \frac{pr}{t} \tag{3.4}$$

where

p Internal pressure (MPa);
r Curvature radius (mm);
t Thickness (mm).

It is assumed that the radius of the corner, r_c, is a constant at a particular moment during forming. Moreover, the curvature radius of the straight wall area in contact with the die of the polygon cross-section, r_f, is infinite. Since the curvature radius is continuous, the curvature radius of the transition area is greater than corner radius, $r_t > r_c$. During the pressurization process, the pressure inside the tube is consistent within the tube. Therefore, according to Eq. (3.4), it can be known that the circumferential stress $\sigma_{\theta t}$ in the transition area is greater than the circumferential stress $\sigma_{\theta c}$ at the corner. Due to such a stress state, materials locate in the transition zone firstly satisfy the yield condition due to the higher circumferential stress in this zone

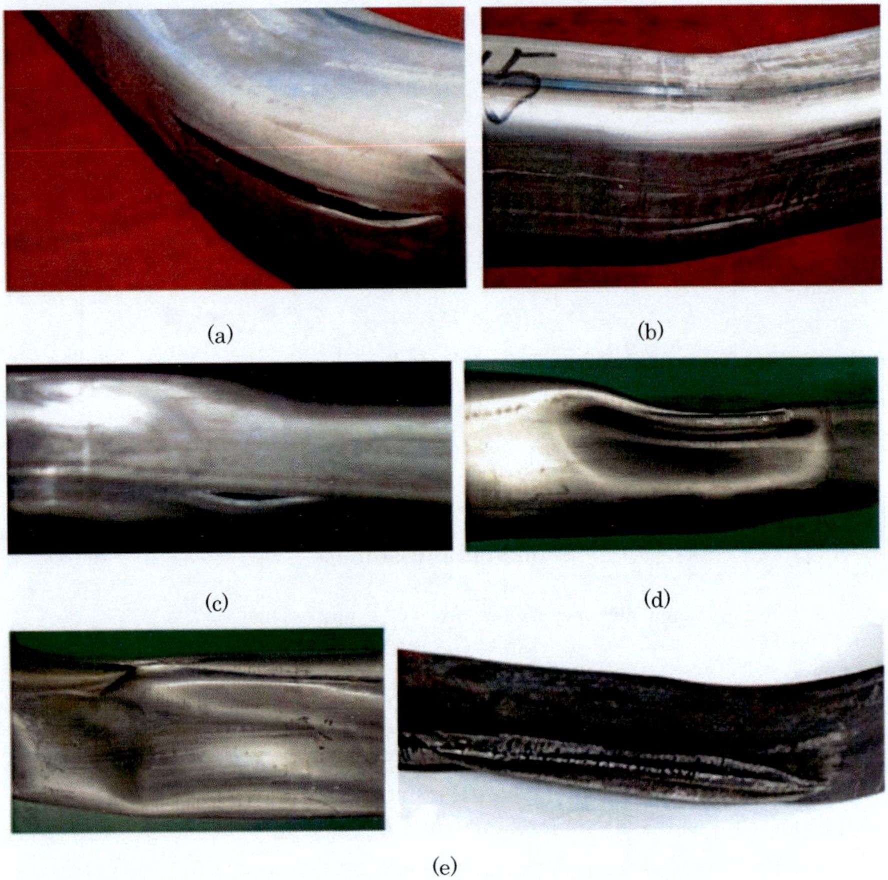

Fig. 3.15 Main defects in hydroforming of tubular parts with irregular cross-sections. **a** Splitting in curved area; **b** splitting in transition area; **c** splitting in welding seam; **d** dead wrinkle; **e** flash

and experiences plastic deformation, which results in the circumferential strain and continuous thinning of wall thickness, leading to the final fracture.

Loading path has an great influence on splitting at the transition area. Therefore, four loading paths shown in Fig. 3.17 were used to investigate the influence of axial feeding effect on hydroforming of part with a square cross-section. In Fig. 3.17, the forming pressures of loading paths 1 and 2 are same, axial feeding is not used in loading path 1. Loading paths 3 and 4 have the same internal pressure while the axial feeding used are different.

For the condition of no axial feeding (loading path 1), when the internal pressure reaches 105 MPa, splitting occurs at the transition zone where the corner area is tangent to the straight wall area, as shown in Fig. 3.18a. In this case, the corner radius is only 13.5 mm, which fails to meet the design requirements. When loading path 2 is used, the folding phenomenon occurs due to the lower forming pressure and fast

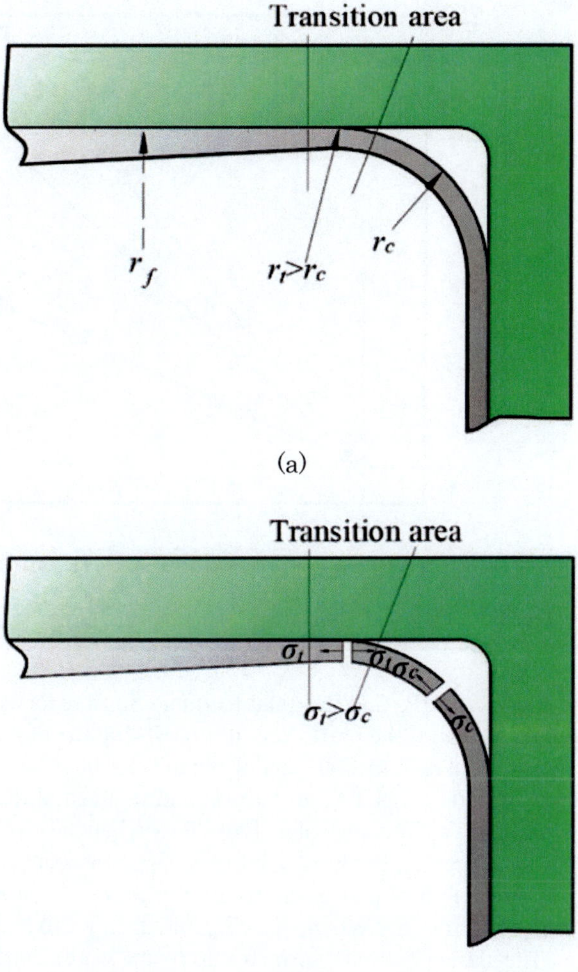

Fig. 3.16 Curvature radius and circumferential stress in transition zone. **a** Curvature radius; **b** circumferential stress

axial feeding (see Fig. 3.18b). In the subsequent calibration stage, the folding cannot be flattened even if a high calibration pressure is used. The increase of pressure is too slow that is not cooperative to the fast material accumulation in the axial direction. Folding is generated as the axial deformation cannot be converted into circumferential deformation in time. If the pressure increases faster and the speed of axial feeding is slower, the axial feeding length is not sufficient to compensate the circumferential deformation, the thickness is reduced. When the internal pressure is too high, splitting occurs (see Fig. 3.18c), which is the result of selecting Loading path 3. The pressure when splitting occurs is higher than that of loading path without axial feeding. If the loading path 4 is adopted, sound parts were formed when the internal pressure

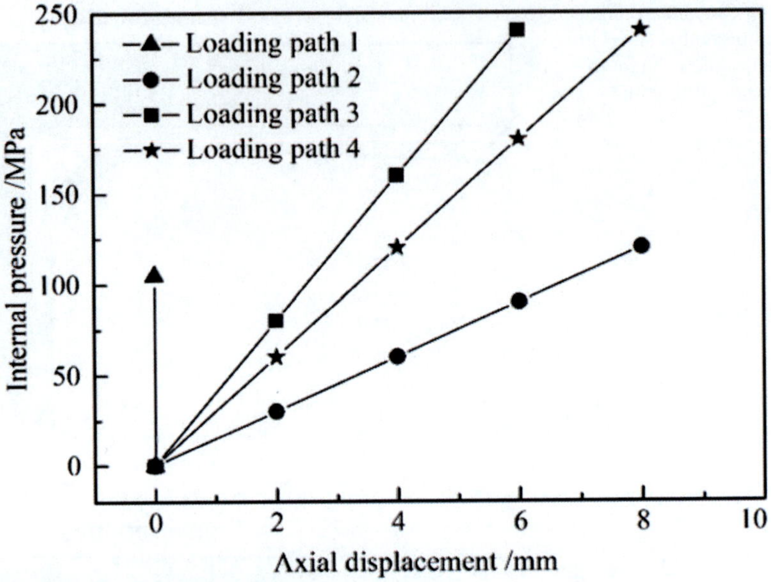

Fig. 3.17 Loading paths for the hydroforming of a square cross-section

reaches 240 MPa and the axial feeding length reaches 16 mm (see Fig. 3.18d). The corner radius of the part is 6.2 mm satisfying the design requirements.

The dead wrinkles are mainly induced when the tube diameter is too large, the cross-section shape of the preform and the design of the parting surface of hydroforming die are unreasonable. Especially when the cross-section shape of the preform is unreasonable, the dead wrinkle is easier to occur. In Fig. 3.19a, for the segment AB, the length of part or die cavity profile is L_0 while the corresponding length of this preform is L_1. When $L_1 > L_0$, the dead wrinkle is generated at this position.

The flash is mainly formed due to the unreasonable parting surface design of hydroforming die and preform cross-section, when a part has a special cross-section shape. In this case, part of the tube in contact with the die is compressed to be extruded from the parting surface before the die is fully closed, then the flash is formed, as shown in Fig. 3.19b. Flash defect can be produced either on one side or on both sides of the part. Flash is a kind of severe defect of hydroforming, that not only wastes the formed part but also damages the dies.

3.5 Thickness Distribution of a Square Cross-Section

Figure 3.20 shows the experimental results of thickness distribution of a part with square cross-sections after hydroforming. The length of square edge is 43.5 mm, and the corner radius is 5.5 mm. The low carbon steel tube with an outer diameter

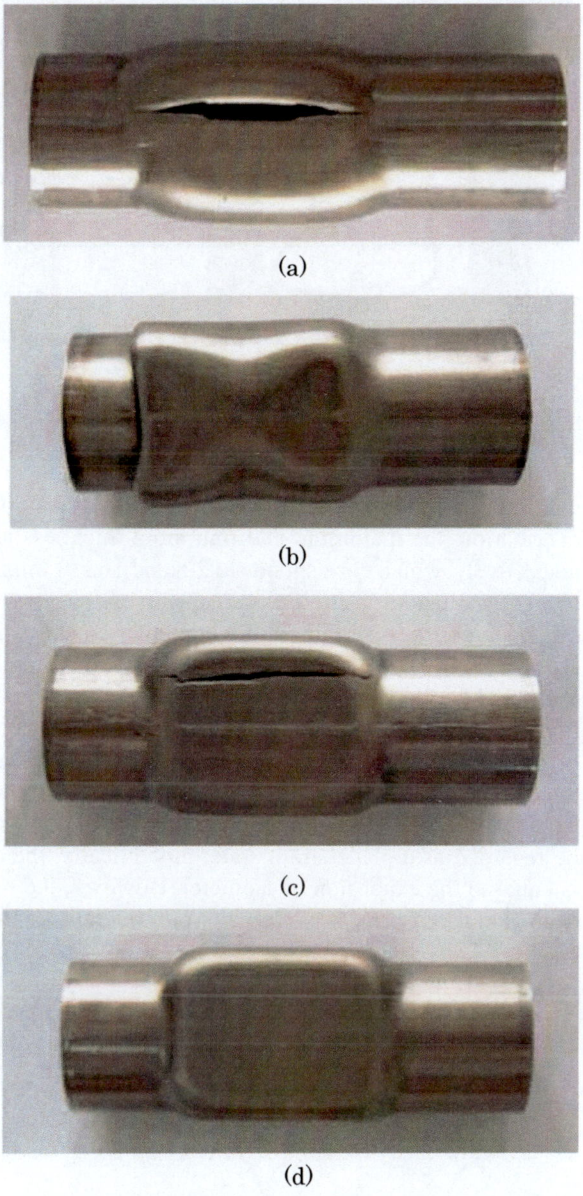

Fig. 3.18 Hydroforming of square cross-section. **a** Loading path 1; **b** loading path 2; **c** loading path 3; **d** loading path 4

of 51 mm and thickness of 1.5 mm was used. Therefore, when the square cross-section of 43.5 mm × 43.5 mm was formed from a tube with a 51 mm diameter, the resulted expansion ratio is 3.5%. The thickness gradually reduces from the midpoint of straight wall to the transition area. Therefore, the maximum thickness is found at the midpoint of the straight wall, which is basically equal to the initial thickness,

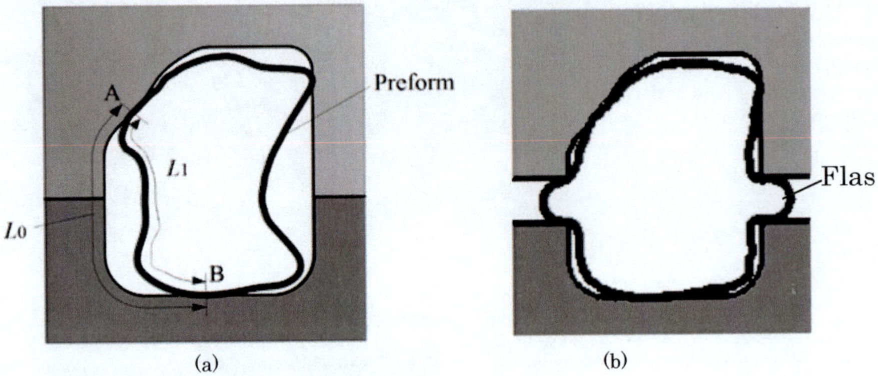

Fig. 3.19 Schematic of the occurrence reason for dead wrinkles and flash. **a** Dead wrinkle; **b** flash

while the minimum thickness locates at the transition area. For such a thickness distribution, the maximum and minimum thickness are 1.462 mm and 1.255 mm, respectively, which corresponds to 2.5 and 16.3% thinning ratios for the midpoint of straight wall and transition area, respectively.

The thickness distribution of rectangular cross-section is similar to that of square cross-section. Only slight difference of minimum thickness at the transition zone is found, when the length–width ratio of rectangular cross-section is different or the transition corners locate in the upper and lower die cavities. The most typical feature of thickness distributions of hydroforming square or rectangular cross-section is that, the thinning at transition zone is greatest, as shown in Fig. 3.20b. When the expansion ratio is 3.5%, since there is no axial feeding, the tube deformation can be regarded as a plane strain state, theoretically, the average thinning of thickness is equal to the expansion of diameter. However, the maximum thinning ratio in the transition zone is 16.3%, which is approximately 4.6 times the average thinning ratio

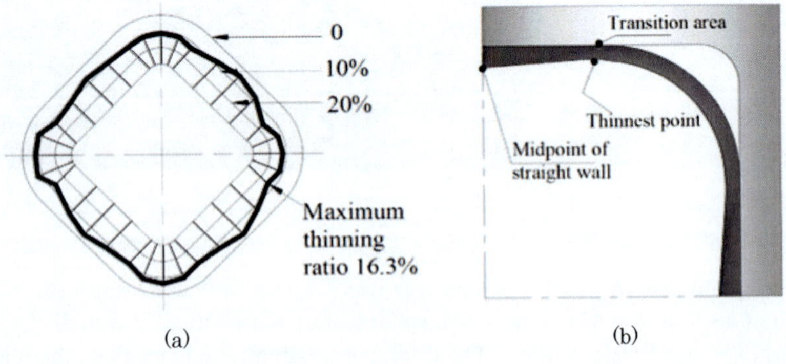

Fig. 3.20 Thickness distribution of a square cross-section. **a** Experimental result; **b** the thinnest point in transition area

Table 3.5 Thickness variations of parts with different expansion ratios

Expansion ratio (%)	Edge length/mm	Corner radius/mm	Thickness at the midpoint of straight wall/mm	Thinning ratio at the midpoint of straight wall/%	Thickness of transition area/mm	Thinning ratio of transition area (%)
3.5	43.5	5.5	1.46	2.5	1.26	16.3
10.0	46.0	6.0	1.43	5.0	1.12	25.5

or expansion ratio. Excessive thinning at the transition zone will lead to splitting during forming. The fatigue performance can be also deteriorated even if no splitting is found. Therefore, controlling the thinning ratio of the transition zone is a key technique for the hydroforming of irregular cross-sectioned parts.

3.5.1 *Effect of Expansion Ratio on Thickness Distribution*

The expansion ratio is one of the main factors influencing the thickness distribution. Thickness variation under different expansion ratios are given in Table 3.5. It can be clearly seen that, with the increase of the expansion ratio, the thickness at the center of the straight wall varies insignificantly, while the thickness in the transition area decreases significantly. When the expansion ratio is 10%, the thickness at the midpoint is 1.43 mm, corresponding to a thinning ratio of 5.0%. While for the transition area, the thickness is 1.12 mm, indicating a 25.5% thinning ratio, which is easy to experience splitting.

3.5.2 *Effect of Friction Coefficient on Thickness Distribution*

Friction also has an important effect on the thickness distribution. For the hydroforming of rectangular cross-sections, as shown in Fig. 3.21, with the increase of friction, the non-uniformity of thickness increases. The higher the friction, the greater the non-uniformity of thickness and the thinner the transition area. When the friction coefficient is 0.05, the minimum thickness of the transition area is 1.72 mm, and the thinning ratio is 14%. While when the friction coefficient is increased to 0.15, the minimum thickness of the transition zone is 1.65 mm and the thinning ratio is 17.5%. Therefore, it is important to use proper lubricant in actual processes to reduce friction and obtain a uniform thickness distribution.

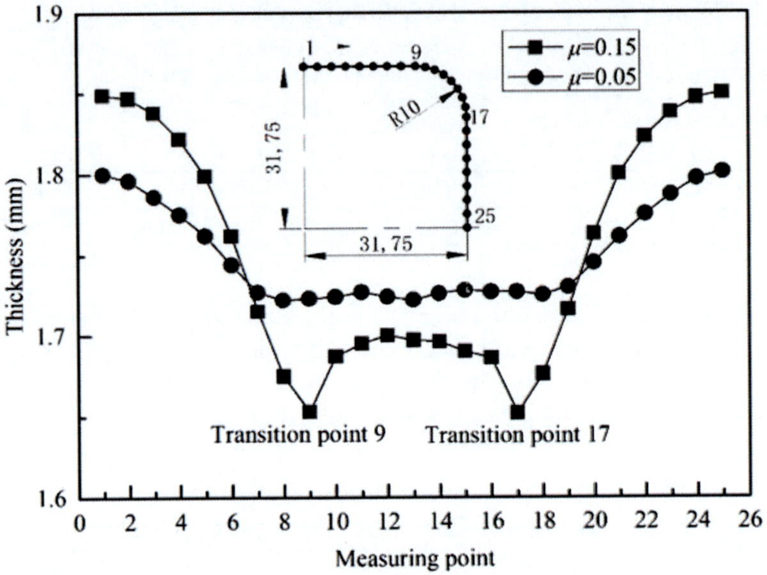

Fig. 3.21 Effect of friction on the thickness distribution

3.5.3 *Effect of Parting Surface on Thickness Distribution*

In the processes of die clamping and hydroforming, if the parting surface is different, the relative movement direction and distance between the tube and die are different. Then, the resulted friction force effect on material flow is different, which has an important effect on the thickness distribution of polygon cross-section.

For the rectangular cross-section, the main parting modes mainly include the following forms (see Fig. 3.22): middle horizontal parting, upper horizontal parting, upper and lower diagonal parting, and middle diagonal parting. Among these four parting modes, if the upper horizontal parting (Fig. 3.22b) is used in the preforming, thinning is greatest and lowest uniformity of thickness distribution is resulted after hydroforming. For the preformed tube using the upper and lower diagonal parting (Fig. 3.22c) mode, the thickness thinning is the smallest resulting in the highest uniformity of thickness after hydroforming. For other parting modes, their thickness distribution uniformities locate between these above two boundaries.

3.5.4 *Effect of Material Mechanical Properties on Thickness Distribution*

For materials with different mechanical properties, their thickness distributions after hydroforming are also different. Particularly, the strain hardening exponent n and

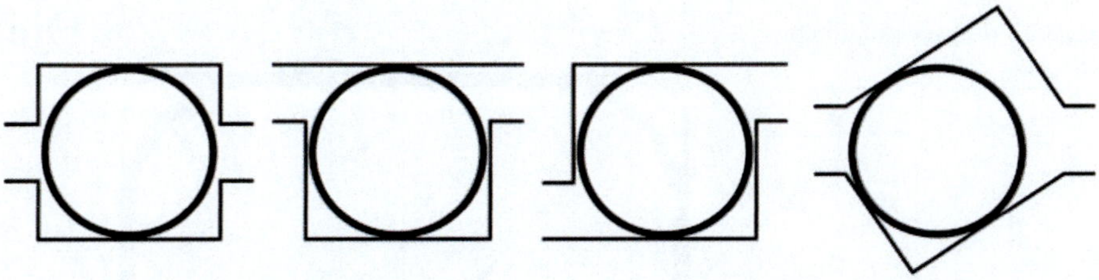

Fig. 3.22 Different modes of parting surface for the hydroforming of rectangular cross-sections. **a** Middle horizontal parting; **b** upper horizontal parting; **c** upper and lower diagonal parting; **d** diagonal parting

the normal anisotropic coefficient r have important effects on the thickness distribution. In general, with the increase of the strain hardening exponent n and the normal anisotropy coefficient r, the thickness thinning tendency of the material decreases, and the uniformity of thickness distribution increases. For example, for a square cross-section, according to numerical simulations, when the strain hardening exponent $n = 0.23$ (equivalent to low carbon steel), the minimum thickness of the transition zone is 1.23 mm, and the thinning ratio is 17.8%. When strain hardening exponent $n = 0.32$ (equivalent to stainless steel), the minimum thickness of the transition zone is 1.32 mm and the thinning ratio is 12.1%.

3.6 Principle and Method for Reducing Calibration Pressure

3.6.1 Principle of Reducing Calibration Pressure by Concave Preformed Cross-Section

According to the calculating formula of calibration pressure in Chap. 2, it is known that the final forming pressure is determined by the minimum corner radius of the cross-section, once the used material and thickness is fixed. The smaller the corner radius, the higher the calibration pressure required. However, if the calibration pressure is too high, a press with large tonnage of die clamping force is needed. In the meantime, the die is subjected to a higher stress due to the large die clamping force, which greatly increases the cost of equipment and die. Therefore, the technique of forming small rounded corners using low internal pressures cooperatively is a key technology of hydroforming of irregular cross-sections.

Through analyzing the force condition in the corner area shown in Fig. 3.23, it can be known that the internal pressure should be able to not only generate the force F to deform the tube plastically, but also overcome the friction force F_f between the tube and the die. Since the magnitude of the friction force F_f is directly proportional

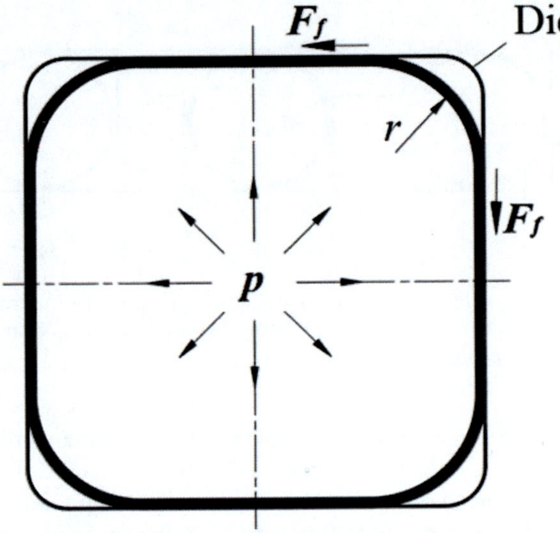

Fig. 3.23 Force condition in the corner area

to the positive pressure (which is approximately equal to the internal pressure), the higher the internal pressure, the greater the friction force. Therefore, the effect of increasing internal pressure to obtain small rounded corners is not significant, and resulted cost is high. If the tube cross-section is compressed into a concave shape by preforming (see Fig. 3.24), the material tends to flow towards the corner during calibration due to the concave shape and reduced contact area with the die. In the meantime, an additional tangential pushing force, F_x, is generated by the internal pressure to overcome friction, F_f, and promote the material flowing towards the corner area. Under such an approach, the calibration pressure is greatly decreased, and small rounded corners can be formed under low internal pressures.

3.6.2 The Correlation Between the Tangential Pushing Force and the Concave Depth

The key of using concave preform to lower calibration pressure is to produce pushing force along tangential direction opposite to the friction force near the transition area. The pushing force along tangential direction is highly related to the depth of concave. For the concave preformed cross-section, the force on the straight wall during pressurization for calibration is shown in Fig. 3.25. For the simplicity of analysis, only one-unit length along the axial direction of part is taken for force analysis. In addition, due to the symmetry of the straight wall area, only the force condition of segment CO is analyzed.

According to the force balance condition in the horizontal direction, the pushing force along tangential direction of the interaction between the straight wall segment

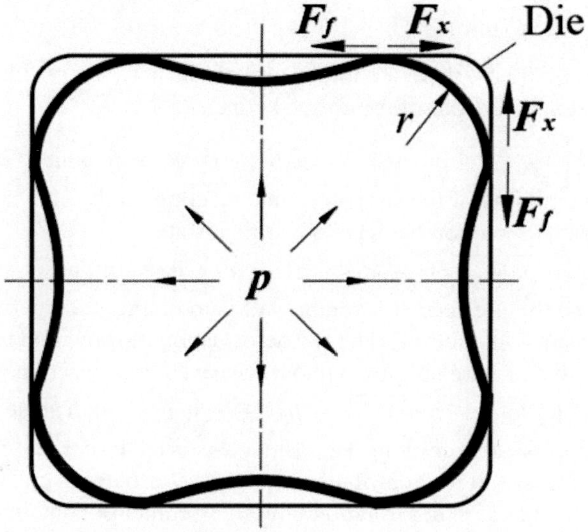

Fig. 3.24 Principle of reducing calibration pressure

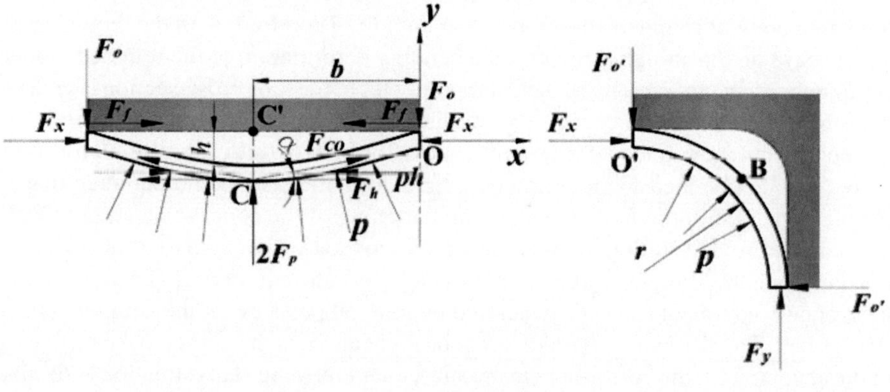

Fig. 3.25 Force analysis of concave preformed cross-section

and the rounded corner segment at point O within the transition zone can be obtained as:

$$F_x = \frac{p_c}{1+\mu}\left[\frac{b^2}{h} - h - \mu(b+r)\right] \tag{3.5}$$

where

p_c Calibration pressure (MPa);
μ Friction coefficient between tube and die;
b 1/2 of the width of the straight wall of the die (mm);

h Depth of concave (mm).

According to Eq. (3.5), there are three correlations between pushing force along tangential direction and depth of pre-formed concave, h:

1. When $h = \frac{1}{2}\left[\sqrt{\mu^2(b+r)^2 + 4b^2} - \mu(b+r)\right]$, $F_x = 0$. which represents there is no force at point O of the straight segment acting on the corner segment. The shape of the cross-section belongs to a critical state.
2. When $h > \frac{1}{2}\left[\sqrt{\mu^2(b+r)^2 + 4b^2} - \mu(b+r)\right]$, $F_x > 0$. That is, when the depth of concave on the pre-formed straight wall is too large, the internal force F_x at point O is in tension. At this point, the deformation of transition rounded corner is similar to that of pure bulging without concaved cross-section.
3. When $h < \frac{1}{2}\left[\sqrt{\mu^2(b+r)^2 + 4b^2} - \mu(b+r)\right]$, $F_x < 0$. The straight segment exerts a compressive force along the positive x-axis on the corner segment at point O. This compressive force contributes to pushing the material into the corner area resulting in the bending deformation of corner segment to obtain a small corner radius. Such a force can be called as "pushing force for calibration".

Figure 3.26 shows the variation of F_x with depth of concave h calculated using Eq. (3.5). As shown in Fig. 3.26, with the decrease of h, the force F_x exerted on the transition point of rounded corner increases rapidly. Therefore, with the flattening of the concave on the straight segment, the bending deformation of the rounded corner continuously occurs. As long as the outer perimeter of the tube cross-section is greater than the perimeter of die cavity, a good die attaching effect between the corner of workpiece and the rounded corner of the die is formed. In this case, a small rounded corner can be obtained without any bulging deformation. It should be noted that if the magnitude of initial h is too small, the flattening of the concave on straight wall area cannot provide sufficient material for the rounded corner area to be attached on the die. Once the straight wall area is formed before the corner area (i.e. $h = 0$), the horizontal component force F_x generated by internal pressure on the straight wall is diminished. The flattening effect does not occur again, which leading to the difficulty of die attaching for the corner area in the subsequent forming. This situation is similar to the forming of small corner radius without concave cross-section.

Figure 3.27 shows the stress distribution in horizontal direction of the concave preformed cross-section obtained by numerical simulation during the calibration. It can be seen that when the concave preformed cross-section is subjected to internal pressure, the stress on the straight wall of the tube along the horizontal direction is mainly compressive, and its absolute value maximizes at the transition point between the rounded corner and straight edge, which proves that the "pushing force for calibration" proposed in the above mechanics analysis is valid.

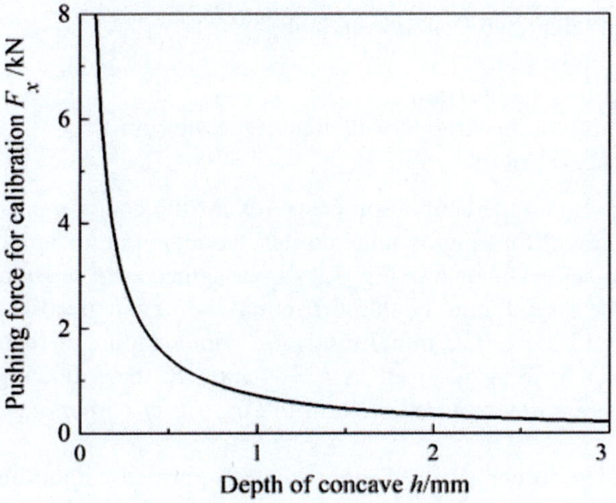

Fig. 3.26 Relationship between pushing force for calibration at transition point and depth of concave

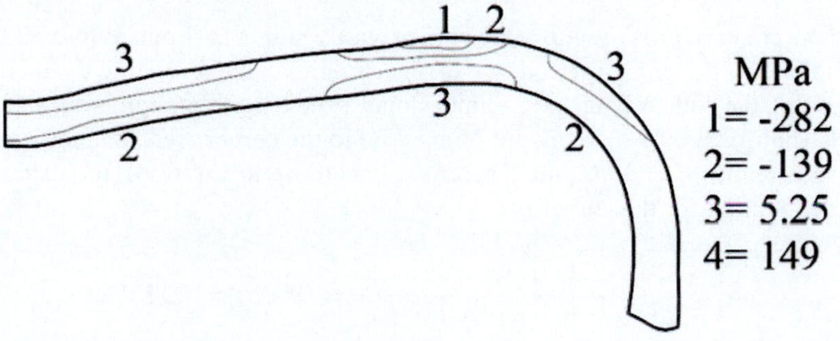

Fig. 3.27 Stress distribution in horizontal direction during the calibration process of concave cross-section

3.6.3 Calculation of Calibration Pressure for Concave Preformed Cross-Section

According to the force analysis and yield condition of the concave cross-section in the calibration process as shown in Fig. 3.25, the calibration pressure at a certain depth of concave h is given as

$$p_c = \frac{t^2 \sigma_x}{b^2/h - h - \mu(b + r_c) + r_c} \tag{3.6}$$

where

μ Friction coefficient between tube and die;
t Thickness of tube (mm);
r_c Transition corner radius (mm);
b 1/2 of the width of the straight wall area of the die (mm);
h Depth of concave (mm).

In order to compare the calibration pressures for the concave pre-formed cross-section and the direct forming of small corner, the main parameters of the concave preformed cross-section shown in Fig. 3.25 are assumed as: $b = 10$ mm, $r = 4$ mm, thickness of tube $t = 1$ mm, coulomb friction coefficient μ=0.1, material yield strength $\sigma_s = 245$ MPa. The initial transition corner radius of rectangular cross-section for direct forming is given as $r = 4$ mm. In the following calculations, the calibration pressures required to form the transition corners to $r = 3$ mm are calculated.

When transition corner of $r = 3$ mm is directly formed without preforming, the required calibration pressure is

$$p_c = \frac{t}{r_c}\sigma_s = 0.33\sigma_s = 81.7\,\text{MPa}$$

For the concave cross-section, the corner with a radius of 4 mm is formed into a 3 mm radius by flattening the concave on the straight wall. Neglecting the thickness variation of the tube, using the 2-dimensional scheme, the straight wall needs to provide additional 0.43 mm amount of material to the corner area, then the depth of concave becomes $h = 2.96$ mm. Therefore, according to Eq. (3.6), the calibration pressure required for the corner is

$$p_c = \frac{t^2\sigma_s}{b^2/h - h - \mu(b + r_c) + r_c} = 0.03\sigma_s = 7.3\,\text{MPa}$$

Thus, it can be seen that when the concave cross-section is used, the calibration pressure is significantly lower than the internal pressure required for the direct forming of transition corner without preforming. According to the above theoretical equation, the calibration pressure is reduced by about 90%. Figure 3.28 shows the comparison of the situations of die contacting during the calibration process using numerical simulation. It can be seen that the calibration pressure for the direct forming without preforming is approximately $0.25\sigma_s$ while it is approximately $0.05\sigma_s$ when the concaved cross-section is adopted. An approximate 80% reduction of calibration pressure can be obtained.

In actual processes, due to the variety of cross-section shapes, it is impossible to obtain an ideal concave cross-section with a single shape, and the reduction in calibration pressure is less than 80–90%. However, the calibration pressure can be also reduced as long as a reasonable concave cross-section is given. In this case, the exact value needs to be determined by experimentation. Figure 3.29 shows the experimental results of the preformed part with a concaved cross-section. In the experiments, the initial tube thickness is 1.5 mm, the diameter is 51 mm, and the edge length of square

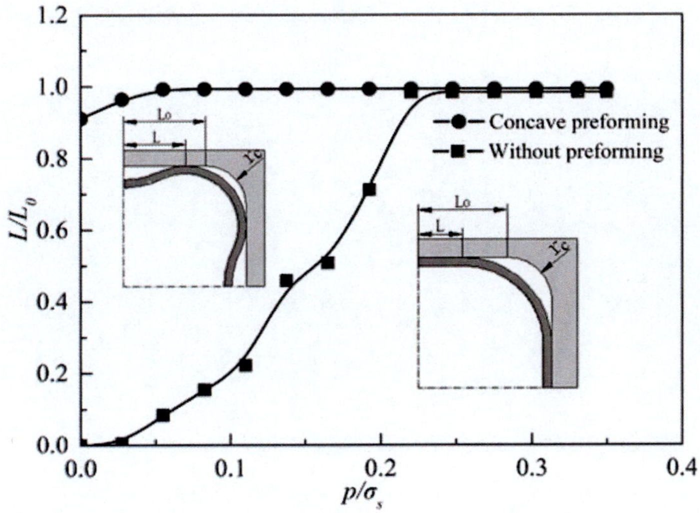

Fig. 3.28 Comparisons of the die contacting situation and calibration pressure between direct hydroforming and using concaved cross-section

cross-section is 43.5 mm. Through the hydroforming of concaved cross-section, the formed part with a transition corner radius of 5.5 mm was successfully formed. The sound corner was formed and the used forming pressure required was very low. However, using the same pressure to directly form the rectangular cross-section with corners, only an approximate 10 mm corner radius can be achieved as shown in Fig. 3.30, and a higher pressure for further deformation is needed.

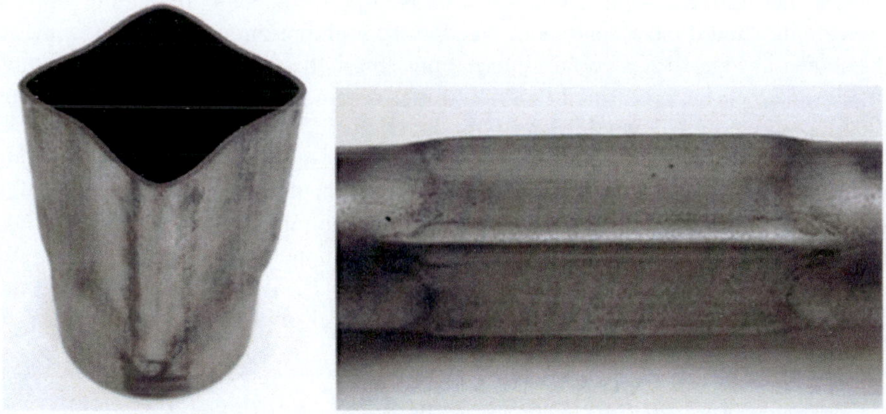

(a) (b)

Fig. 3.29 The tubular preform with concaved cross-section. **a** Preform with concaved cross-sections; **b** tubular part

Fig. 3.30 The tubular part obtained by direct hydroforming

3.6.4 Shape Optimization Design of Concave Preform

As mentioned above, the shapes of typical cross-sections in curved parts are square, rectangle, trapezoid, polygon, ellipse and other irregular shapes. The shapes of concave cross-section corresponding to different part cross-section is also different, as shown in Fig. 3.31. In general, the shapes of the concaved cross-section required for square and rectangular cross-sections are relatively simple. However, the shapes of concave cross-section required by trapezoid and polygon are complex, which is difficult to obtain practically. In the following, quadrilateral cross-section is taken as an example to explain how to obtain a reasonable shape of concave preform through optimization.

Considering the complexity of the concaved preform shape, the equipment for preforming should have degrees of freedom in multiple directions for movement. As shown in Fig. 3.32, a certain internal pressure is firstly exerted within the tube inside, and then the preformed cross-section is obtained by compressing using four die blocks in the up, down, left and right directions.

A reasonable shape of preform was obtained by optimizing the displacements of the up, down, left and right die blocks, as well as the internal pressure. In this process, the optimization objective is to increase the thickness uniformity, and the process variables are the feeding displacements of upper and lower die blocks X_1 and X_2, the feeding displacements of left and right die blocks X_3 and X_4, and the internal pressure p_j. For the above forming process example, the objective function is obtained as

$$\begin{cases} minimize \quad F(x_i, p_j)(i = 1, 2 \ldots 4, j = 1) \\ subject\,to \quad M \le x_i \le N(i = 1, 2 \ldots 4) \\ \qquad\qquad q \le p_j \le s(j = 1) \end{cases} \tag{3.7}$$

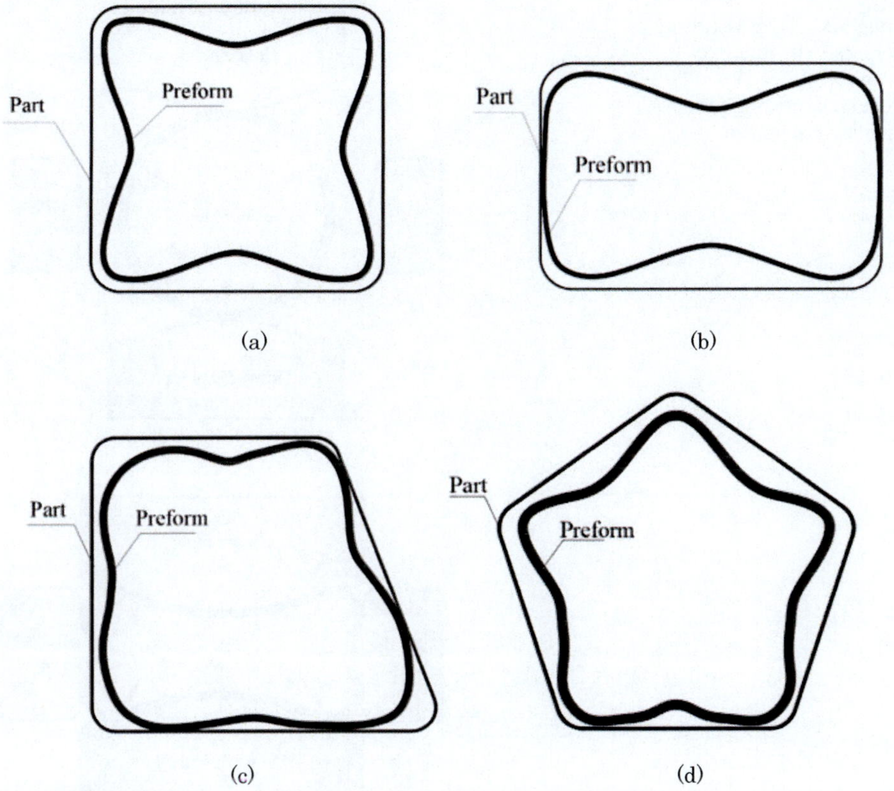

Fig. 3.31 Shape of preform with typical cross-section. **a** Square; **b** rectangle; **c** trapezoid; **d** pentagon

The specific constraint condition is a variable within a certain range, and the boundary condition constants are M, N, q and s. In the meantime, considering the structure symmetry, there are $X_1 = -X_2$ and $X_3 = -X_4$.

The optimal Latin super cube design was adopted in the optimal design process. Two hundred sets of experimental data were designed and analyzed within the range of process variables. Then the quadratic regression model was established according to the sample points. The primary, secondary and interaction effect of each parameter can be expressed by the coefficient of normalized model, which can reasonably indicate the contribution of input variables to the response. Through optimizing the process parameters, the matching scheme of 20 groups of process parameters satisfying the requirements of objective function was obtained. Three groups were randomly selected to form the optimization and comparison schemes with the scheme 1 randomly selected within the variable range of process parameters, as shown in Table 3.6. The optimized preform shape is shown in Fig. 3.32.

The maximum thickness thinning ratios of the final hydroformed parts of each scheme were compared and analyzed. In scheme 1 (random), severe thinning was

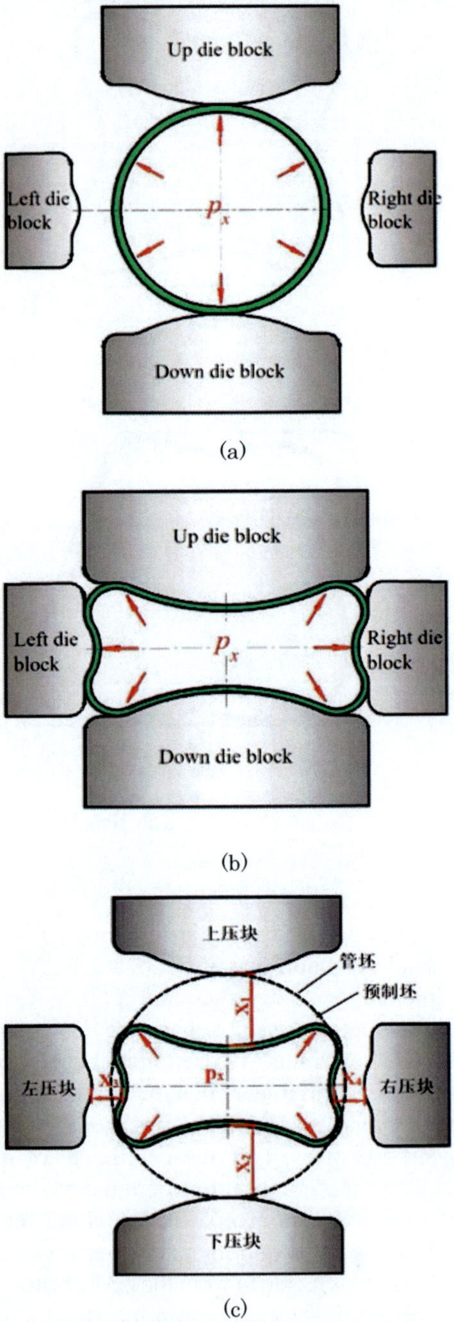

Fig. 3.32 Hydro-pressing process with multiple degrees of freedom. **a** Filling liquid; **b** pressing forming; **c** process parameters

Table 3.6 Optimization and comparison schemes

Scheme number	Upper feeding displacement X_1 (mm)	Lower feeding displacement X_2 (mm)	Left feeding displacement X_3 (mm)	Right feeding displacement X_4 (mm)
Scheme 1 (random)	−24.00	24.00	11.00	−11.00
Scheme 2 (optimization)	−19.05	19.05	15.77	−15.77
Scheme 3 (optimization)	−21.10	21.10	10.60	−10.60
Scheme 4 (optimization)	−22.44	22.44	13.01	−13.01

observed in the transition area of long-side, with a maximum thinning ratio of 16.8%. In optimized scheme 12, the maximum thinning ratio is 11.4%, and the thinning ratio of the transition area between the long and the short side is quite different. In scheme 3, the thinning ratio of the transition area between the long and the short side is the minimum, and the maximum thinning ratio is 13.5%. In scheme 4, severe thinning was observed in the transition area of long-side, with a maximum thinning ratio of 14.6%. By comparison, it can be seen that the maximum thickness reduction ratio of the final formed specimen in scheme 2 is the minimum, while that of the final formed part using scheme 1 is the maximum (Fig. 3.33).

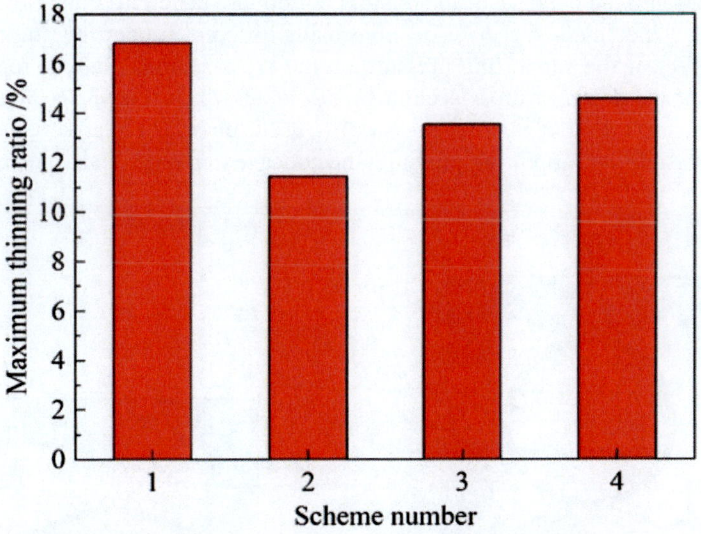

Fig. 3.33 Comparison of maximum thinning ratio of hydroformed parts with different optimization schemes

3.6.5 Hydro-Pressing Process for the Preform

The shape of preform obtained through optimization is often complex. However, it is difficult to obtain the desired shape by conventional mechanical pressing process. Complex preform shapes can be obtained by hydro-pressing process. As shown in Fig. 3.34, the hydro-pressing process can be divided into two stages:

1. Filling and pressurization of liquid. The tube is placed into the die and filled with liquid medium, then both ends of the tube is sealed, and the liquid pressure is increased to the required supporting pressure;
2. Hydro-pressing. When the die moves downwards, the tube is deformed to the desired shape under the combined action of mechanical pressing of the die and liquid pressure in the tube. During the pressing process, the liquid pressure in the tube can be adjusted to ensure that the supporting pressure is within a reasonable range.

The main process parameters of hydro-pressing include: ultimate internal supporting pressure, pressing force and liquid reaction force at tube ends.

Supporting pressure is an important parameter that affects the shape of cross-section. When supporting pressure (conventional pressing) is not given, the upper and lower sides of the cross-section collapses obviously. Sharp circular arcs on the two sides are found resulting in an unqualified formation of ellipse shape. When the supporting pressure is in the range of 3–15 MPa, the shape of cross-section is a long ellipse. When the supporting pressures are different, the shape of the elliptic curve at two sides and the concave of the top and bottom are different. Figure 3.35 shows the obtained specimens and cross-sections using different supporting pressures.

The effect of the supporting pressure on the cross-section shape is represented by the concave depth of cross-section (h_1-h), as shown in Fig. 3.36. When traditional pressing (without internal pressure) is used, obvious collapse would occur. With increasing the supporting pressure, the concave depth gradually decreases and

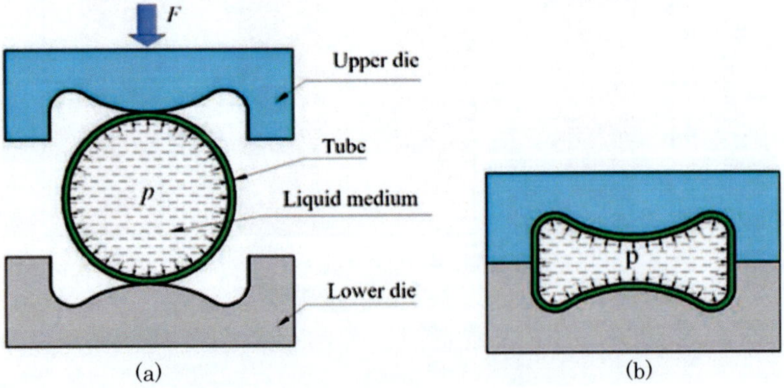

Fig. 3.34 Schematic diagram of tube hydro-pressing process. **a** initial state; **b** pressing

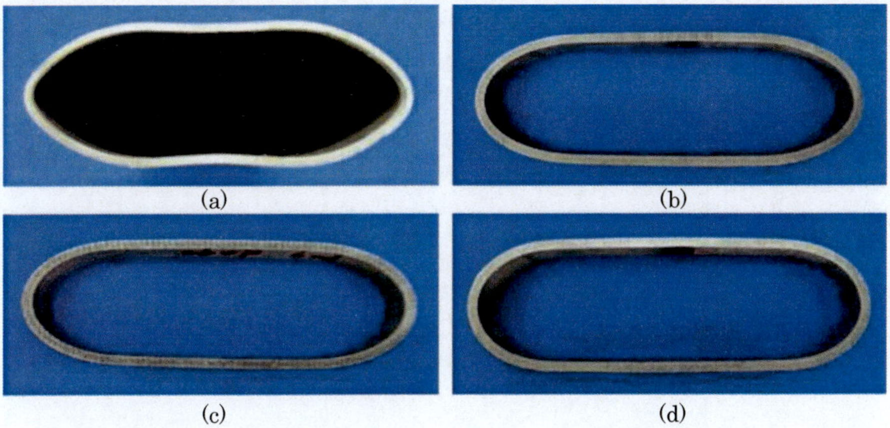

Fig. 3.35 Specimens and cross-section shapes under different supporting pressures. **a** $p = 0$ MPa; **b** $p = 3$ MPa; **c** $p = 9$ MPa; **d** $p = 15$ MPa

diminishes when the pressure is greater than 6 MPa. When the pressure reaches 15 MPa, external convex defects appear. This is mainly because when the support pressure is high, springback of the middle section occurs, resulting in a slight convex phenomenon. Therefore, hydro-pressing can effectively inhibit the occurrence of concave defects in the elliptical cross-section, and when the supporting pressure increases to a certain value, the defect will disappear. However, if additional pressure is applied, not only does it have no effect on the defect, but also too high pressure causes the cross-section springback, which could lead to the external convex phenomenon.

After hydro-pressing, the thickness is basically unchanged with excellent uniformity. However, with increasing the supporting pressure, the maximum thinning ratio of thickness increases. Fortunately, the variation of thickness is not significant.

The height reduction also has an important effect on the cross-section shape. Figure 3.37 shows the relationship between the depth of concave and height reduction. It can be found that for traditional pressing, the greater the height reduction, the higher the depth of concave. The correlation between the concave depth and height reduction is nearly a linearly increasing function. When the height reduction is 50 mm, the concave depth is 2.2 mm. When the supporting pressure is 6 MPa, the concave is almost diminished. The above results indicate that the hydro-pressing is able to completely eliminate the concave defects.

For conventional mechanical pressing, collapse occurs at the top and bottom, and severe undercut appeared on both sides of the tube. While in the hydro-pressing process, the two sides are arcs with very small curvature, there are basically no straight segments when the pressure is 3 MPa. When the pressure is greater than 6 MPa, the arcs on both sides start to become straight lines. With increasing the supporting pressure, the length of straight line keeps increasing, and the desired shape of the concave rectangular cross-section is obtained, as shown in Fig. 3.38.

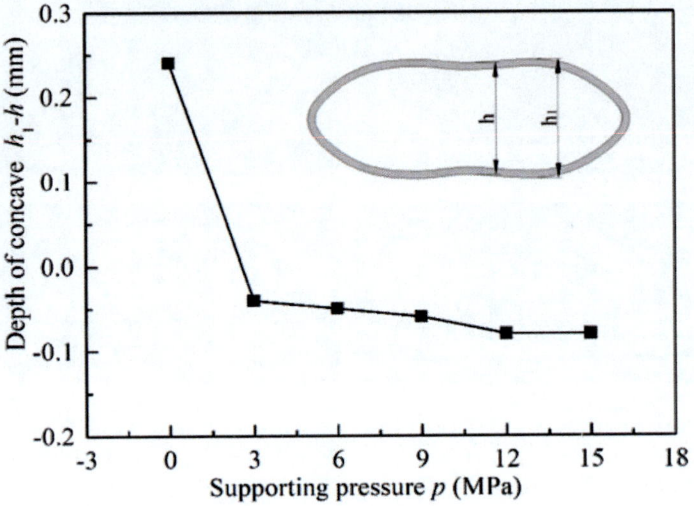

Fig. 3.36 The change curve of concave depth with supporting pressure

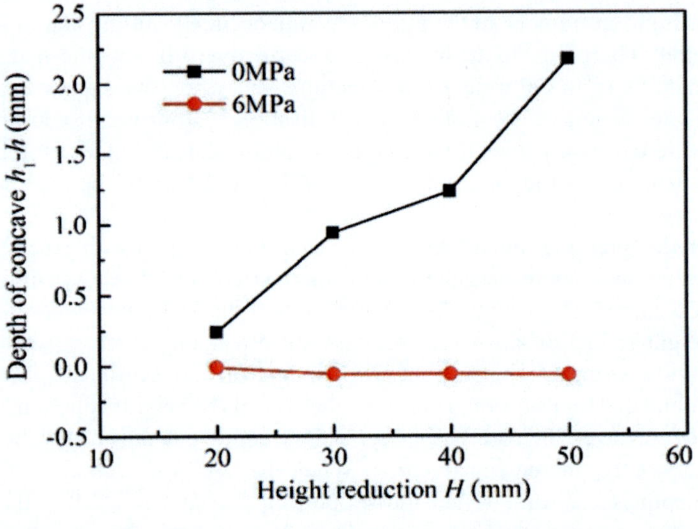

Fig. 3.37 The relationship between depth of concave and pressure

When the supporting pressure is different, the thickness distribution of the obtained concave rectangle preform is also different. Figure 3.39 shows the thickness distributions of a concave rectangular preform under different pressures. It can be found that the thickness of the preform is basically unchanged when the supporting pressure is equal to 3 MPa. When the supporting pressure is greater than 6 MPa, the maximum thinning point occurs in the middle point of the vertical symmetry

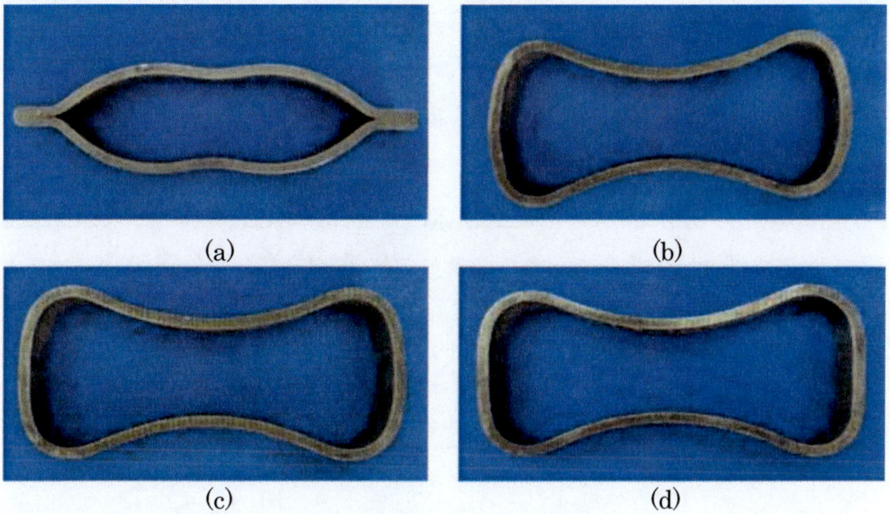

Fig. 3.38 Shapes of preformed cross-sections under different supporting pressures. **a** Without internal pressure; **b** $p = 3$ MPa; **c** $p = 9$ MPa; **d** $p = 15$ MPa

plane (points 1 and 13), rather than the corner area (points 4, 5 and 6). Under the same pressure, the thickness distribution is symmetrically distributed on both top and bottom sides. With the increase of supporting pressure, the maximum thinning ratio increases. For example, when the pressure reaches 15 MPa, the maximum thinning ratio of the cross-section (points 1 and 3) is 4.62%, and the thinning ratio of the corner area is 1.5%. In summary, the thickness thinning in the corner area of the rectangular cross-section after hydro-pressing is smaller and the thickness uniformity is good.

3.7 Hydroforming of Typical Parts with Curved Axis

3.7.1 Hydroforming of the Engine Cradle

Figure 3.40 schematically shows the model of a subframe tubular part for a passenger car. The axis of this part is a three-dimensional curve, and the change of cross-section shapes along the axis is complicated. In total, there are 18 cross-sections with different shapes and sizes, including rectangular, trapezoid, and polygon cross-section etc. Subsequently, the deformation of cross-sections along the circumferential direction is not uniform during hydroforming, which is easy to induce wrinkling or splitting. Such a nonuniformity becomes severely at the corner positions with complex shapes, as the thickness thinning has occurred during the bending in the previous operation. Splitting is prone to occur at the outside of tubular part.

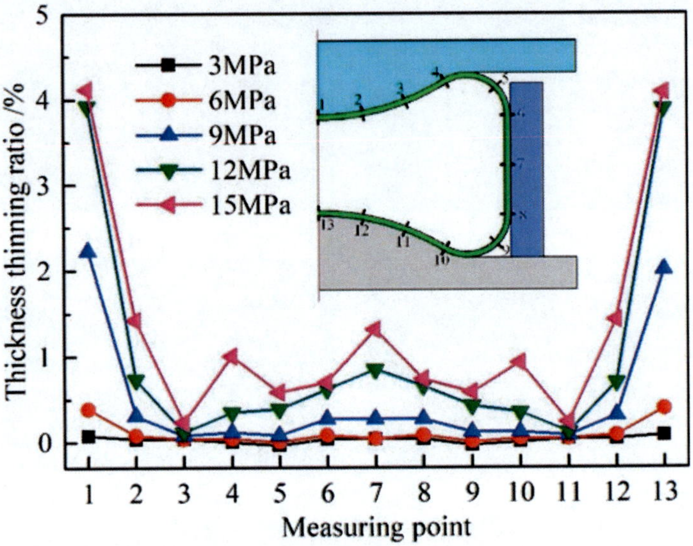

Fig. 3.39 Thickness distribution (thinning ratio) of concave cross-section

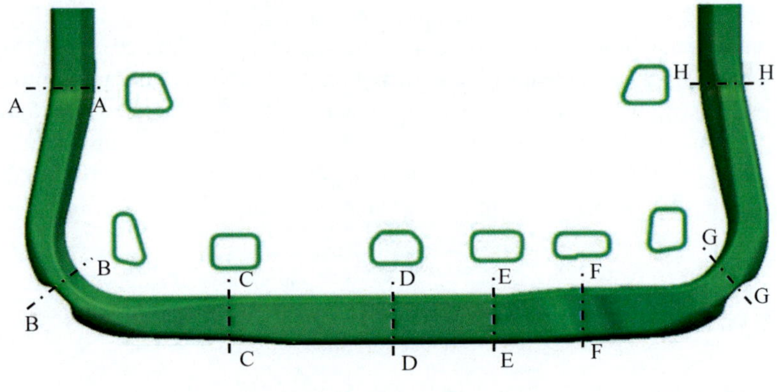

Fig. 3.40 Geometrical model of an engine cradle

For the case in this book, the low alloy high strength steel tube with an outer diameter of 63.5 mm and thickness of 2 mm is used for this subframe. The initial tube length is 1835 mm, and the corresponding volume is 5.1 L. The yield strength and tensile strength of this tube are 390 MPa and 480 MPa, respectively. The elongation is 40%. The main process parameters for hydroforming are defined as that the calibration pressure is 150 MPa, the clamping force is 16,600 KN, and the axial force is 428 KN. Figure 3.41 shows the thickness distribution in the whole process including bending, preforming and hydroforming obtained by multi-step numerical simulation.

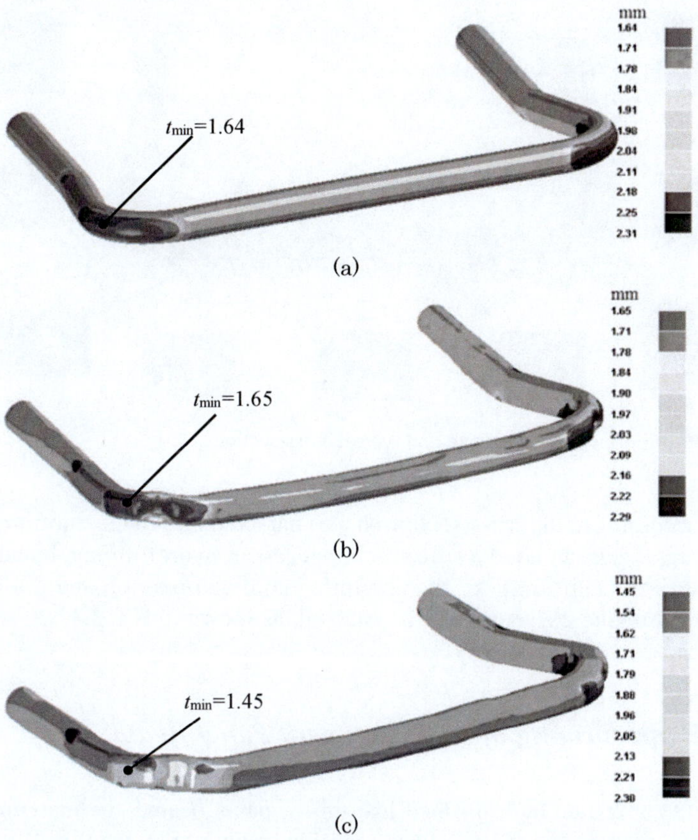

Fig. 3.41 Hydroforming process and the thickness distribution. **a** Bending; **b** preforming; **c** hydroforming

It can be found from Fig. 3.41 that after bending the minimum thickness on the outside of bend is 1.64 mm, and the corresponding thinning ratio is 18%. In the meantime, thickening occurs on the inside of bend, and the maximum thickness is 2.31 mm, which corresponds to a thickening ratio of 15.5%. After preforming, the minimum thickness is nearly unchanged. The thickness distribution after hydroforming is summarized as follows: viewing from the axial direction of the part, the minimum thickness locates in concave area of curved corner and the transition area of the straight wall; viewing from the circumferential direction of the part, the minimum thickness is found at the transition area between the corners and straight wall of the cross-section. Moreover, the minimum thickness is 1.45 mm which locates on outside of the left corner, and the maximum thinning ratio is 27.5%. The maximum thickness is 2.30 mm corresponding to a thickening ratio of 15%.

Based on the principle of lowering the calibration pressure using the concave preformed cross-section, the preform with different concave cross-sections were

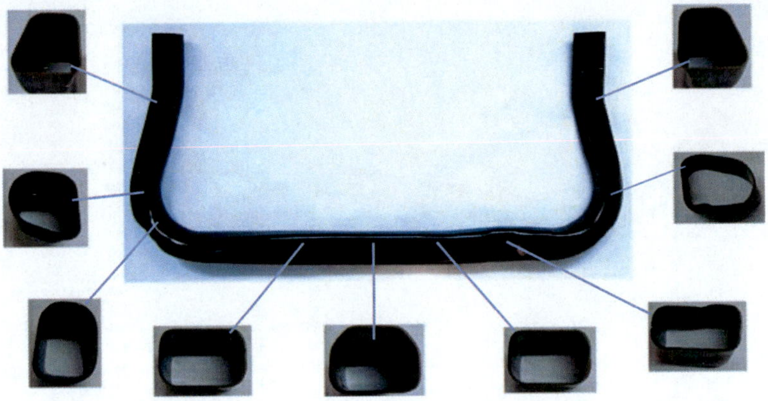

Fig. 3.42 Hydroformed subframe part and typical cross-sections

designed according to the cross-section shape characteristics of this subframe tubular part. This approach can not only ensure the success of hydroforming, but also effectively control the uniformity of the circumferential deformation and the thickness distribution. Finally, the sound part is obtained, as shown in Fig. 3.42.

3.7.2 Hydroforming of an Instrument Panel Beam

Figure 3.43 shows the hydroformed instrument panel beam. The material for this beam is low-carbon steel. The yield and tensile strength of material is 205 MPa and 260 MPa, respectively. The elongation is 36%. The forming difficulties of this part are:

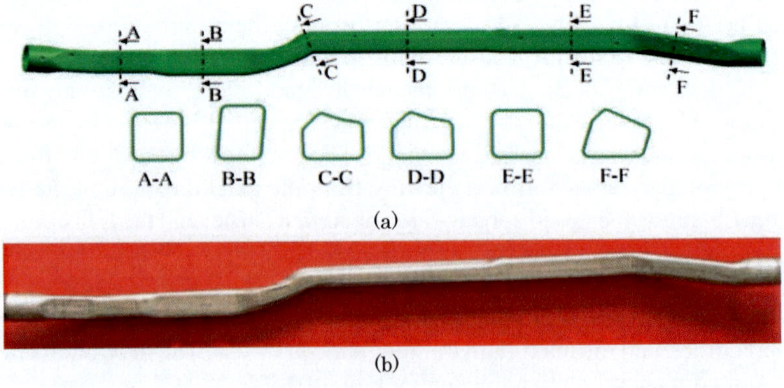

Fig. 3.43 Instrument panel beam. **a** Part shape; **b** hydroformed part

1. the axis is a space curve with a length of 1600 mm, so the size of press bed needs to be larger.
2. there are a variety of different polygonal cross-sections along the axis, including rectangles, parallelograms, pentagons and irregular cross-sections, etc. Due to the complex transition shapes between each cross-section, the preforming is very difficult.
3. the tube thickness is only 1.75 mm, which brings great difficulty in achieving the sealing of tube ends.
4. the asymmetry along the axial direction leads to non-uniform deformation of the material.
5. the welded tube leads to inhomogeneous mechanical properties, as the mechanical properties of the weld and heat-affected zone are significantly poorer than that of the base material.

The typical process of forming an instrument panel beam includes bending, preforming, and hydroforming. During bending, the tube inside bend is prone to experience wrinkling due to the small bending radius. To avoid such a defect, reasonable structural mandrel and anti-wrinkle block can be used. Preforming is the key stage for forming the instrument panel support. Through the appropriate pre-pressing operation, the reasonable distribution of materials is realized in advance, the cooperative circumferential deformation during hydroforming is ensured, which can avoid the defects of local wrinkling and cracking. As a result, the qualified parts are formed, as shown in Fig. 3.43b. It is seen clearly that the corner dimensions of the parts meet the design requirements, and the surface is free from wrinkles.

3.7.3 Hydroforming of an Aluminum Alloy Tubular Part

Figure 3.44 is a schematic diagram of an aluminum alloy tubular part. Its typical cross-section shapes include trapezoid and ellipse. The material used for this part is rust-proof aluminum alloy. The forming difficulties are:

1. the material is aluminum alloy, which has poor ductility resulting in a high tendency of splitting during forming;
2. after deformation, coarse grains and the appearance of orange peel are easy to generate, which will result in the deterioration of performance;
3. during forming, the deformation of such an asymmetric structure is not uniform. As a result, the local deformation is too large leading to the splitting defect;
4. the bending radius of the axis is too small to be bent using CNC bending.

Through the processes of bending, preforming and hydroforming, the aluminum alloy tubular part with irregular cross-sections that meets the design requirements is formed, as shown in Fig. 3.45. In addition, Fig. 3.46 shows the thickness distribution in each forming process obtained through numerical simulations. The initial thickness is 1.2 mm. It can be found from Fig. 3.46 that, during bending, the thickness

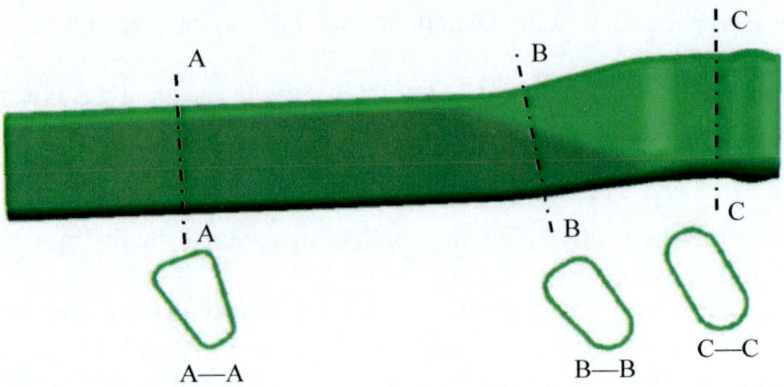

Fig. 3.44 Geometrical model of an aluminum alloy tubular part

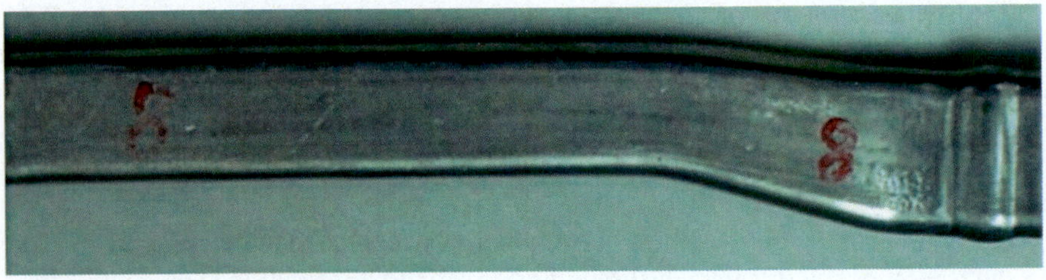

Fig. 3.45 Aluminum alloy tubular part with irregular cross-sections

on the outside of the bended tube is reduced to 1.17 mm with a thinning ratio of 2.5%, while the thickness on the inside correspondingly is increased to 1.27 mm with a thickening ratio of 5.8%. In the preforming process, the thickness remains basically the same. During hydroforming, when the pressure reaches a certain value, the minimum thickness is 1.11 mm with a thinning ratio of 7.5%, while the maximum thickness is 1.30 mm with a thickening ratio of 8.3%.

3.7.4 Hydroforming of an Aluminum Alloy Subframe Tubular Part

Figure 3.47 shows the shape and typical cross-sections of an aluminum alloy tubular part for passenger car subframe. This part has a three-dimensional space axis. Its cross-section is mainly circular, and the tube end has a track-shaped cross-section. Among all typical cross-sections, section A-A has the largest expansion ratio of 3.8%, while section G-G has the smallest value of 2.54%. The mechanical properties of this aluminum alloy tube were measured by hydraulic bulging test. The yield and tensile strength are 61 MPa and 170 MPa, respectively. The n value is 0.26.

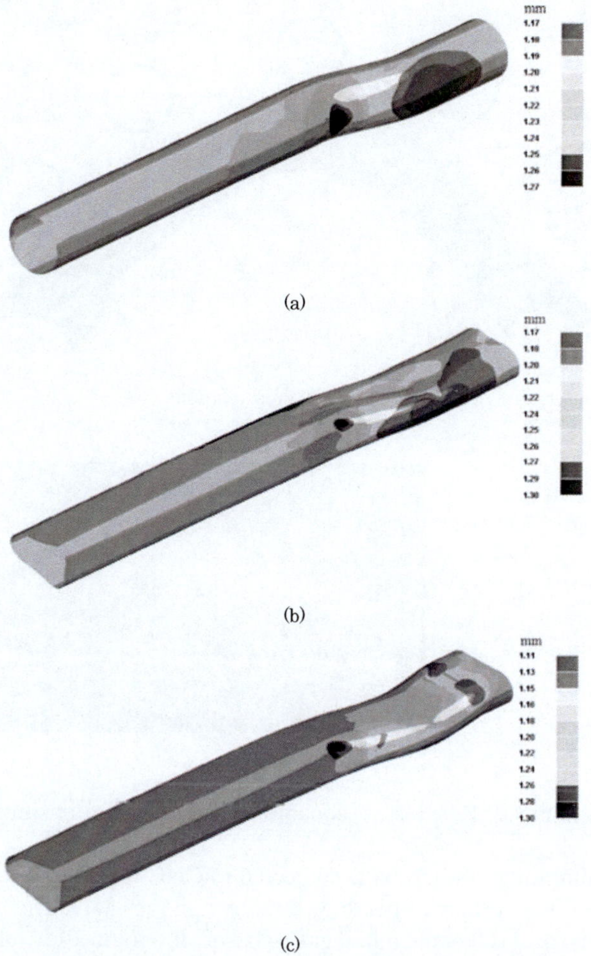

Fig. 3.46 Forming procedure and thickness distributions of aluminum alloy tubular part. **a** Press bending; **b** preforming; **c** hydroforming

Different defects are possibly found in the process of CNC bending and hydroforming of aluminum alloy. In the CNC bending process, the main defects are wrinkling and orange peel. However, splitting is the dominant defect for hydroforming.

Wrinkling is one of the defects in CNC bending of aluminum alloy tubes, as shown in Fig. 3.48. The main factors that affect the wrinkling of aluminum alloy tubes during CNC bending are gap between guide die and tube, size of mandril, lateral and axial pushing forces, etc. In order to avoid the wrinkling, a series of solutions can be adopted. For example, the position of the wrinkle-proof plate can be adjusted so that it effectively constrains the material on the inside of tube to flow back. Moreover, a suitable gap between guide die, anti-wrinkle plate, mandrel and

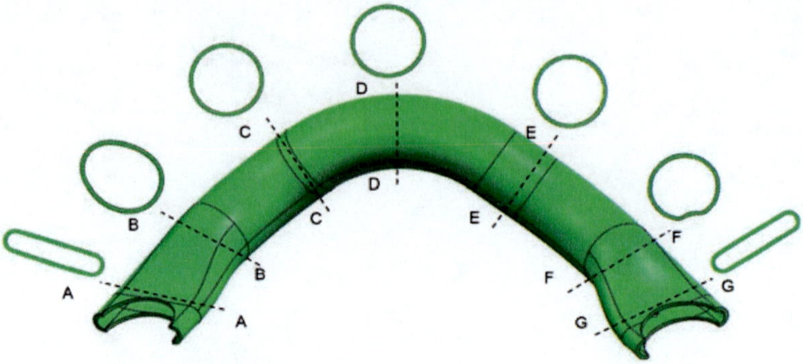

Fig. 3.47 Aluminum alloy subframe and typical cross-sections

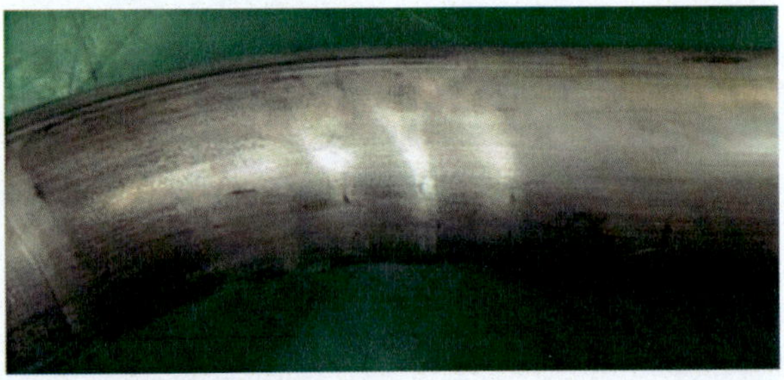

Fig. 3.48 Wrinkling in bending process

tube can be guaranteed. Besides, reasonable lateral and axial pushing force can be also adopted.

When the aluminum alloy tube is subjected to CNC bending, orange peel tends to occur on the outer surface of bending, as shown in Fig. 3.49. The main reason of occurring orange peel is that the initial grain size of aluminum alloy tube is large, and when the non-uniform plastic deformation occurs, local dent and bulge are formed on the surface. The factors influencing the peel structure include the following aspects: initial grain size, microscopic texture, plastic deformation, and loading mode, among which the initial grain size is the main reason influencing the orange peel. Therefore, in order to effectively control the formation of orange peel structure, the initial grain size of the tube must be strictly controlled.

Splitting is the most common defect in the hydroforming of aluminum alloy tubes. It generally occurs on the outer side of the bent tube, the transition area with dramatic changes in shape, and positions with large deformation, as shown in Fig. 3.50. The splitting on the outside of the bending is mainly caused by excessive thinning during bending, which leads to splitting in the subsequent hydroforming process due to insufficient ductility. Moreover, for the position with a dramatic shape variation, after die clamping, the gap between the tubular part and the die is too large due to the obvious shape difference. Therefore, the local expansion ratio is too large which would lead to splitting in the hydroformed part. In addition, for the tubular part with

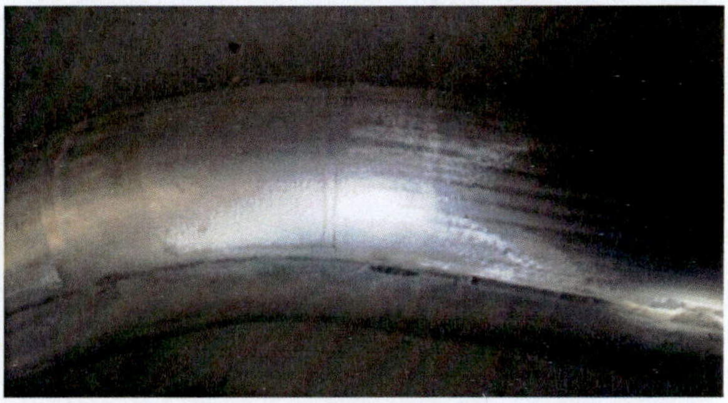

Fig. 3.49 Orange peel on the outside surface

polygonal cross-sections, it is easy to crack in the transition area between corner area and straight wall area, which is consistent with the deformation behavior of steel and other materials. This is mainly caused by the excessive thinning of the transition area due to friction. In order to avoid the splitting, one approach is to increase the axial pushing force and improve the surface quality of the bending die. Subsequently, the thickness thinning on the outside can be controlled, and the surface quality of the bending parts can be improved. The other is to optimize the preform shape, and adjust the gap between the preformed tube and hydroforming die, so that the shape of the tube axis is approaching to the shape of the die cavity as much as possible. In the meantime, the gap between tube and die is able to be evenly distributed in the circumferential direction.

The thickness distributions of hydroforming aluminum alloy subframes have an important influence on their operational performance. The thickness distribution during bending is the main factor that affects the thickness distribution of the final hydroforming. Therefore, if the thickness of bending parts decreases seriously, it is easy to lead to splitting defects in the subsequent hydroforming process. Figure 3.51 shows the thickness thinning ratio during hydroforming of aluminum alloy tubes. It can be seen that the maximum thinning ratio of the tube is 15.8% after bending, which locates on the outside of the D-D cross-section, and the maximum thickening ratio is 17.4%, which locates on the inside of the D-D cross-section. After hydroforming, the

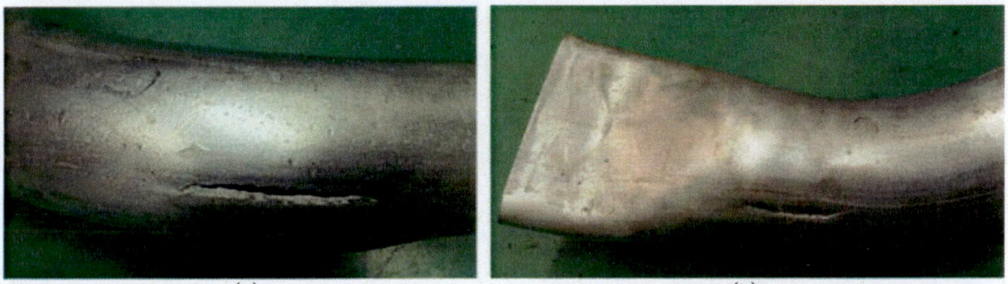

(a) (b)

Fig. 3.50 Splitting defect of aluminum alloy tube. **a** Splitting on the outside of bending; **b** splitting at the transition area

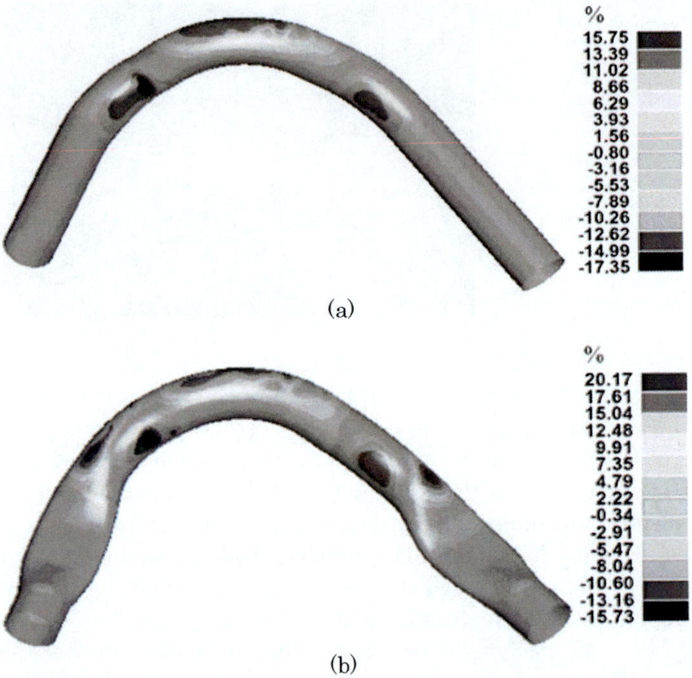

Fig. 3.51 Thickness thinning ratio during hydroforming of aluminum alloy tubes. **a** Bending tube; **b** hydroformed part

maximum thinning ratio is 20.2%, locates on the outside of the F-F cross-section, while the maximum thickening ratio is 15.7%, locates on the inside of the D-D cross-section.

The forming process of the aluminum alloy subframe is given as follows: tube material → CNC bending → preforming → hydroforming. To increase the final mechanical properties, the artificial aging process at the condition of 175 °C × 8 h is needed. The forming dies for this aluminum alloy subframe include the CNC bending die, the preforming die and the hydroforming die. Because the bending radius of the part is small, and there is no transition between each bending segment, the bending die is designed into a complicated three-layered forming die,. The hydroforming die is composed of upper die, lower die and sealing device. A special-shaped sealing is adopted in the sealing device. Figure 3.52 shows the parts of each forming process for the aluminum alloy subframe, including CNC bending part, preformed part and hydroformed part.

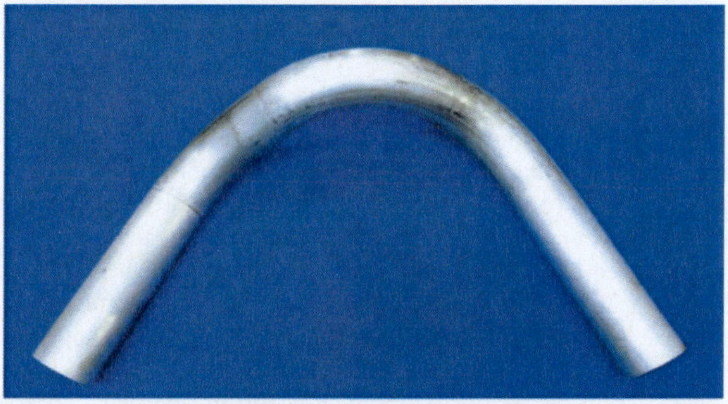

(a)

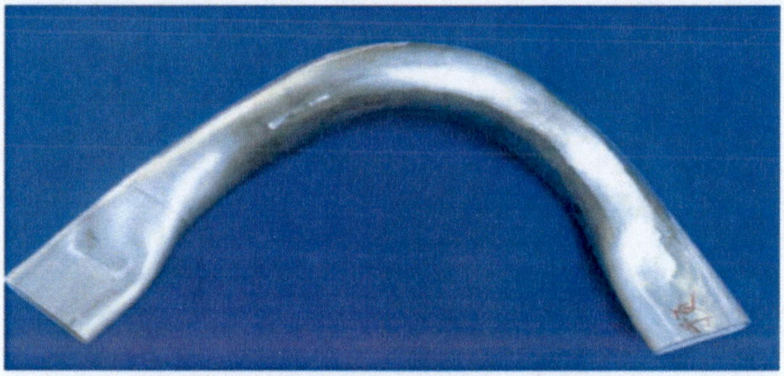

(b)

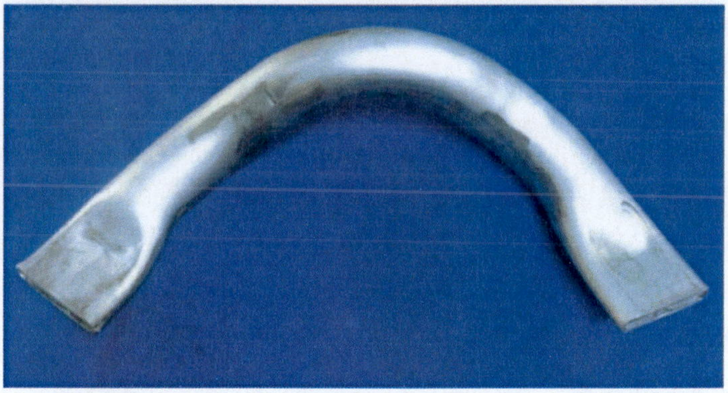

(c)

Fig. 3.52 Parts of each forming process for the aluminum alloy subframe. **a** Bending part; **b** preformed part; **c** hydroformed part

3.7.5 Hydroforming of MPV Passenger Car Subframe

The tubular part for a MPV subframe is a typical hollow structure with variable cross-sections and a three-dimensional axis, as shown in Fig. 3.53. The perimeter of cross-section varies along the axis. The maximum cross-section perimeter is 245.4 mm (D-D cross-section), and the minimum value is 234.2 mm (C-C cross-section). In addition, there are 10 holes with a diameter of 10 mm on the part.

The forming process of the subframe mainly includes CNC bending, preforming and hydroforming. If preforming is not used, the bent tube is directly put into the hydroforming die followed by die closing, then the undercut defect appears in several positions (Fig. 3.54a) and the subsequent hydroforming process cannot be completed. Therefore, preforming process is essential. First of all, the bent tubes should be preformed, and then put into the hydroforming die followed by die clamping, which can successfully complete the die clamping process (Fig. 3.54b).

Figure 3.55 shows the thickness distributions of part in each process for the subframe forming. It can be found that after bending the minimum thickness on the outside of the bend is 2.23 mm corresponding to a thinning ratio of 14.2%, while the maximum thickness is 3.04 mm, which locates on the inside of the bend corresponding to a thickening ratio of 16.9%. After preforming, the thickness remains basically unchanged. After hydroforming, the minimum thickness is 1.97 mm, corresponding to a thinning ratio of 24.2%, while the maximum thickness is 3.23 mm, corresponding to a thickening ratio of 24.3%.

Finally, the parts in each operation for the subframe of MPV passenger car are obtained, including the CNC bent part, the preformed and the hydroformed part, as shown in Fig. 3.56. Moreover, there are 10 holes on the hydroformed tubular part. These holes are all formed by the hydro-piercing process integrated with the hydroforming.

For the tubular parts having the welding seam on the inside, no defects appear in the positions of welding seams and heat-affected zones when the workpiece is completely attached to the die during hydroforming. Good experimentation consistency has been achieved by repeated experimental verifications. Although the welding seam is

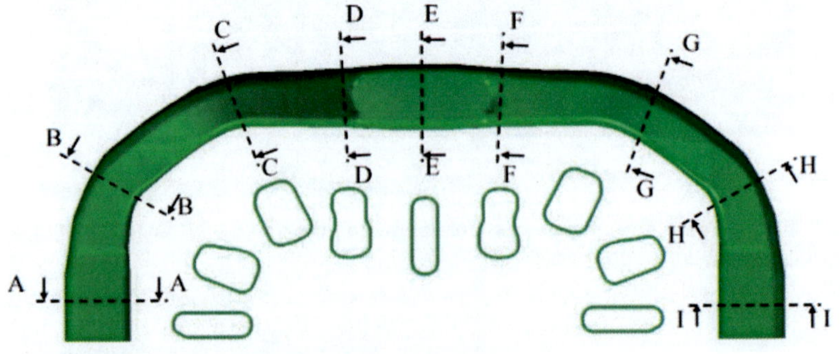

Fig. 3.53 Geometrical model of a MPV passenger car subframe

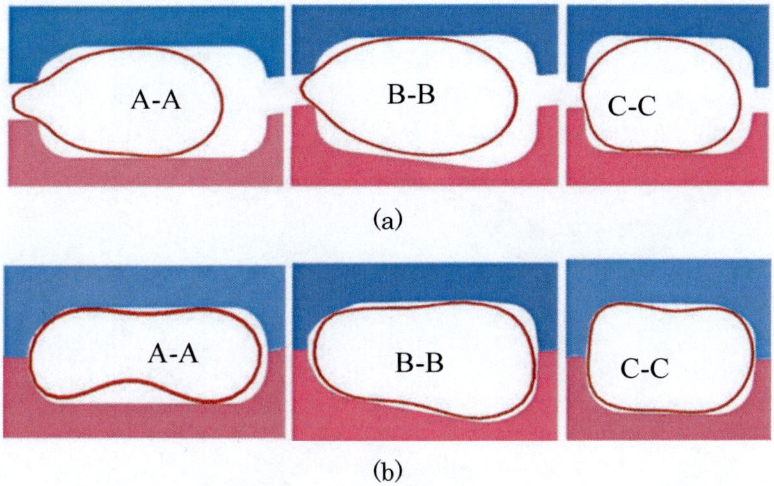

Fig. 3.54 Die clamping process. **a** Die clamping without preforming; **b** die clamping of a concave preform

prone to experience splitting, there is no splitting defect on the welding seam in the calibration process even if the calibration pressure is very high. This is due to that the welding seam is far away from the large deformation corner area.

In order to obtain the mechanical properties of the part after hydroforming, samples were machined from different positions to test the tensile properties and hardness. The mechanical properties obtained are shown in Table 3.7. It is clearly seen that the strength of the weld is much higher than that of the base metal, while the elongation is lower than that of the base metal.

3.7.6 Hydroforming of a DP590 Trailing Arm

The axis of the trailing arm tubular part is a three-dimensional space curve with different bending radii. Moreover, there are small bending radii at local locations sustaining large bending deformation. Figure 3.57 shows the part drawing of this control arm. It can be seen that the typical cross-sections of the control arm are rectangular and trapezoid, which has been dramatically deformed along the axis. The maximum width of the cross-section is 82 mm, which locates at cross-section C-C. This cross-section is not only the location where bending deformation occurs, but also the location with the largest width. Due to the double accumulation of thinning by bending and large expansion, this area is the weakest position and the most difficult area for forming. The minimum width of the cross-section is 65 mm, located at cross-section A-A and E-E, as shown in Fig. 3.57. Moreover, the material

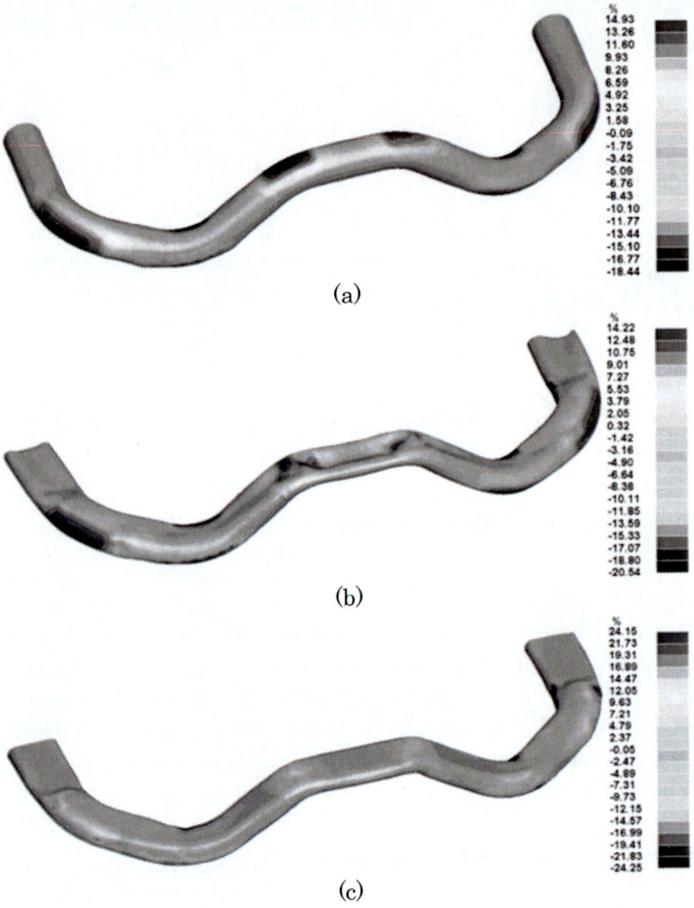

Fig. 3.55 Thickness variation during the hydroforming process. **a** Bent tube; **b** preform; **c** hydroformed part

for this control arm is duplex high strength steel DP590. The yield and tensile strength are 390 MPa and 590 MPa, respectively, and the elongation is 21.6%.

The springback of high strength steel is severe after bending. The bending angle of the tube is one of the main factors that influences the springback. The degree of springback has a roughly linear correlation with the bending angle. As shown in Fig. 3.58, with the increase of bending angle, the springback angle of the tube also increases. The larger the bending angle, the larger the length of the deformation zone. In this case, the ratio of the total plastic deformation to the elastic deformation increases correspondingly, which results in a larger springback angle.

The effect of springback must be considered in CNC bending process. Generally, the compensation method of springback is used to compensate the springback in the process of CNC bending. For the tubular part without springback compensation,

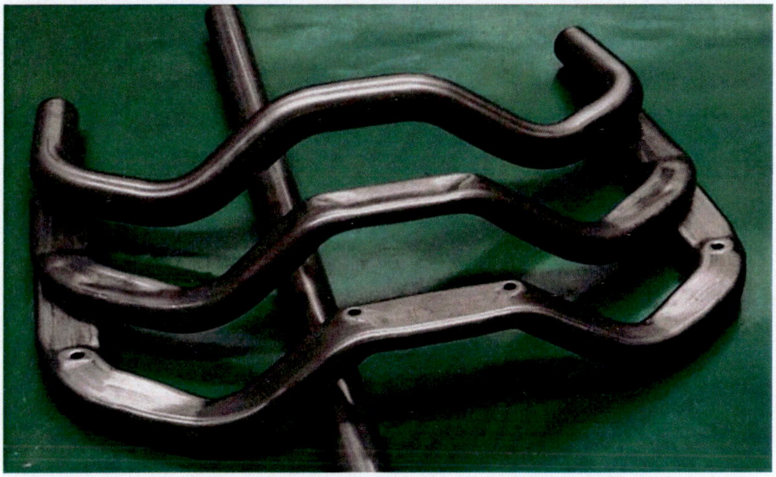

Fig. 3.56 Parts in each operation for the subframe of MPV passenger car

Table 3.7 Mechanical properties of the hydroformed parts

Section	Materials	Yield strength (MPa)	Tensile strength (MPa)	Elongation (%)
A-A	Welding seam	443.47	517.47	15.43
	Base metal	359.27	437.53	27.14
D-D	Welding seam	441.96	514.89	16.57
	Base metal	372.14	454.25	25.14
E-E	Welding seam	439.44	513.45	17.51
	Base metal	374.65	451.43	24.71
F-F	Welding seam	437.78	510.24	17.58
	Base metal	371.58	448.47	25.29
I-I	Welding seam	428.81	498.85	17.65
	Base metal	368.30	443.70	26.14

the actual bending angle is less than the designed value. Therefore, the undercut defect is easy to occur in the preforming process. If the springback is considered and a certain amount of compensation is given in the bending process, the bending angle of the obtained tubular part is consistent with the design angle. In this case, the subsequent preforming process can be successfully completed, and the qualified preformed tubular part can be obtained.

The location of welding seam has an obvious influence on the splitting defect in the hydroforming process. Figure 3.59 shows the effect of different weld positions on hydroforming. The welding seams are located in the neutral layer of bending, the outside and inside of neutral layers. When the welding seam is located in the neutral layer and the outside of bending, splitting defects occur in both conditions. However,

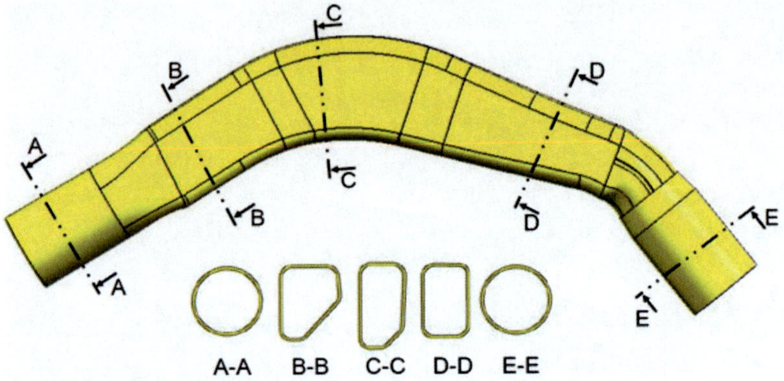

Fig. 3.57 Part drawing of the trailing arm

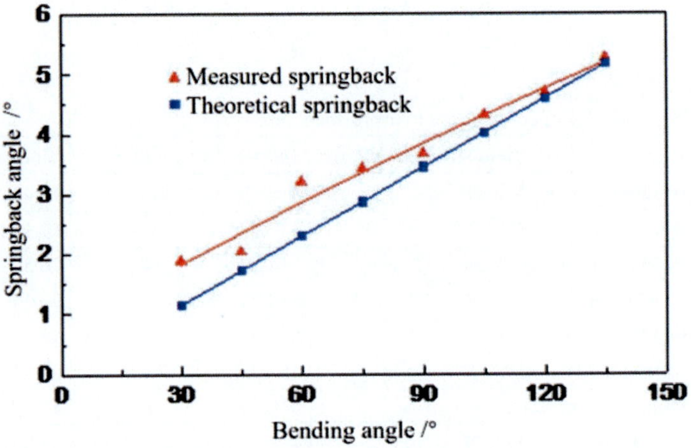

Fig. 3.58 The relationship between springback and bending angle

when the weld is located in the inner side of the bending, no splitting occurs, and a qualified hydroformed part is obtained. Therefore, during the hydroforming of high strength tubular parts with a curved axis, placing the welding seam in the inside position of tube blank is beneficial to suppress the generation of splitting defects.

Moreover, the axial feeding length is also an important process parameter that affects forming. If the axial feeding lengths are different, two typical defects of splitting and wrinkling will appear, respectively. When the axial feeding length is very small, splitting occurs at low pressures. However, in the case of excessive axial feeding, wrinkling cannot be eliminated even though the pressure is very high. Therefore, only when a reasonable axial feeding length is given, neither splitting nor wrinkling occurs, as shown in Fig. 3.60.

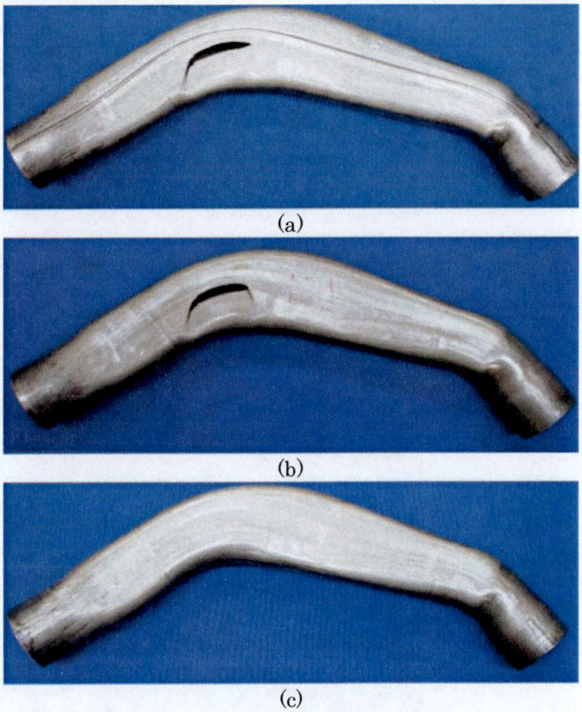

Fig. 3.59 Effect of weld position on forming. **a** Neutral layer of bending; **b** outside of the neutral layer; **c** inside of the neutral layer

3.7.7 *Hydroforming of Torsion Beam of a High Strength Steel (780 MPa)*

Figure 3.61 shows the 3D diagram of a torsion beam and its typical section shapes. The tubular part is a structure with V-shaped cross-section, which has a complex variation along the axis. Moreover, the perimeter of cross-section varies significantly along the axis. The width of the cross-section is analyzed along the axis direction, and the shape of the cross-section changes from the end to the middle. The maximum width of the cross-section is 120 mm, located at the end region (cross-section I-I), and the minimum width is 90 mm, located in the middle region (cross-section A-A). The tube material used for the torsion beam is the advanced high strength steel with a tensile strength of 780 MPa. The strain hardening exponent n is 0.10 and strength coefficient K is 1050 MPa.

In the hydroforming process, the loading path is usually composed by three aspects:

1. The curve of die clamping force with time;
2. The curve of internal pressure with time;
3. Relationship between axial feeding length and internal pressure. Figure 3.62 shows the loading paths for the hydroforming process of the torsion beam, where the initial internal pressure is used to avoid buckling caused by the axial force.

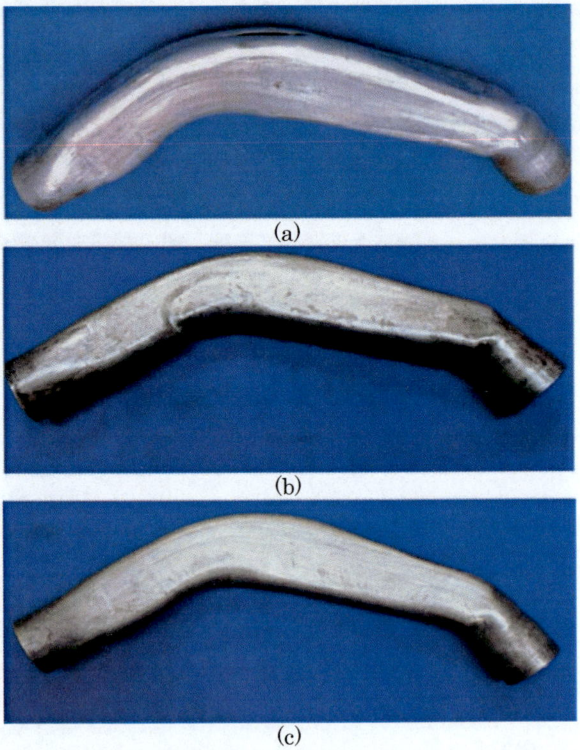

Fig. 3.60 Effect of axial feeding length on hydroforming. **a** Too small axial feeding length; **b** too large axial feeding length; **c** reasonable axial feeding length

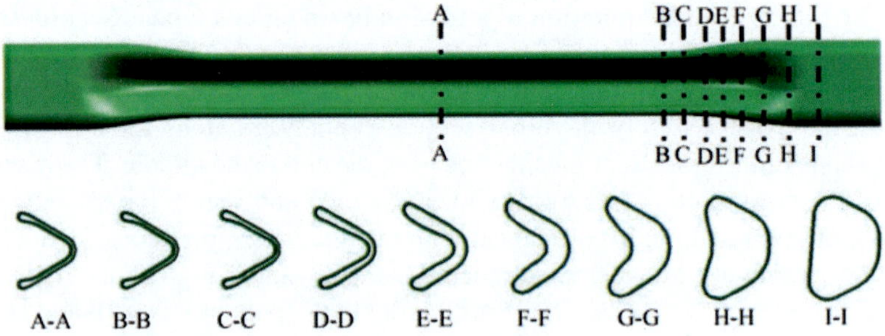

Fig. 3.61 Geometrical model and typical cross-section shape of torsion beam

The axial feeding length has an important influence on the thickness distribution and forming accuracy of the hydroformed torsion beam. Figure 3.63 shows the numerical simulation results of the thickness distribution when the relative axial feeding lengths (the ratio of axial feeding length to part length) are 4%, 8% and 12%. It can be seen that when the relative axial feeding is 12%, buckling occurs

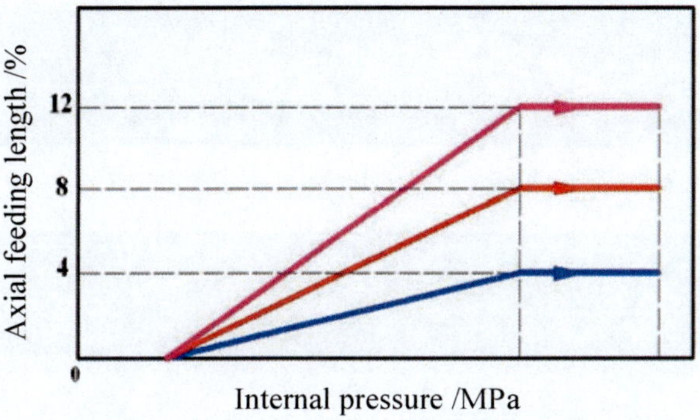

Fig. 3.62 Loading paths for the hydroforming process of the torsion beam

at the tube ends and the transition area of the intermediate V-shaped cross-section. When the relative axial feeding is between 4 and 8%, the optimum forming effect can be obtained. Therefore, the reasonable loading path is critical for the successful forming of the torsion beam free of splitting defect.

When the shape of preform is not reasonable, flash defect appears at the transition area of the V-shaped cross-section in the middle of the tubular part and the end cross-section during the die clamping process, as shown in Fig. 3.64a. However,

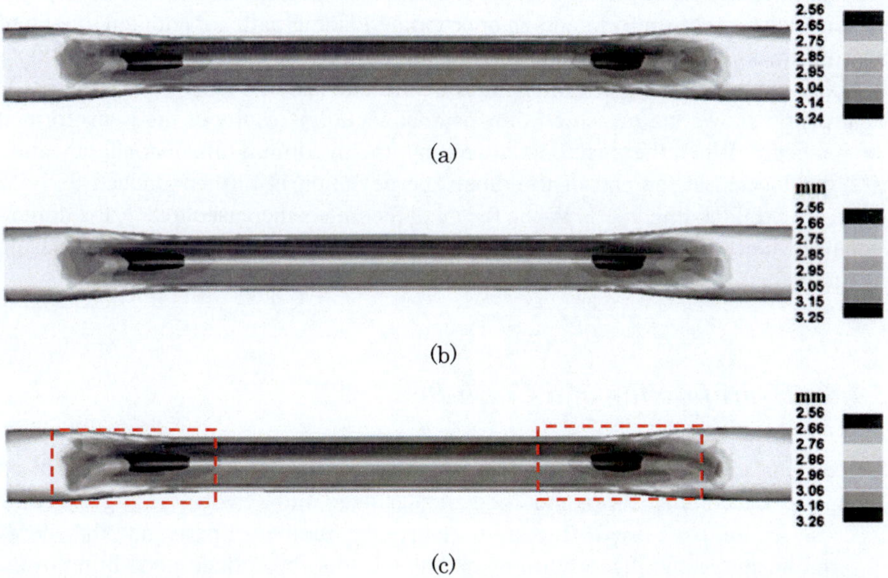

Fig. 3.63 Thickness variation of torsion beam during hydroforming process with different axial feeding length. **a** 4%; **b** 8%; **c** 12%

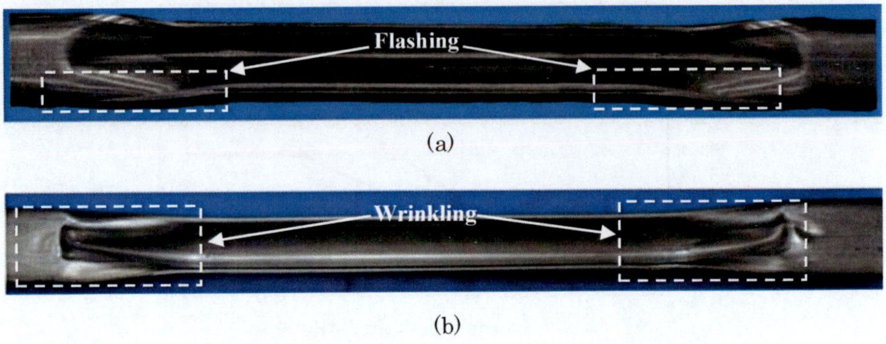

Fig. 3.64 Defects in the hydroforming of torsion beams. **a** Flashing; **b** wrinkling

Fig. 3.65 Sound torsion beam

wrinkling defect appears in the transition area of the tube end in the hydroforming process when the relative axial feeding is more than 8%. Even with high calibration pressure, wrinkles are difficult to be eliminated, as shown in Fig. 3.64b. Only when a reasonable shape of preform and an appropriate loading path are adopted, qualified hydroformed torsion beam can be formed successfully, as shown in Fig. 3.65.

Axial feeding can significantly improve the dimensional accuracy of the parts. Figure 3.66 shows the measured dimensional accuracy results of the hydroformed torsion beam. When there is no axial feeding, the maximum dimensional deviation is 2.7%, located at the end of the tube. The deviation is further reduced to 1.5% when the axial feeding is 4%. When the axial feeding is increased to 8%, the dimensional deviation of the whole torsion beam is within 0.5%, satisfying the design requirements.

3.7.8 Hydroforming of a Crash Box

The original design structure of the crash box for cars is a structure formed by stamping and welding. If it is modified and designed into a hydroforming structure, its cross-section is a closed structure to reduce the number of parts, and the energy absorption can be significantly improved also. Under the collision condition of the front rigid wall, the energy absorption of the front bumper assembly increases by about 8% for the hydroformed structure compared to the stamping-welding structure. During the collision deformation, the crash box collapsed gradually. For these two

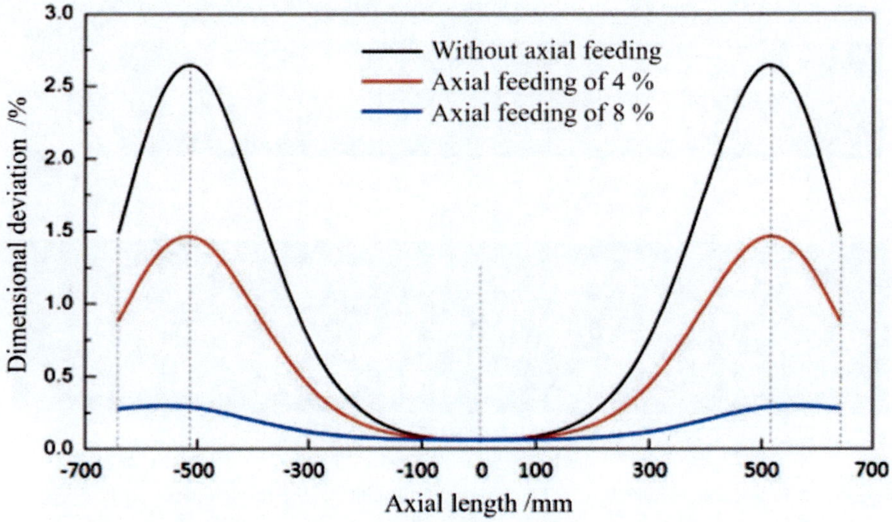

Fig. 3.66 Dimensional deviation of the torsion beam

structures, there is no bending in the middle cross-section of the bumper beam. Moreover, under the condition of 40% offset of the side, the hydroformed front bumper assembly absorbs more energy in the first 25 ms. Similarly, in the process of collision deformation, the crash box collapsed gradually, and there is no bending in the middle cross-section of the bumper beam.

Figure 3.67a shows the numerical results of thinning ratio in hydroforming. It can be seen that the maximum thinning ratio after forming is 13.89% and the maximum thickening ratio is 1.78%. Figure 3.67b shows the final hydroformed crash box. The width of the parts is 82.5 ± 0.2 mm. The height dimension of the parts is 91.7 ± 0.2 mm. All the dimensions meet the design requirements.

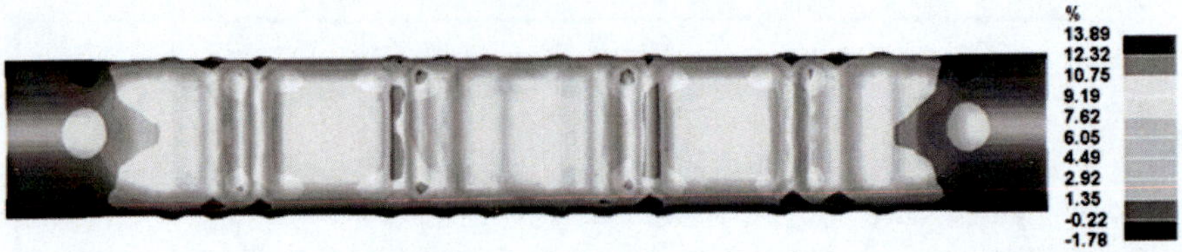

(a)

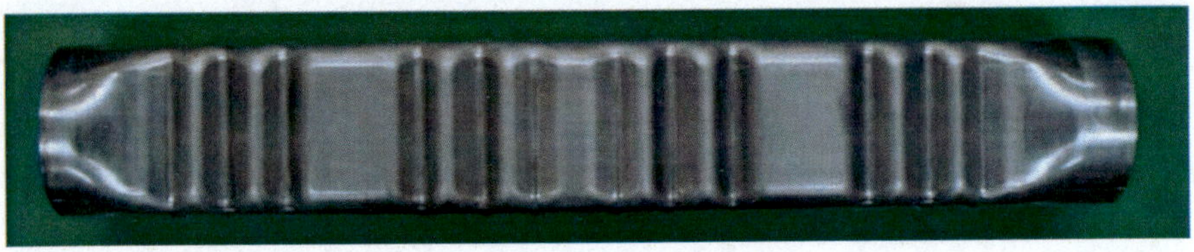

(b)

Fig. 3.67 Hydroformed crash box. **a** Thinning ratio; **b** photo of the formed part

Chapter 4
Hydroforming of Multi-branch Tubular Parts

4.1 Classifications and Processes of Multi-branch Tubes

Multi-branch tubular parts have various types, which can be classified using different criteria as:

(1) According to the specific part profile, there are straight three-branch (T-typed), oblique three-branch (Y-shaped), U-shaped, X-shaped four-branch and multi-branch (greater than 5) tubular parts, as shown in Fig. 4.1;
(2) According to the diameters of main and branch tubes, there are equal and unequal diameter multi-branch tubular parts;
(3) According to the axis shape of part, there are straight and curved multi-branch tubular parts;
(4) Symmetrical and unsymmetrical multi-branch tubular parts can be classified based on the symmetrical feature part;
(5) Multi-branch thick-walled and thin-walled types are classified, while the thin-walled characteristic referring the tube thickness is between 0.5 and 2 mm. Among the above various types of multi-branch tubular parts, T-typed and Y-typed three-branch tubes are the most commonly seen structures.

Three-branch tubes are difficult to form which are traditionally manufactured using two welding-assisted processes, as shown in Fig. 4.2. Figure 4.2a shows the welded structure using two straight tubes with axis intersected, while alternatively, it can be manufactured by welding two stamped half tube segments together to compose a complete tube profile. However, using welding exhibits main problems for manufacturing multi-branch tubular parts, those are:

(1) Deformation resulted from welding induces a high rate of unqualified parts;
(2) Less reliability due to the welding beam and residual stress compared to that using integrated structure;
(3) Unsmoothed inner surface, which results in a high resistance of liquid flowing. In addition, regarding T-typed three-branch part with a low height of branch

S. Yuan, *Modern Hydroforming Technology*,
https://doi.org/10.1007/978-981-19-5775-8_4

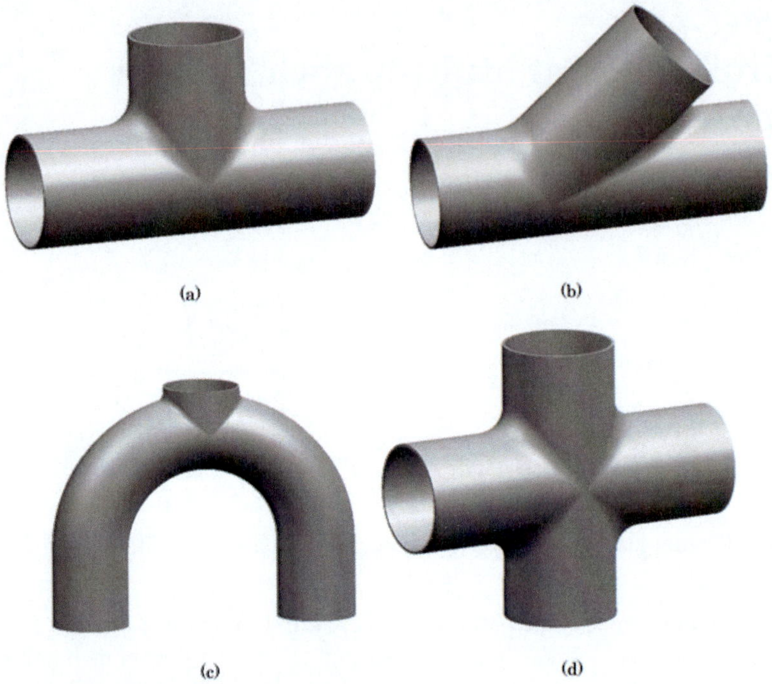

Fig. 4.1 Shapes of typical multi-branch tubular parts

tube, the part can be selectively manufactured using piercing and rigid ball flanging.

Tube hydroforming is a robust and cut-edge technology to manufacture integrated multi-branch tubular part using a single blank within one operation. Figure 4.3 shows the principle of hydroforming a Y-typed tubular part. The tooling is composed by the upper and lower dies, left and right punches, and an additional counter punch that is normally mounted in the vertical center line of die set. Firstly, a tube is positioned on the lower die, followed by a clamping of upper die to fix the tube vertically. Then, the tube is filled with liquid medium, and sealed at the ends by horizontal punches. Finally, the horizontal punches moves inwards to feed, and the tube is deformed at a certain pressure within the tube.

The complete process of hydroforming Y-typed tube can be divided into three stages:

(1) Initial stage (shown in Fig. 4.3a), both the left and right punches perform axial feeding with the counter punch fixed simultaneously. As the top of the branch tube is not in contact with the counter punch still, which results in the biaxial bulging state of material due to inner pressure exerted inside the tube;
(2) Intermediate stage (shown in Fig. 4.3b), this stage initiates when the top of branch begins to contact the counter punch. With further increasing the inner

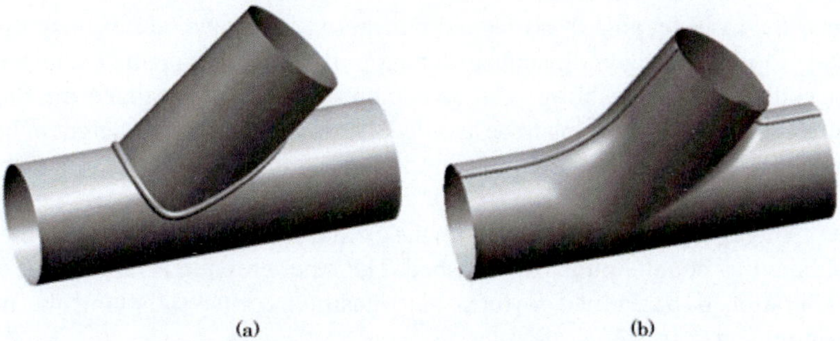

Fig. 4.2 Traditional manufacturing processes of three-branch tubes. **a** Welding two straight tubes; **b** welding stamped half-tube segments

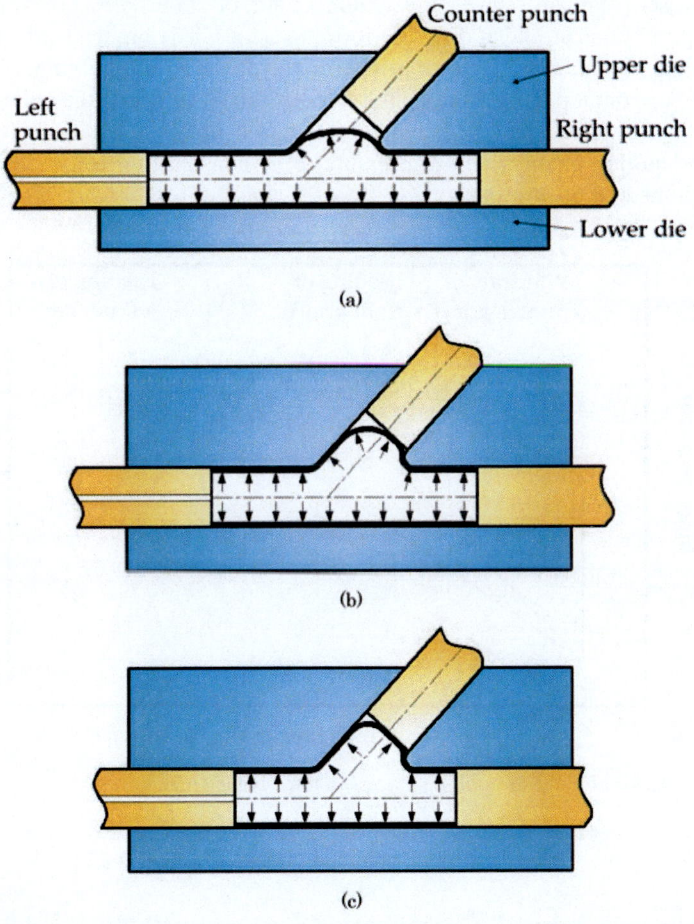

Fig. 4.3 Principle of hydroforming three-branch tube. **a** Initial stage (free bulging); **b** intermediate stage (forming of branch tube), **c** final stage (calibration)

pressure, axial feeding of horizontal punches is continued while counter punch starts to withdraw simultaneously. The top of branch maintains a contact with the counter punch, enabling a certain compressive stress to be given and splitting defect due to over-thinning to be avoided. During this stage, the height of branch should be formed, while the transition corner at the top of branch is not fully obtained yet;

(3) Final stage (shown in Fig. 4.3c), both the axial feeding of horizontal punches and withdraw of counter punch are finished. The inner pressure is rapidly increased. Calibration is performed to form the transition corner satisfying the design requirements.

During the hydroforming of three-branch tubular part, proper combinations between inner pressure and punch stroke are of vitally importance for different stages. Figure 4.4 schematically gives the loading path of inner pressure varied with proceeding of punch stroke at different forming stages. The branch shape is mainly accomplished by the axial feeding of horizontal punches and bulging. Therefore, feeding of horizontal punches dominates at the initial stage, while the counter punch starts to withdraw once the tube is bulged freely to contact the counter punch. Such a withdraw motion enables a back pressure to be constantly given by the counter punch until the end of forming.

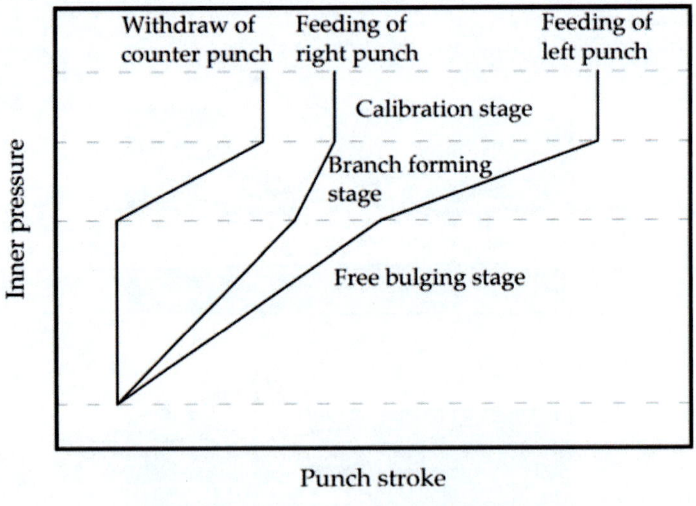

Fig. 4.4 Schematic of the hydroforming process of three-branch tube

4.2 Defect and Limit Height of Branch

4.2.1 Typical Defects

During hydroforming of multi-branch tubular parts, different kinds of defects may occur due to the improper combination between axial feeding of horizontal punches and the withdraw amount of counter punch. The main defects of hydroforming T-typed three branch tube are the splitting at the top of branch and wrinkling of main tube, as shown in Fig. 4.5. Additionally, a concave defect at the transition zone of branch shown in Fig. 4.6, may be experienced due to unsymmetrical feature of Y-typed tube structure.

T-typed tube has a typical unsymmetrical structure vertically due to the branch on one side, while the structure is symmetrical horizontally and the feeding amount of left and right ends are same during hydroforming. The splitting defect is believed to be caused by the improper control of parameters and material properties. For instance, During the initial free bulging stage, the top of branch tube sustains a bi-axial tensile stress state. The inner pressure is too high while the axial feeding is

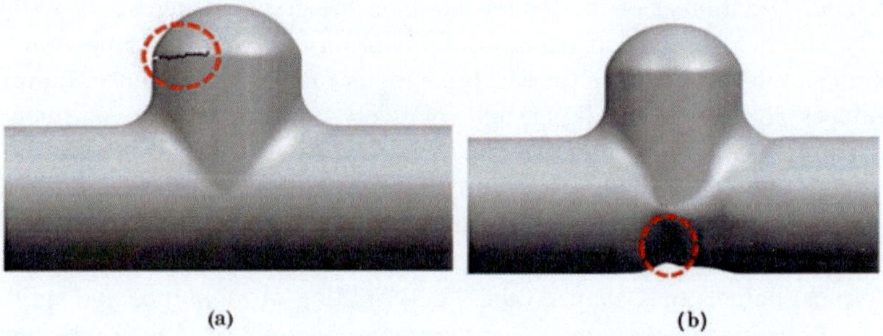

Fig. 4.5 Typical defects of T-typed tube. **a** Splitting of the branch tube; **b** wrinkling of main tube

Fig. 4.6 The concave defect at the transition zone of Y-typed tube

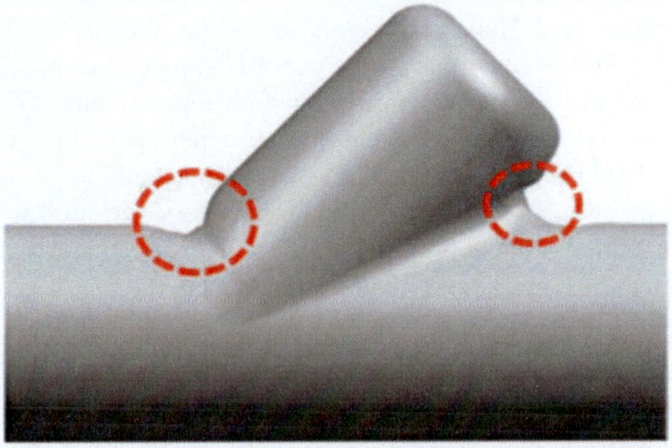

too slow; During the intermediate stage of hydroforming, when the counter punch withdraws too fast, the absence of contact between top of branch and counter punch results in the lack of back pressure support and splitting occurs. In addition, mechanical properties and lubrication also significantly affect the forming, the smaller the strain hardening component, n, the severer the thickness reduction and tendency of experiencing splitting; the smaller the thickness anisotropic parameter, r, the greater the tendency of splitting at the branch tube. The friction mainly influences on the free flow of material at the bulging zone and the success of tube end feeding. When the lubrication is poor and resulted friction is relatively large, the bulging pressure is subsequently increased to form a branch tube feature with a small height, which induces a greater biaxial tensile stress. Splitting is more likely to generate. Therefore, the interfacial friction between tube and die surface is required to be reduced for the sake of successful forming. The interfacial friction can be reduced by, either increasing surface roughness of die cavity, e.g. surface hardening treatments using CVD and PVD, or giving lubricant at the locations with relative movement between the tube and die.

Wrinkling defect of main tube mainly occurs before calibration stage. Either a too fast axial feeding or too low inner pressure could result in wrinkles on the main tube as the axially fed material is unable to flow to compensate the deformation of branch tube. The thinner or the longer the tube, the greater tendency of wrinkling. Similarly, a higher friction due to the poor lubrication could constrain material to flow to the branch tube. Then, once the axial feeding is further given, constraint material flow induces compressive bulking, and relatively uniform wrinkles are generated subsequently.

Y-typed three-branch tube belongs to an unsymmetrical structure both horizontally and vertically. Among the various multi-branch tube, forming of Y-typed tube is the most difficult and its defects are typical. During the hydroforming, besides the above typical defects of T-shaped tube, i.e. wrinkling of main tube and splitting at the top of branch tube, the concave at the transition zone of branch tube resulting from improper allocation of left and right feeding amount, may be also found. During the final calibration stage, neither left nor right side feeding is necessary theoretically, calibration is accomplished by increasing inner pressure with reliable sealing maintained. However, in some occasions, the unsatisfying control accuracy of hydroforming system or improper punch sealing could cause the punch displacement larger than the optimal magnitude, which induces the concave at the transition zone of branch tube. Such a defect may be experienced on either left, right or both sides.

Reasoning from the above onset mechanisms of defects in hydroforming multi-branch tubes, the loading paths must be controlled within the processing windows, meaning that a proper combination between inner pressure, axial feeding and withdraw stroke of counter punch is the key factor governing a sound forming.

4.2.2 Limit Height of Branch

The achievable limit height of branch tube, H, is the most critical factor of hydroforming multi-branch tubes. The higher the designed branch tube requires, the more difficult the forming. The limit height of branch tube in a multi-branch tube is related to the geometry, material, thickness, loading path and lubrication of multi-branch tube.

Figure 4.7 shows the limit heights of branch tube for some typical multi-branch tubes those can be obtained using hydroforming. For T-shaped three branch tube, the limit height can reach 1 time of original tube diameter; while 0.85 time for Y-typed and bent-typed tubes. In comparison, the limit height of U-shaped tube is much lower, which is only 0.15 times of the original tube diameter. The reasons of above correlations are:

(1) T-typed tube has an equal fed amount on both sides resulting in a greater limit height;

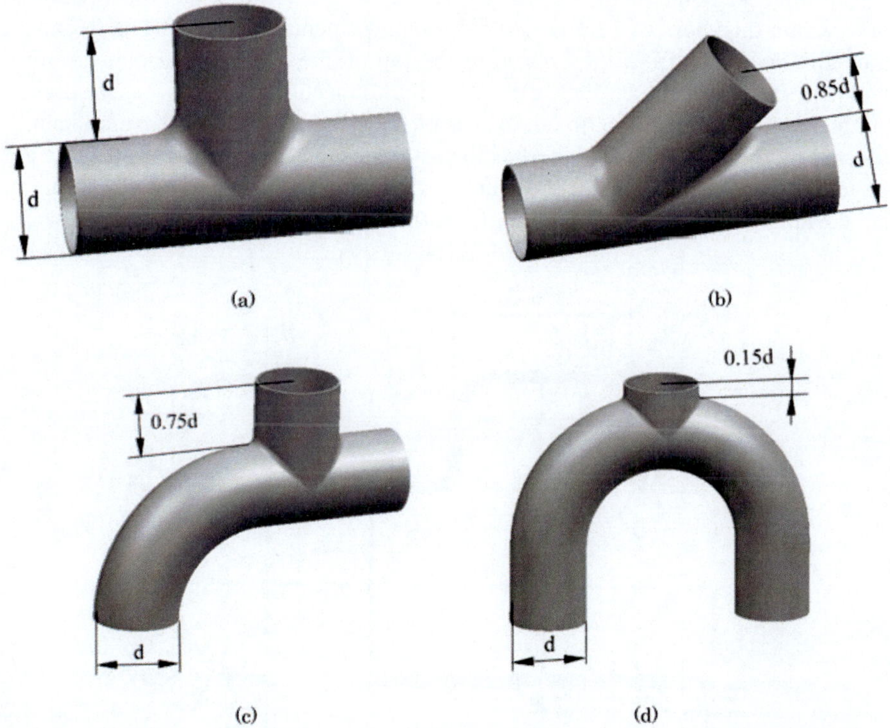

Fig. 4.7 Limit heights of branch for typical multi-branch tubes. **a** T-shaped (H = d); **b** Y-shaped (H = 0.85d); **c** bent-typed (H = 0.75d); **d** U-shaped (H = 0.15d)

(2) Bent-type or U-shaped three branch tubes are difficult to perform feeding, as the feeding zone is bent tube, hence, the limit height of branch tube is relatively smaller.

4.3 Thickness Distribution of Hydroformed Three-Branch Tubular Parts

The thickness distribution varies with different locations on the T-typed tubular part, which is thinning at the top of branch tube, and thickening for the majority of main tube. The thinnest and thickest points are the vertex of branch tube and fed zones of main tube, respectively. Considering the horizontally symmetrical feature of T-typed structure, the thickness distributions are same for the left and right side. Figure 4.8 shows the thickness distribution of a typical hydroformed T-typed tubular part. The initial thickness 2 mm, the material was stainless steel, and the forming pressure utilized was 68.6 MPa.

The thinnest point of this T-typed part locates at the center of top of branch tube, with a thickness of 1.56 mm and thinning percentage of 22%. The thickness gradually decreases from the bottom to the top on the symmetrical plane, with a thickness unchanged point roughly locates at the connection point between the branch and main tubes. In addition, the thickness of three-branch tube is also significantly affected by the mechanical properties of material. In general, the greater the strain hardening part, n, and the thickness anisotropic parameter, r, the less the thinning and deviation of thickness, a more uniform thickness distribution can be obtained.

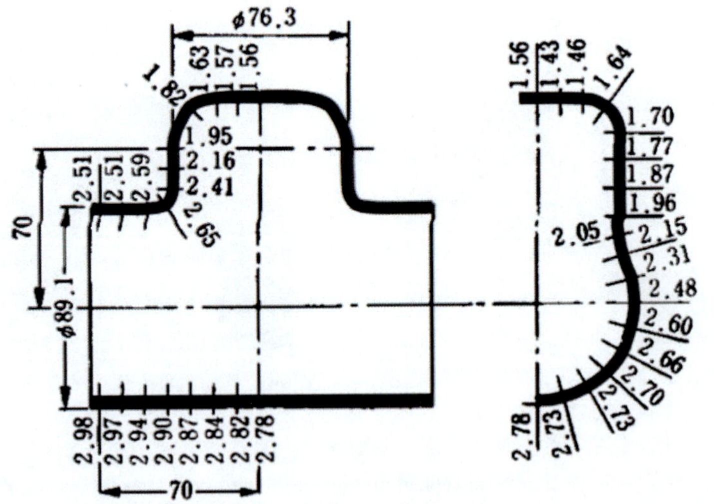

Fig. 4.8 The thickness(mm) distribution of T-typed tubular part

T-typed tubular part belongs to the horizontally symmetrical structure, with the same axial feeding at the two ends of tube during hydroforming, which results in a symmetrical thickness distribution on both sides of the three-branch tube. In comparison, Y-typed tubular part is unsymmetrical in both horizontal and vertical direction resulting in a much more complex thickness distribution than that of T-typed part. Figure 4.9 shows the thickness distribution and location of thickness unchanged line on the Y-typed three-branch part. The initial thickness was 2 mm, and tube material was the stainless steel. The inclined angle of branch tube was 45°. As can be seen in this figure, after hydroforming, a significant thickening can be observed at the transition zone of both sides of formed part. Following the path from the transition zone corner to the top of branch tube, thickness gradually decreases. The locus of thickness unvaried line is V-shaped, locating between the middle and bottom of this branch tube. Thinning mainly occurs on the top region of branch tube with the thinnest point approaching the left corner of top of branch tube. The rest region of this part experiences thickening, with the thickest point locates at the point A on the corner of left transition zone. The thickness is 3.2 mm and the thickening percentage is 60%. The thinnest point locates at the point C on top of branch tube with a thickness of 1.16 mm and a maximum thinning percentage is 38%.

In order to further illustrate the thickness variation of Y-typed tubular part during the three different forming stages, FE simulations were performed to analyze the hydroforming. The computed thickness distributions at different forming stages are shown in Fig. 4.10.

Figure 4.10a shows the thickness distribution at the initial forming stage (inner pressure is 0.25 σ_s), there exists obvious thinning at the top of branch tube with a minimal thickness of 1.81 mm and thinning percentage of 9.5%. In comparison, the feeding zone of main tube, and the corners of left and right transition regions exhibit different degree of thickening. The most severest thickening locates at the left end of main tube, where the maximum thickness reaches 2.42 mm and thickening percentage is 21%. Figure 4.10b shows the thickness distribution during the intermediate forming state when the inner pressure is equal to 0.45 σ_s. In this stage, axial regions of the tube continues to become thicken, while thinning is experienced for the top of branch tube. The most thick location is unchanged and still at left end of main tube, where the maximum thickness reaches 2.6 mm and thickening percentage is 30%; The thinnest location is at the top of branch tube, where the minimal thickness is 1.4 mm, and thinning percentage is 30%. Figure 4.10c shows the thickness distribution after calibration stage when the inner pressure is equal to 0.55 σ_s. For the corner of left transition zone, the maximum thickness is 3.67 mm, and corresponding maximum thickening percentage is 83.5%; For the corner of right transition zone, the thickness is 3.39 mm, and corresponding thickening percentage is 69.5%. The top of branch tube experiences the severest thinning, with a minimal thickness of 1.34 mm and corresponding thinning percentage of 33%.

Based on above simulation results, counter punch plays an important role in the evolution of thickness. Initially (inner pressure is lower than 0.45 σ_s), the absence of contact with counter punch results in a biaxial free bulging state of the branch tube and a quicker thinning rate of branch top. In comparison, the thickening of corners

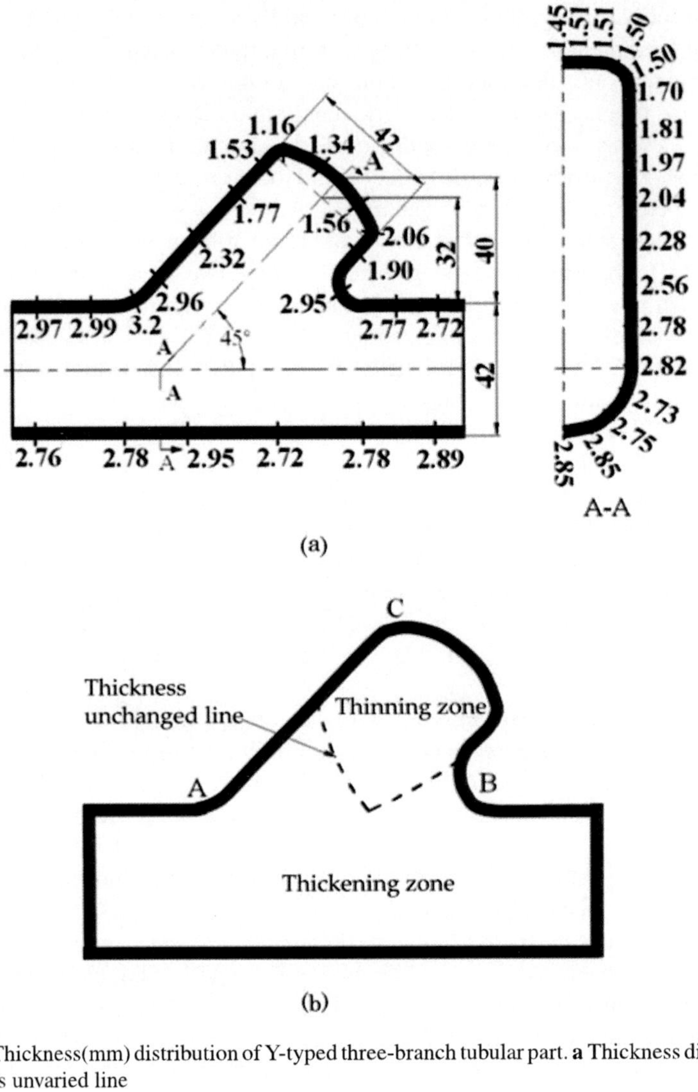

Fig. 4.9 Thickness(mm) distribution of Y-typed three-branch tubular part. **a** Thickness distribution; **b** thickness unvaried line

of left and right transition zones are relatively slow. During the increase of branch tube height (inner pressure is lower than 0.45 σ_s), resulting from the counter push force of counter punch, thinning of branch tube becomes slow, while thickening of corners of left and right transition zones becomes relatively quick.

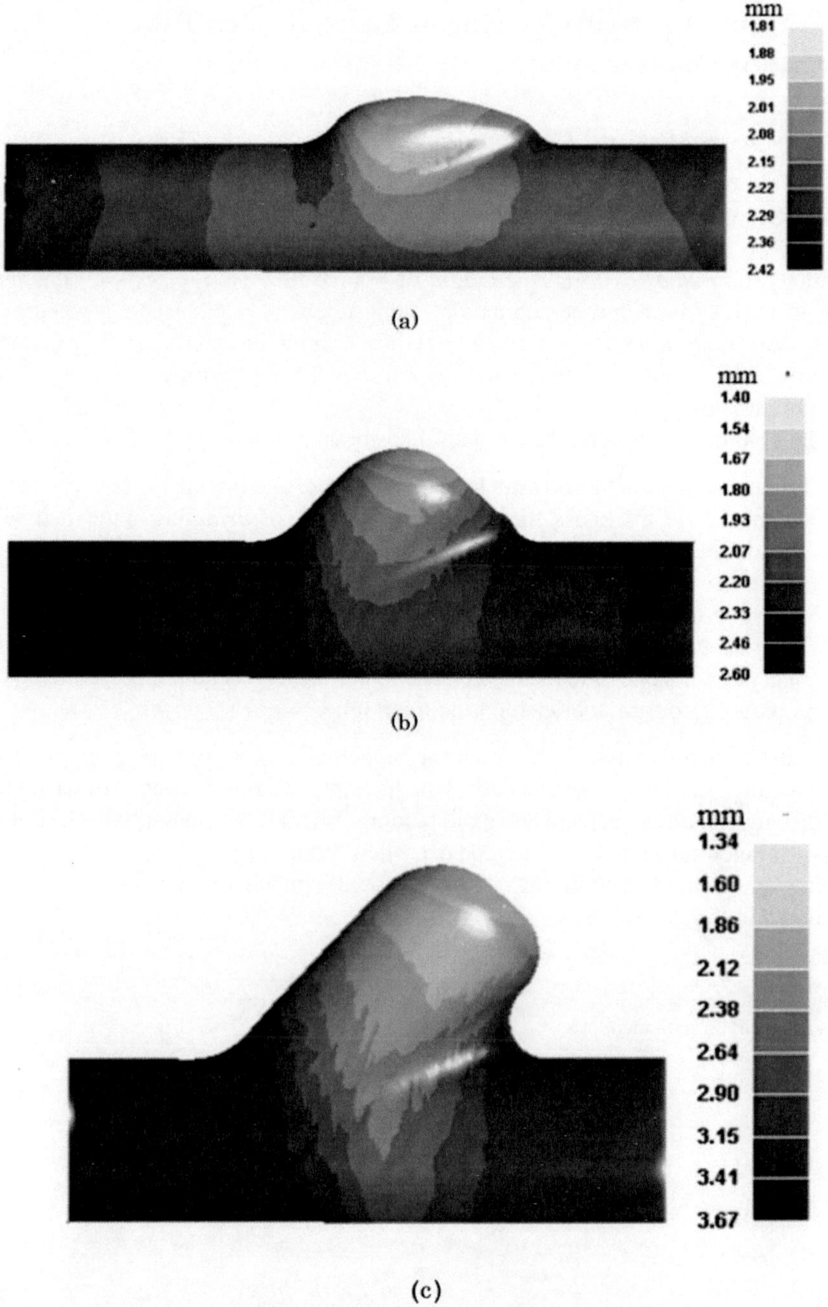

Fig. 4.10 FE results of thickness distribution at different stages during hydroforming Y-typed tubular part. **a** Initial stage; **b** intermediate stage and **c** calibration stage

4.4 Multi-step Hydroforming of Three-Branch Tube of Aluminium Alloys

4.4.1 Geometry and Challenges

Figure 4.11 is the schematic diagram of Y-typed tubular part of aluminium alloy with main geometrical dimensions. The diameter of main tube is $D = 40t$, the diameter of branch tube is $d = 0.9D$, the height of branch tube is $H = 0.95D$, the angle between axis of branch tube and axis of main tube is $\alpha = 62°$, such geometry and dimensions determine the part as a typical three-branch tubular part. The diameter of initial tube is 100 mm, the initial thickness is 2 mm, and the material is 5A03 aluminium alloy.

The dominant difficulties in forming three-branch tubular part are:

(1) Relative small thickness (ratio between diameter and thickness, $D/t = 40$ mm), the thinner the thickness, the greater the tendency of wrinkling of the main tube;
(2) The diameter of branch tube is smaller than the main branch, such a feature results in a greater shrinking of branch tube, the compressive type of deformation is easy to induce wrinkling defect;
(3) The structure is unsymmetrical. Considering the Y-typed part is an unsymmetrical part with one-side locally convex outwards, a great amount of axial feeding is required to form a branch on the main tube.

In addition, there is an angle between the branch and main tube requiring different axial feeding amount of the tube ends, which results in a robust control of the combination between inner pressure and axial feeding. Wrinkling (main tube) and splitting (top of branch tube) defects may coexist when the inner pressure, horizontal axial feeding amount and the withdrawn stroke of counter punch are controlled improperly, as shown in Fig. 4.12.

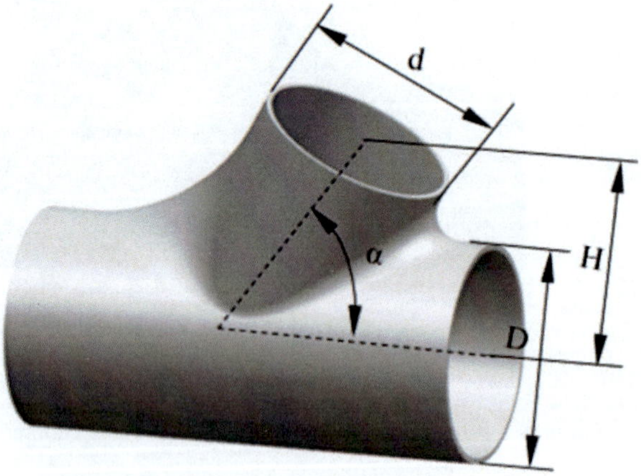

Fig. 4.11 Geometry and dimensions of three-branch thin-walled tube of aluminium alloys

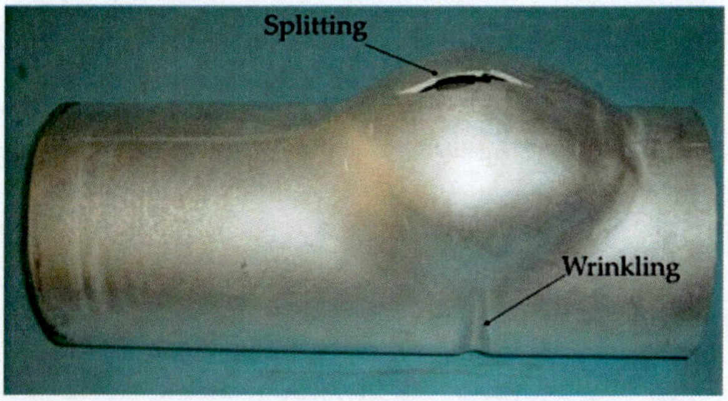

Fig. 4.12 Typical defects of hydroforming Y-typed tubular part of aluminium alloys

The main reason of main tube wrinkling is that, inner pressure is relatively lower, while the axial feeding is too fast during the initial and intermediate forming stages. Then, the material fed axially is unable to flow to compensate the deformation of branch tube in time. When the tube thickness is thinner and initial length is greater, the greater tendency of wrinkling of main tube occurred. The longer the feeding zone, the greater the induced friction force. Subsequently, the material flow is constraint to flow to the branch tube. Compressive buckling is induced with further proceeding of axial feeding. In terms of such a thin-walled Y-typed tubular part, a higher branch tube is normally required, which is difficult to be satisfied using bulging. The axial feeding is the dominant approach to form the branch tube. However, if there is no back pressure supplied by the counter punch, thinning is very likely to be experienced at the top of branch tube, and further developed to fracture.

To solve the difficulty of coexistence of wrinkling and fracturing defects simultaneously regarding three-branch thin-walled tube of aluminium alloys, multi-step forming process is developed with a preforming stage introduced before final forming stage. The preforming can be also multi-step, which contributes to, not only avoiding splitting at the top of branch tube, but also adjusting inner pressure to suppress wrinkling. Greater axial feeding amount can be obtained, enabling more material to flow towards die cavity. The forming limit of such a process is increased finally.

4.4.2 *Simulations of Multi-step Hydroforming of Three-Branch Tubular Part*

With regard to the multi-step forming processes, the number of preforming is determined by the needed axial feeding amount. Under the premise of avoiding wrinkling and splitting, the number of preforming step is required to be minimized. Finite element simulations were performed to analysis the loading path effect on hydroforming 5A03 three-branch tube, which enables to identify the pre-forming process. Counter punch with different structures were designed according to the specific process of each step. The counter punch used in the pre-forming requires to be designed owning an end surface that has a reasonable angle with the axis of main tube. A 90° angle was selected for the counter punch to accomplish the final forming. Through changing the angle of counter punch top end, the distance between top of branch tube and counter punch is able to be reduced in each step, subsequently, the location deformation degree of top region can be reduced enabling the avoidance of splitting.

Figure 4.13 shows the computed thickness distribution between preformed and final parts. To clearly show the variation, the contour legend utilizes the thinning rate range, with positive and negative values indicating thinning and thickening, respectively. During preforming, due to the decreased end surface angle of counter punch in the middle, the maximum thinning rate at the top of branch tube is only 7.57%. A relatively sufficient filling is accomplished for the top of branch tube, which exhibits a positive effect of material accumulation for next step forming. Due to the effect of axial feeding, Both the sidewall of main tube and the transition zone of branch tube occur relatively large compressive deformation. The maximum thickening rate is 20.52%. In the meantime, arising from the increase of branch tube height, tensile deformation occurs in the hoop direction, which is highly likely to induce wrinkling defect. The inner pressure is required to be controlled to suppress the wrinkling. During the final forming, the top of branch tube is the thinnest with a maximum thinning rate of 25.77%. The end facing the obtuse angle and the region neighboring branch tube, of the main tube are thickest with a maximum thickening rate of 49.29%. The top of branch tube is needed to be cut off before using, and the thinning rate of this part is required to be controlled within 18%.

Figure 4.14 gives the stress-state variations and their corresponding stress locus on the yield ellipse of (a) one-operation forming and (b) multi-step forming. During one-operation forming, biaxial tensile stress state (Fig. 4.14a) is experienced by the deformation zone of branch tube top. The great magnitude of tensile stress is easy to induce splitting defect; In comparison, the feeding zone sustains one tension and one compression stress state (Fig. 4.14b). The great magnitude of compressive stress is easy to induce wrinkling defect. While during multi-step forming, resulting from the decreased distance between top of branch tube and counter punch in each step, the local deformation at the top of branch tube is reduced. Consequently, the biaxial tensile stress and axial compressive stress sustained by the top of branch tube and feeding zone, respectively, are reduced significantly, enabling the difficulty in simultaneous avoidance of splitting and wrinkling defects to be successfully addressed.

4.4.3 Experiment on Multi-step Forming of Three-Branch Tube

Figure 4.15 shows the preform and final part of hydroforming thin-walled three-branch tubular part. Annealing treatment is needed by the preform to eliminate the work-hardening influence on subsequent forming. Due to the utilization of a variety

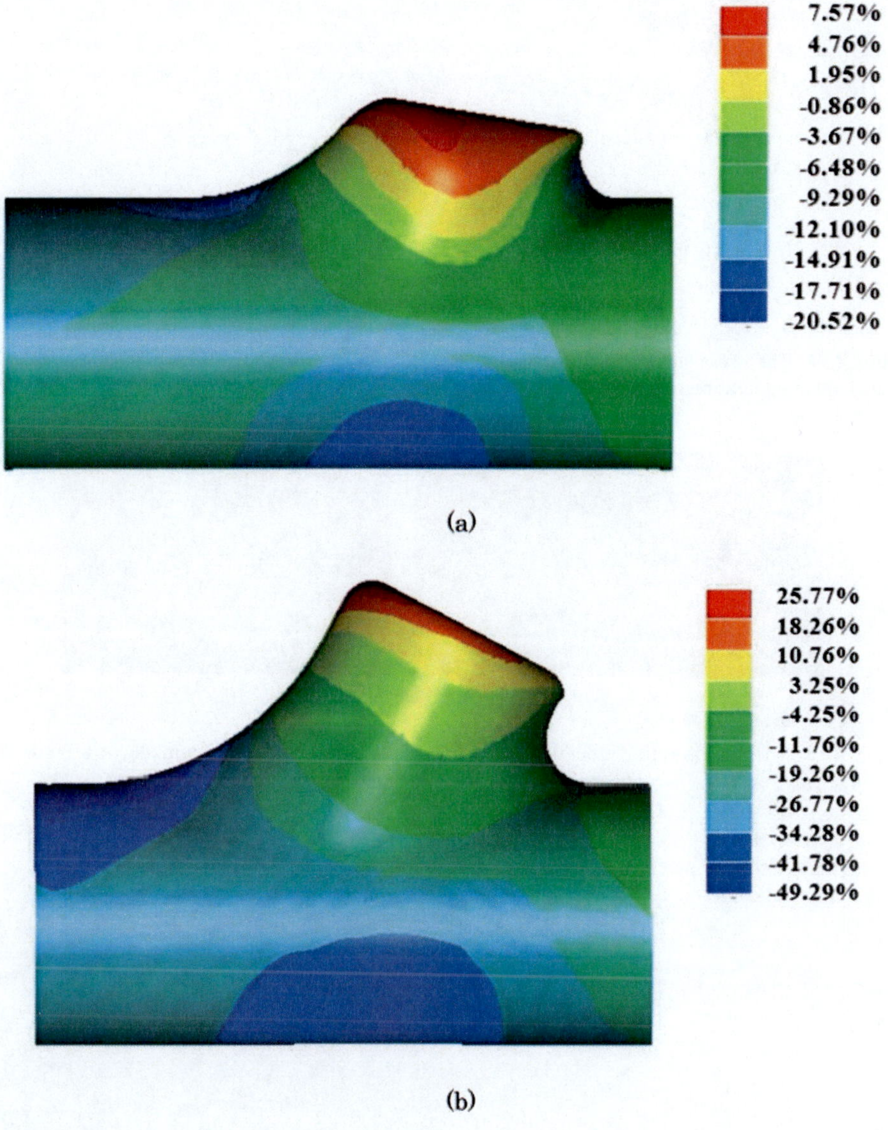

Fig. 4.13 The thinning rate of thickness distribution of three-branch aluminium alloy tubular part. **a** Preforming and **b** final forming

of counter punches with different end angles, the height of branch tube gradually increased. Through preforming step, the thin-walled three-branch tube of aluminium alloys can be successfully formed. Figure 4.16 shows the final part after cutting off the unnecessary top of branch tube (processing segment), where the height of branch tube was up to 0.98 times of the main tube diameter.

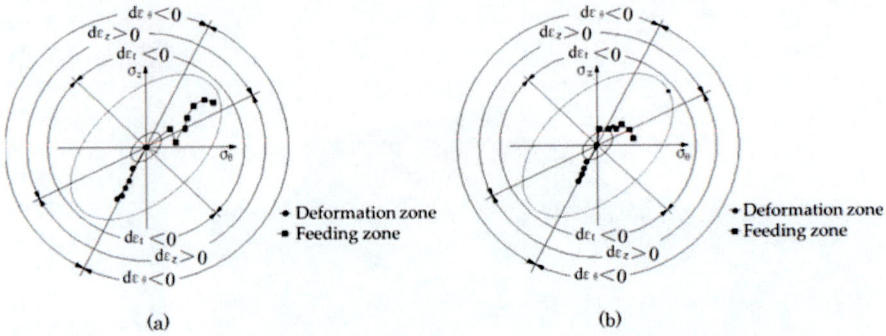

Fig. 4.14 The stress-states and corresponding locus during hydroforming three branch thin-walled tube. **a** One-operation forming and **b** multi-step forming

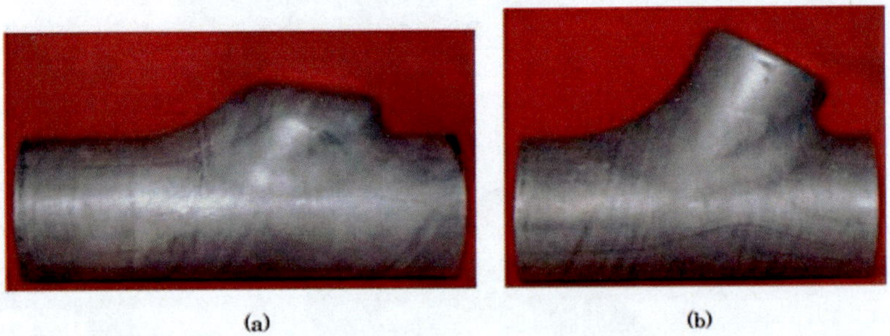

Fig. 4.15 The hydroforming process of three-branch tubular part of aluminium alloys. **a** Preform; **b** final part

Fig. 4.16 Y-typed three-branch tubular part of aluminium alloy

Figure 4.17 shows the experimental results of thickness distribution of a thin-walled Y-typed three-branch tube of aluminium alloys. The general trend of thickness distribution of such three-branch tube is thickening of main tube and thinning of top of branch tube. The thickening region is larger than the thinning region. As the formation of branch tube is mainly caused by the mutual contribution between axial feeding and bulging of the top of branch tube, the thickness variation of branch tube gradually changes from thickening to thinning, with the location varies from the bottom near main tube to the top. As the part of branch tube is higher on the obtuse angle side, the required axial feeding amount is greater, which results in greater thickening and thinning percentages compared to those on the acute angle side. Regarding the material attaching to the end surface of counter punch, the experienced bulging deformation degree is greater. Consequently, thinning of the peak point of the branch tube is the severest with a thinning rate of 23.2%. In comparison, the lowest point of the branch tube is relatively smaller with a thinning rate of 4.8%. The Y-typed three-branch tube is a symmetrical structure both in vertical and horizontal directions. Material at the bottom of main tube is difficult to flow to the branch tube resulting in an obvious thickening. The maximum thickness is 4.19 mm and the thickening rate is 67.6%. In comparison, the flow of material at the top of main tube approaching the branch tube is relatively easy. The closer the distance to the branch tube, the smaller the thickening rate of main tube.

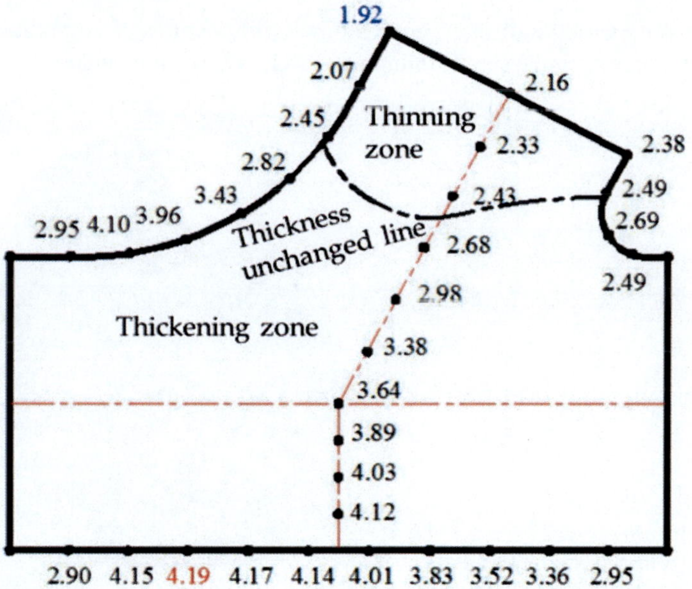

Fig. 4.17 The thickness distribution of thin-walled three-branch tube of aluminium alloy

4.5 Hydroforming of Ultra-Thin Three-Branch Tube of Stainless Steel

4.5.1 Geometry and Challenges

Figure 4.18 shows the geometry of a Y-typed ultra-thin three-branch tube with a large diameter. The diameter of main tube is 220 mm, while the diameter of branch tube is 100 mm. The inclined angle between axis of branch tube and main tube is 62°. This part requires to use 1Cr18Ni9Ti stainless steel tube with a thickness of 1.2 mm to manufacture, which is a typical large-diameter three-branch ultra-thin tube.

The difficulties in forming such kind of three-branch tube are:

(1) relatively smaller thickness, the ratio between diameter and thickness reaches 183. The structure is ultra-thin which is highly likely to occur wrinkling during axial feeding;
(2) the ratio between diameters of branch and main tubes is smaller than 0.5, the material is difficult to flow to branch tube, subsequently, the thickness of branch tube is difficult to be guaranteed within safety range resulting in splitting occurs;
(3) the whole structure is unsymmetrical. Utilizing the optimization of loading paths only is unable to eliminate the defects during hydroforming, as shown in Fig. 4.19.

Therefore, for such kind of three branch tubes with features of large diameter and ultra-thin thickness, multi-step forming process is still recommended.

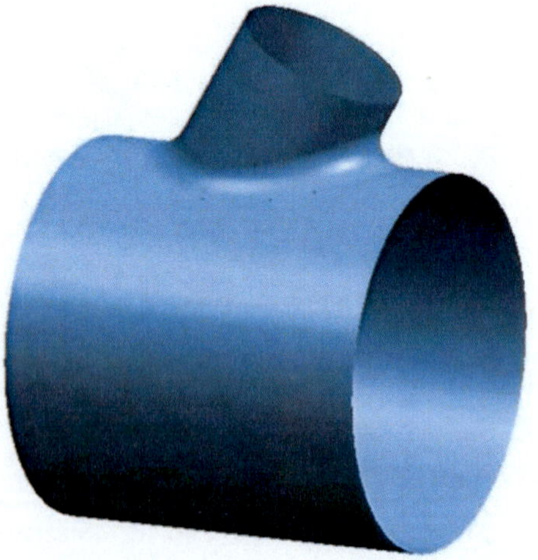

Fig. 4.18 Geometry and dimensions of three-branch ultra-thin tube

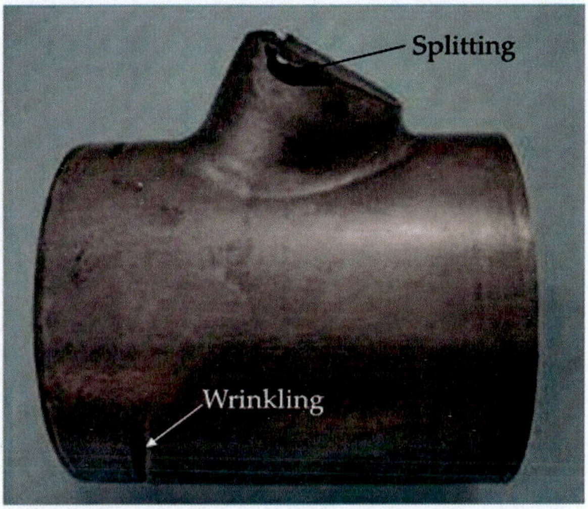

Fig. 4.19 Typical defects of large-size ultra-thin three-branch tube of stainless steel

4.5.2 Finite Element Simulations of Multi-step Forming of Three-Branch Ultra-Thin Three-Branch Tube

Considering the angle between branch tube end and axis of main tube is relatively large, a greater bulging of top of branch tube is required to make this region contact with the counter punch. For the sake of decreasing the bugling percentage of top of branch tube, the angle between branch tube end and axis of main tube should be minimized during the multi-step forming, to achieve an earlier contact between the top of branch tube and counter punch. After a full contact, the branch tube is further deformed under the back pressure enabling the excessive thinning to be avoided. Then, the angle is gradually increased to the target value in the final forming, which is consistent with geometry of product. The increment of angle between each forming step is mainly determined by the thickness distribution of branch tube and forming limit of tube material. The larger the increment, the severer the thinning and the less the uniformity of thickness.

Figure 4.20 shows the computed thickness distributions of preform and final part. The contour value is thinning rate with positive and negative values indicating thinning and thickening, respectively. In general, the thickness variation is thickening of main tube and thinning of branch tube, and the extent of thickening is greater than that of thinning.

During preforming, due to the decreased angle of counter punch end surface, the top of branch tube is able to contact early with the counter punch, which contributes to suppressing thinning effectively through the back pressure and accumulating material for next forming step. The filling of top of branch tube is sufficient with a thinning rate of 12.5% only. In addition, as the diameter of branch tube is obviously smaller than that of main tube, material flow towards the branch tube is restricted due to the

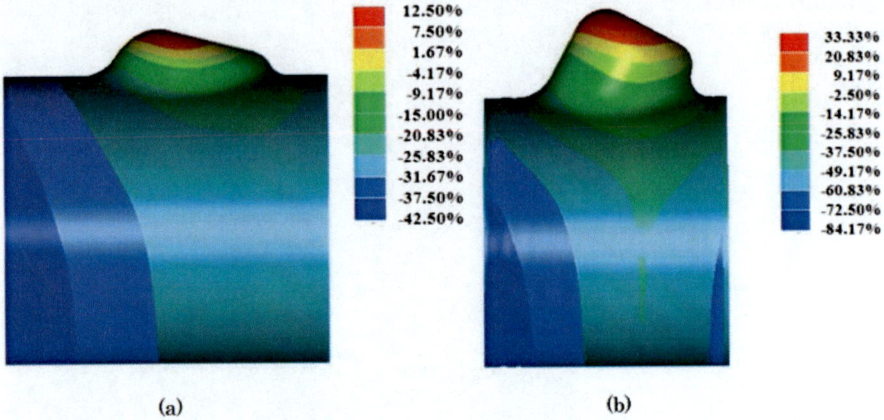

Fig. 4.20 The thickness distribution (thinning rate) of three-branch ultra- thin tube. **a** Preform; **b** final part

friction, which results in an obvious thickening of the main tube. The greater the degree of axial feeding, the larger the thickening.

During final forming, the angle between counter punch end and axis of main tube has reaches the target value. At this moment, the thickness of the top of branch tube is smallest with a thinning rate of 33.3% only; The main tube end facing the obtuse angle side has the greatest thickness. Therefore, reasoning from above findings, such kind of large-scale ultra-thin three branch tube can be successfully formed, through controlling the axial feeding amount and variation of branch tube reasonably in each step.

4.5.3 Experiment on Multi-step Forming of Three-Branch Ultra-Thin Tube

Figure 4.21 shows the experimentally formed three-branch ultra-thin tube of stainless steel. The process was multi-step using a series of counter punches with different end angles. From the preform (a) to the final part (b), the height of branch tube is gradually increased until the final sound three-branch tube.

Through the proper combination between axial feeding and inner pressure, a sound three-branch ultra-thin tubular part was successfully formed without the defects of wrinkling and fracture. The difficulty in forming such kind of three-branch tube was overcame. Figure 4.22 shows the qualified part after trimming the top of branch tube.

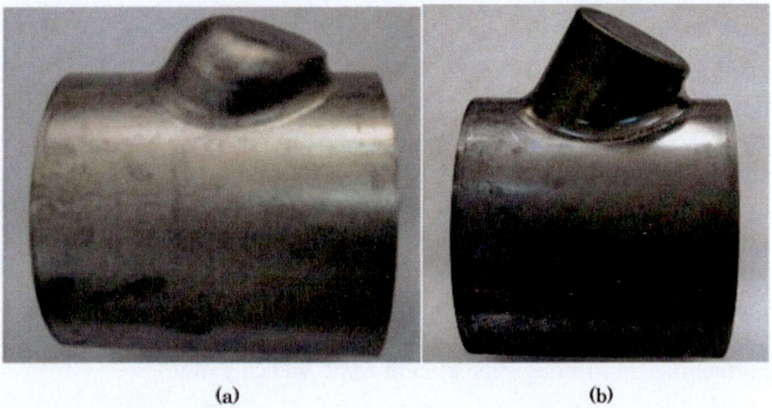

Fig. 4.21 The hydroforming process of three-branch ultra-thin d tube of stainless steel. **a** Preform; **b** final part

Fig. 4.22 Large-size ultra-thin three-branch tubular part of stainless steel

4.6 Hydroforming of a Three-Branch Thin-Walled Tube of Ni-Based Superalloys

4.6.1 The Geometry of Ni-Based Superalloy Three-Branch Tube and Forming Difficulties

Figure 4.23 shows the schematic diagram of three-branch tube with a spherical bottom of Ni-based superalloy. Such a part has an unsymmetrical structure both in the vertical and horizontal directions. For an easy description, the part is divided into a series of zones, including the main tube, varied-diameter tube, branch tube, and bottom half spherical.

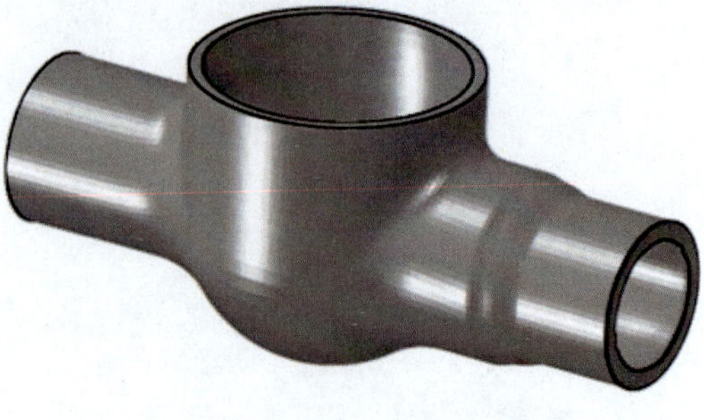

Fig. 4.23 Schematic diagram of three-branch tube with a spherical bottom of Ni-based superalloy

The difficulties in forming such kind of three-branch part are:

(1) The diameter of main tube is smaller than that of branch tube, which is only 58% of the branch tube diameter. Significant axial feeding is required which is likely to induce splitting;
(2) The whole structure is unsymmetrical. The material flow is very complex during axial feeding. Wrinkling of main tube and splitting of the top of branch is easy to be induced;
(3) The formability of Ni-based superalloy at room temperature is poor resulting in the high tendency of occurring splitting.

Figure 4.24 gives the typical defects of hydroforming Ni-based superalloy three-branch tube, which are wrinkling of main tube and splitting at the deformation zone. In some occasions, there also exists the wrinkling at transition zone and the splitting at the corner radius of branch tube. To address the difficulty of coexistence of wrinkling and splitting of Ni-based superalloy three-branch tube, multi-step forming process with optimized loading paths was used.

4.6.2 *Experimentation of Multi-step Forming of Ni-Based Superalloy Three-Branch Tube*

Figure 4.25 shows the preform and final part of a Ni-based superalloy three-branch tube. During the multi-step forming, the preform was annealed to eliminate the work-hardening effect on the subsequent forming. As the multi-step forming scheme was utilized, the main tube was preformed to a part with various diameters, which also results in a certain extent of material accumulation at the branch tube zone simultaneously. Then, the height of branch tube gradually increased with the exerted back pressure.

Using the preforming step, such a three-branch Ni-based superalloy tube was successfully formed. A final qualified part with top of branch tube cut off partially is shown in Fig. 4.26. The structure of this three-branch Ni-based superalloy tube is unsymmetrical, which requires the material flow to be strictly controlled. As can be seen from the pre-etched grinds on the surface, the tube experienced a relatively

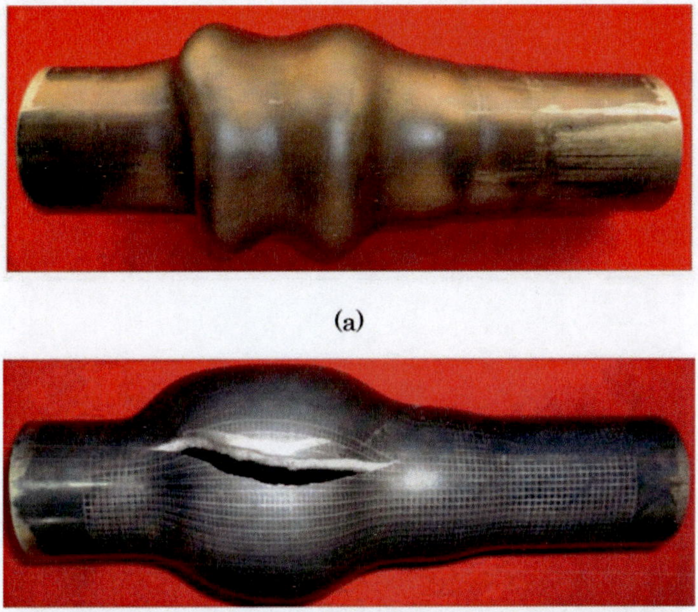

(a)

(b)

Fig. 4.24 Typical defects of hydroforming three-branch of Ni-based superalloy. **a** Wrinkling of main tube; **b** Splitting at the deformation zone

large plastic deformation to flow to the top of branch tube. Therefore, it can be seen that, the thickness distribution of formed part is significantly affected by the axial feeding and material flow.

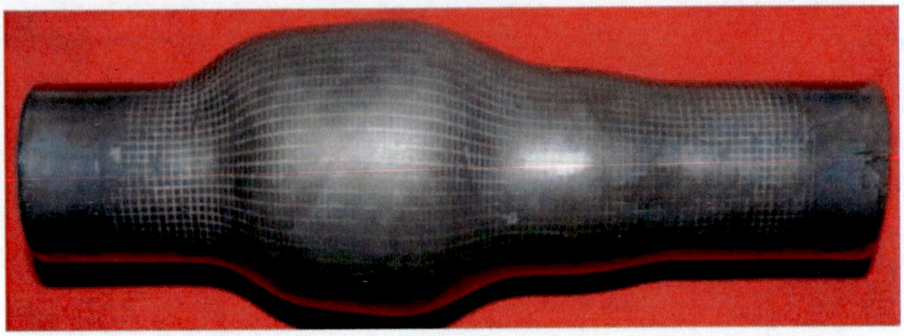

(a)

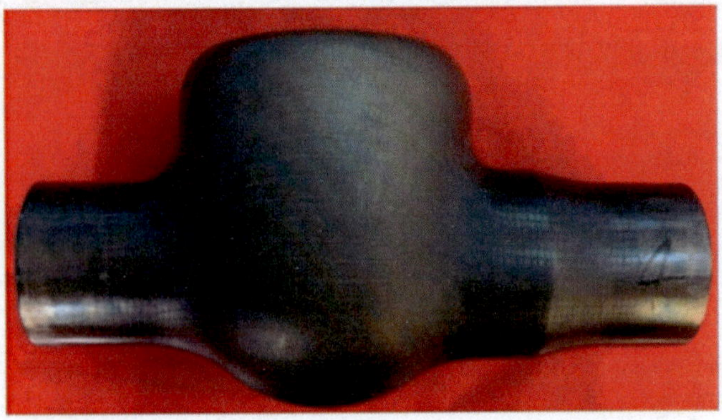

(b)

Fig. 4.25 The hydroforming process of three-branch tube of Ni-based superalloy. **a** Preform; **b** final part

Fig. 4.26 A sound three-branch superalloy tube

Chapter 5
Stress–Strain Analysis for Tube Hydroforming

5.1 The Stress–Strain States and the Yield Locus During Hydroforming of Tubes

When the tube is assumed as a thin-wall structure, the inner pressure p that exerted on the inner-wall of tube can be neglected. The plane stress state is assumed as the tube only sustains axial stress σ_z and hoop stress σ_θ. Then, substituting such a stress state into the Mises yield criterion, the yield function is rearranged as $\sigma_\theta^2 - \sigma_\theta\sigma_Z + \sigma_Z^2 = \sigma_s^2$ to be utilised for calculating yielding in hydroforming tubes with variable diameters. With the proceeding of deformation, stress states of the tube at, either different locations at a specific moment or same location at different moments, are differently significantly. Theoretically, all the possible stress-states experienced by the deformed tube should locate on the yield ellipse (yield locus) moving following the path from point A to point B, and ultimately to point C.

As previously described in Chap. 2, the tube blank can be mainly dominantly zoned as feeding and forming regions during hydroforming, with an additional transition zone exists between them. According to the sustained stress states and loading characteristic, the hydroforming process can be divided into three stages: (1) initial filling; (2) forming and (3) calibration. The stress states and deformation conditions of the tube vary with specific zones on the tube and the hydroforming stages, which are comprehensively illustrated in the following sections using Fig. 5.1.

5.1.1 Initial Filling Stage

With regard to the initial filling stage, the axial punches at the two ends of tube move towards the die cavity. The tube can be sealed through the deformation of tube end material once the axial punches are in contact with the tube and further deform it. A certain axial thrust on the tube ends is exerted by the punch to accomplish the sealing operation. Once a closed space is established, liquid medium is filled in the tube, and

S. Yuan, *Modern Hydroforming Technology*,
https://doi.org/10.1007/978-981-19-5775-8_5

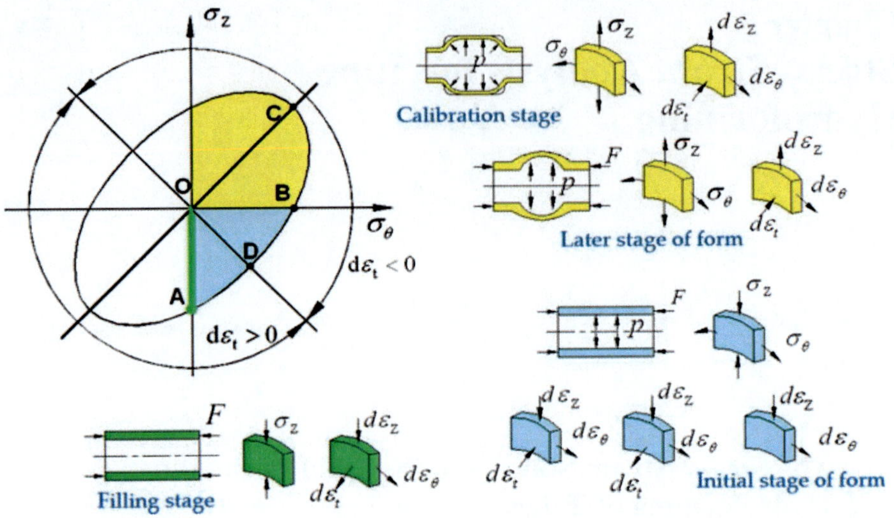

Fig. 5.1 The locations of stress states on the yield locus during hydroforming of tubes with variable diameters

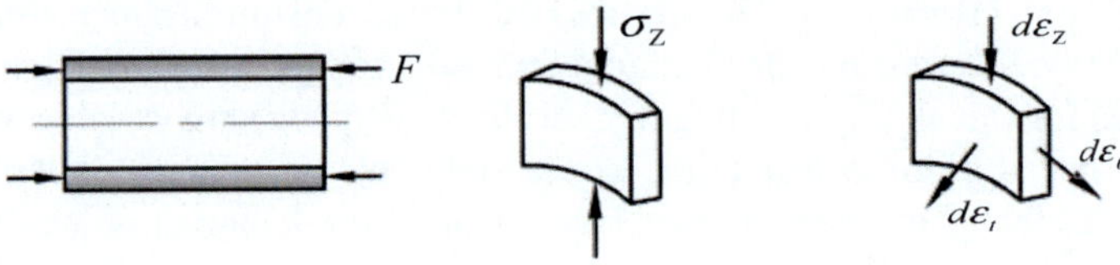

Fig. 5.2 The stress–strain states of tube in the initial filling stage

the resulted internal pressure is relatively small initially. Therefore, the whole tube belongs to a uniaxial compressive stress state corresponding to point A in Fig. 5.1. Then, the strain states are axial compression, hoop expansion and thickness thicken as shown in Fig. 5.2, while all the degrees of strain are relatively small. In this initial filling stage, if the tube is too long, the buckling defect of the whole tube could be induced when the axial compressive stress is too large.

5.1.2 Forming Stage

With regard to the forming stage, the stress and strain states at feeding and forming zones are different, which are described in detail in this section. Figure 5.3 shows the stress and strain states of tube in the feeding zone, which corresponds to point A on the yield ellipse. For the tube material in this zone, although liquid pressure is exerted on the tube inner wall, the normal contact stress σ_N is approximately equal to the inner

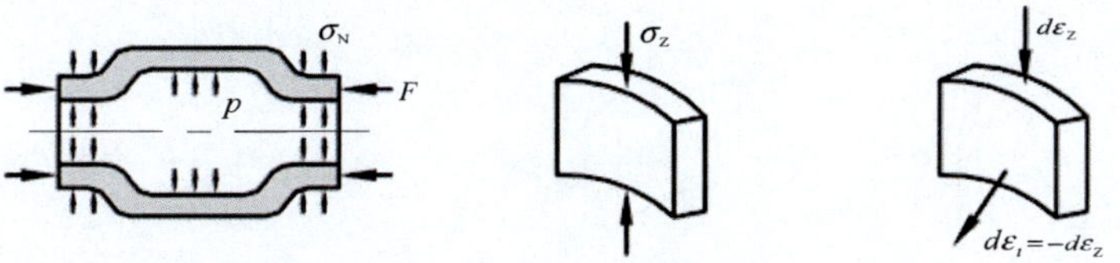

Fig. 5.3 The stress–strain states of tube in the forming stage

pressure p, the hoop stress is zero. Only axial stress exists in the feeding zone. For the strain states, due to the die constraint, the hoop strain becomes zero resulting in the plane strain state experienced in the feeding zone. Considering the axial strain is compressive, the strain in the thickness direction is positive resulting in the thickening of feeding zone in compliance with the volume consistency condition. In addition, arising from the friction at the interface between tube and die, the magnitude of axial stress decreases from the tube end towards the die cavity, which induces the most severe thickening at the tube end.

The stress states of forming zone vary between the beginning and end stage of forming. The tube keeps straight still at the beginning of forming. Figure 5.4 shows the corresponding stress state. The stress state of tube material is one-dimensional tension and one-dimensional compression, which is tensile and compressive stress in the hoop and axial direction, respectively. Such a state locates within the yield locus between point A and B. The strain state is determined by the magnitudes of hoop and axial stresses. As shown in Fig. 5.5, according to the plastic constitutive equation $d\varepsilon_t = -(d\varepsilon_i/2\sigma_i)(\sigma_\theta - \sigma_Z)$; when $\sigma_\theta > |\sigma_Z|$, corresponding to the yield locus between point B and D, thinning occurs as $d\varepsilon_t < 0$; when $\sigma_\theta < |\sigma_Z|$, corresponding to the yield locus between point D and A, thickening occurs as $d\varepsilon_t > 0$; when $\sigma_\theta = |\sigma_Z|$, corresponding to point D on the yield locus, thickness is unvaried as $d\varepsilon_t = 0$, and the tube deformation is plane strain.

With the proceeding of deformation, the tube within the forming zone becomes non-straight with a convex bulging deformation. The tube within this zone sustains a biaxial tensile stress state, as shown in Fig. 5.6. The representation on the yield ellipse exhibits the movement along the path from point B towards point C.

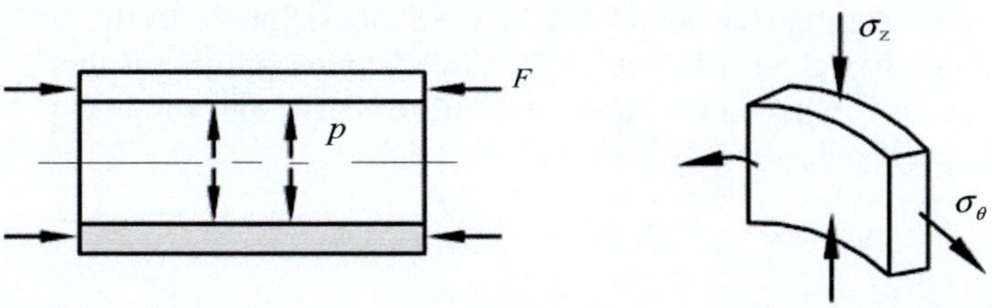

Fig. 5.4 The stress state of forming zone at the beginning of hydroforming. **a** Force acting on the tube, **b** stress state

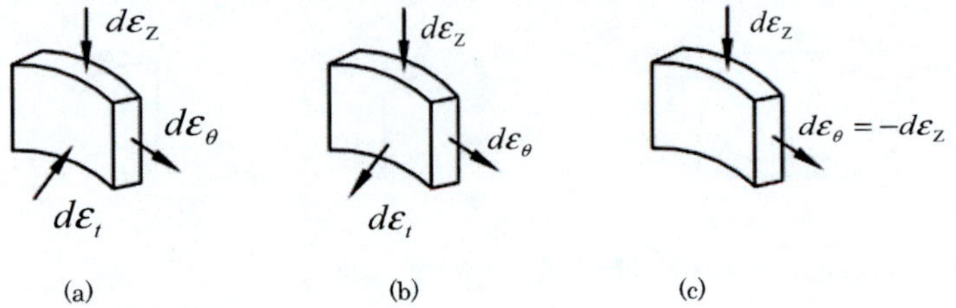

Fig. 5.5 The strain state of forming zone at the beginning of hydroforming. **a** Thinning, **b** thickening, **c** unvaried thickness

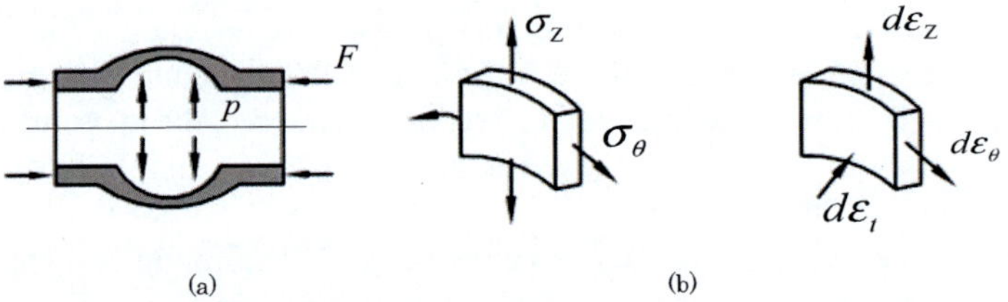

Fig. 5.6 The stress–strain states at the late stage in the forming zone. **a** Force acting on the tube, **b** stress state

In this stage, $\sigma_\theta > 0$, $\sigma_Z > 0$, normally $\sigma_\theta > \sigma_Z$, tensile deformation is always experienced for both hoop and axial directions, while thinning occurs in the thickness depends on the magnitudes of axial and hoop stresses. It should be noted that, the relatively ratio between hoop tensile stress, σ_θ, and axial tensile stress, σ_Z, is also determined by the relatively length of deformation zone.

A particular case of the forming stage is the absence of axial feeding. The tube only undergoes the inner pressure, and the deformation is free bulging. Initially, the tube keeps straight only sustaining hoop stress resulted from the inner pressure. The axial stress equals to zero, $\sigma_Z = 0$. Such a stress state corresponds to point B on the yield locus. Figure 5.7 shows the stress and strain states at this initial stage. With the increase of inner pressure, obvious bulging deformation occurs in the forming zone, the stress and strain states at this moment are consistent with those shown in Fig. 5.6, locating around the point C on the yield locus. Subsequently, the stress state changes to the biaxial tension state, which may induce splitting defect of the tube. The above stress–strain comparison reveals the reason of lower limit expansion ratio of free bulging compared to hydroforming.

5.1.3 Calibration Stage

With regard to the calibration stage, i.e. end of hydroforming, most of the material of the forming zone are already in contact with the die cavity, only local regions of

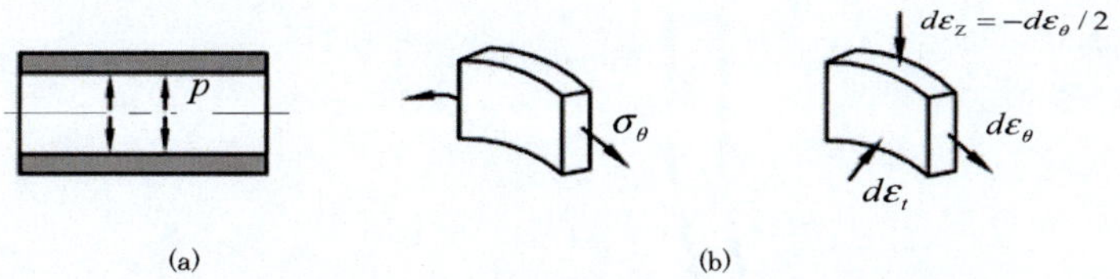

Fig. 5.7 The stress–strain states at the beginning of bulging. **a** Force acting on the tube, **b** stress state

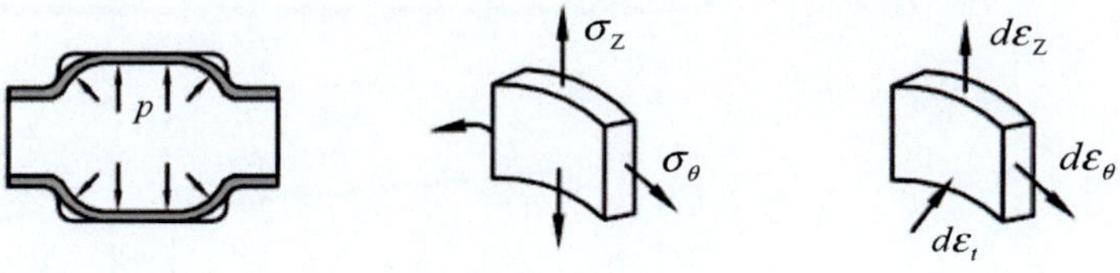

Fig. 5.8 The stress–strain states at the transition zone

the transition corner zone are not fully contact to the die. The objective of calibration stage is to form tight corner radius by increasing forming pressure to force material at the transition zone deformed approaching to the die inner surface. The stress state of transition zone is equivalent to a ring shell sustaining inner pressure, as shown in Fig. 5.8. Tensile deformation occurs in both the hoop and tangential direction, resulting in the thinning of tube thickness. The representation on the yield locus locates between point B and point C.

5.2 The Stress–Strain States During Hydroforming of Curved Tubes and Three-Branch Tubes

5.2.1 The Stress–Strain States During Hydroforming of Curved Tubes

Assuming a curved circular tube is directly formed by the internal pressure to obtain required cross-section geometries of formed part without pre-bending and pre-forming, the part can be zoned as tube end region, bent region and middle region as shown in Fig. 5.9, according to the sustained loading condition.

The tube end refers to the region from end surfaces of tube to an inwards extended distance. The tube ends are required to be sealed reliably by giving a certain axial displacement and force during hydroforming, therefore, the stress state of this zone is similar to that shown in Fig. 5.4, which is compressive stress in the axial direction and tensile stress in the hoop direction. Resulting from the friction effect, the magnitude of axial compressive stress decreases gradually inwards from the tube ends. Then,

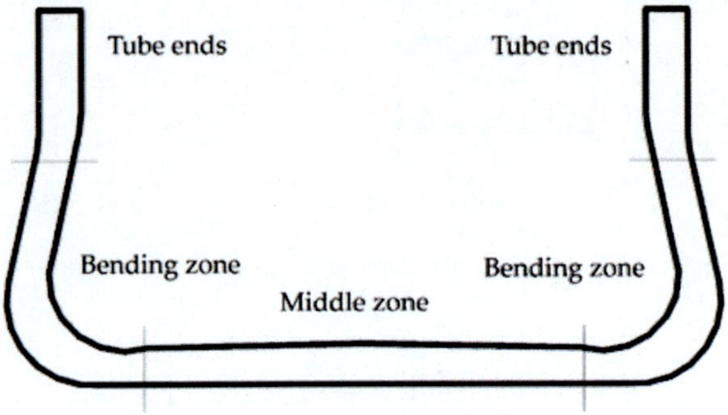

Fig. 5.9 Zoning of the hydroformed curved tube

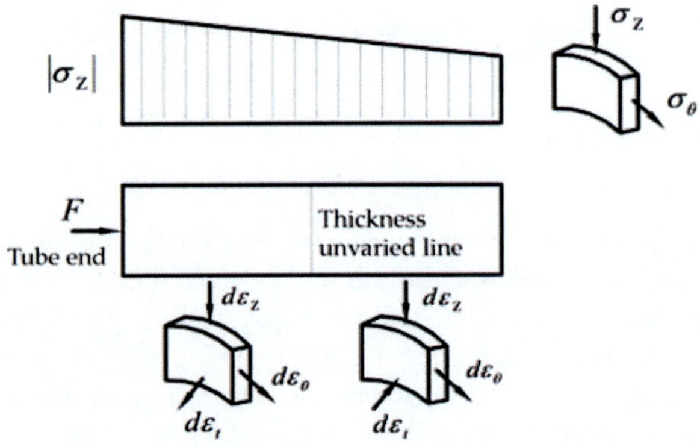

Fig. 5.10 Stress–strain states at the tube end zone

the strain states within this region may be different. Normally, the average thickness increases approaching the tube end referring the strain state shown in Fig. 5.5b; while the thickness decreases with a further distance to the tube end referring the strain state shown in Fig. 5.5a; Subsequently, a boundary with unvaried thickness exists between the two zones, as illustrated in Fig. 5.10.

For the bent zone, the stress state is equivalent to a circular ring shell sustaining an inner pressure, as illustrated in Fig. 5.11. Under the premise of an ideal ring shell and uniform inner pressure, for a part of the shell determined by ϕ, an internal force equilibrium condition in the normal direction as well as a general force equilibrium equation of the shell, can be established and used to work out the hoop stress σ_θ and axial stress σ_Z as follows:

$$\sigma_\theta = \frac{pd}{4t}\left(1 + \frac{R_b}{r'}\right) \tag{5.1}$$

$$\sigma_Z = \frac{pd}{4t} \tag{5.2}$$

where d represents the diameter of tube, R_b is the neutral diameter of bent axis, and r' is the distance between the point on the shell ring to the symmetrical axis.

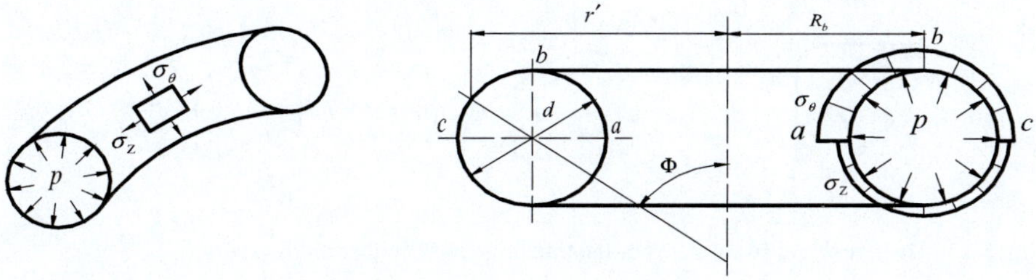

Fig. 5.11 The stress states in the curved zone

As can be seen from Eqs. (5.1) and (5.2), the axial stress σ_Z of the ideal ring shell is a constant, while the hoop stress σ_θ varies with the change of radiusr', maintaining $\sigma_\theta > \sigma_Z$. The hoop stress σ_θ has the minimal value, $\frac{pd}{4t}\left(1+\frac{2R_b}{2R_b+d}\right)$,at the outside point c, while peaks at the inside point a, $\frac{pd}{4t}\left(1+\frac{2R_b}{2R_b-d}\right)$. A value $\frac{pd}{2t}$ is found at point b corresponding to the inside/outside separation point on the top. Due to stress in the thickness direction is always the minimal principle stress, thinning always occurs. Figure 5.12 illustrates the deformation characteristics in terms of different locations within the bent zone, with regard to the inner surface of bent zone where $r' < R_b$, $\sigma_\theta > 2\sigma_Z$, the deformation is circumferential elongation and axial shorten; With regard to the separation point where $r' = R_b$, $\sigma_\theta = 2\sigma_Z$, the deformation is circumferential elongation and axial constant; With regard to the outside surface of bent zone where $r' > R_b$, $\sigma_\theta < 2\sigma_Z$, elongation occurs in both the circumferential and axial directions.

The middle zone refers to the region between two bent zones without the function of axial force. During the calibration stage with an increased pressure, the deformation mainly concentrates on the transition radius of the cross-section, and the stress–strain states are shown in Fig. 5.13. For the middle region of this radius, the uniaxial tensile stress state is found in the hoop direction.

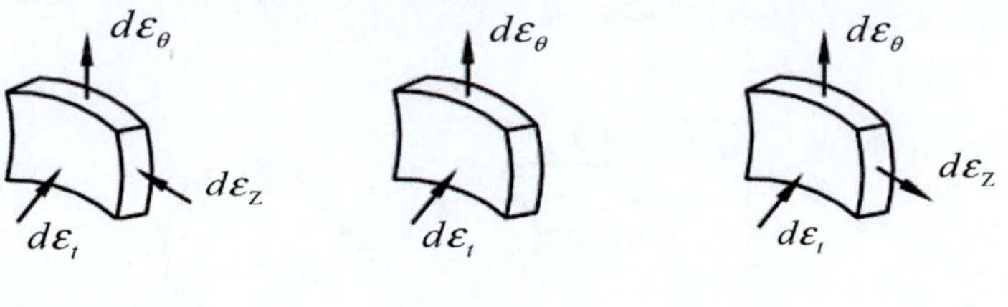

Fig. 5.12 The strain states in the curved zone

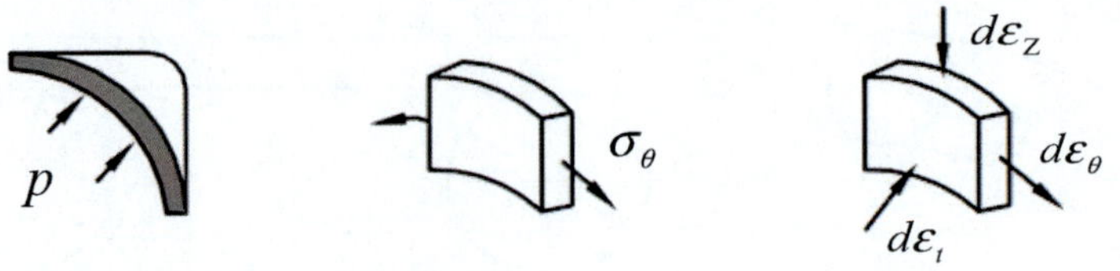

Fig. 5.13 The stress and strain states of the transition zone in the middle zone

5.2.2 The Stress–Strain States During Hydroforming of Y-shaped Tubes

Unsymmetrical geometry is the dominant characteristics of Y-shaped tubes, which brings greater difficulty in forming, compared to hydroforming of T-shaped tubes. Therefore, this section selects the Y-shaped tube to illustrate the stress–strain states variation during hydroforming, which is schematically shown in Fig. 5.14. Three typical locations are identified on the Y- shaped three-branch tube, that are the transition radius on the left, vertex of branch tube and middle point on the sidewall of main tube.

The stress-states of both transition radius and middle point on the sidewall of main tube during forming is one-tension and one-compression, while that of vertex of branch tube is bi-axial tension. Figure 5.15 illustrates the corresponding strain-states. For the thickening zone in the main tube, the strain state is compression in the axial direction and tension in hoop direction; For the thinning zone in the branch tube, bi-axial tensile deformation is experienced; Then, regarding the thickness unvaried line between thickening and thinning zones, plane strain state is found. When the axial compressive stresses of transition zone and middle zone of main tube sidewall are great, inwards concave deformation may generate, and wrinkling defect occurs when such kind of concave deformation becomes severe. In comparison, the vertex of branch tube always sustains bi-axial tensile stress state, resulting in a two-dimensional tensile straining and thinning. Once the thickness is severely reduced, splitting defect will happen on the vertex of branch tube.

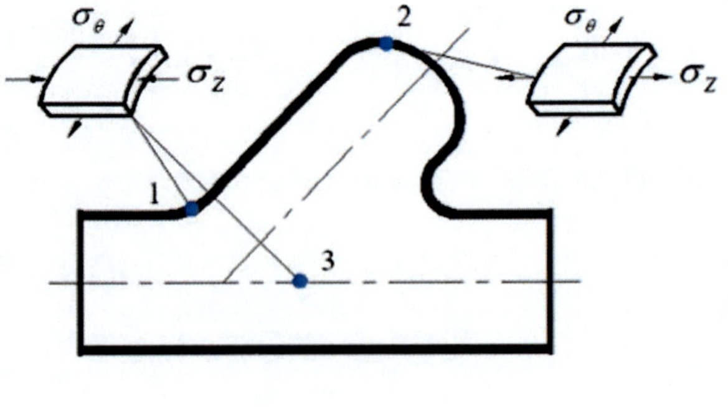

Fig. 5.14 Stress states of typical points on the Y-typed tube

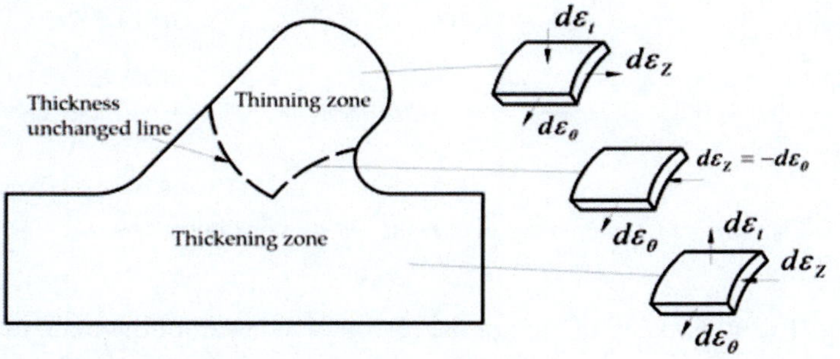

Fig. 5.15 The strain states of Y-typed branch tube

5.3 The Stress Locus of Tube Hydroforming

As described above, significant differences exist in the stress states of either different points at same stage or same point at different stages, on the hydroformed tube. In terms of Y-shaped tube, due to the unsymmetrical geometry and complex loading path, the stress-states of different points on the tube vary greatly, and that of individual point varies continuously during hydroforming. The common defects include wrinkling at transition zone of branch tube, splitting at the vertex of branch tube and the wrinkling at the main tube. Therefore, in this section, using the typical points on the zones prior to failure shown in Fig. 5.14, the stress locus and strain states of hydroforming Y-shaped tube under different loading paths are analyzed.

Considering work-hardening occurs in the plastic deformation of alloys, and assuming the isotropic hardening rule, yielding locus of individual points during the complete forming process can be approximately represented by Fig. 5.1. In particular, Fig. 5.16 shows the variations of stress locus on the yielding ellipse of the three typical points during a successful forming of Y- shaped tube under a given loading path.

As can be seen in Fig. 5.16, the corner of transition zone on the left (point 1 in Fig. 5.14) always sustains bi-axial compressive stresses during forming, resulting in

Fig. 5.16 Stress locus of typical points during tube hydroforming

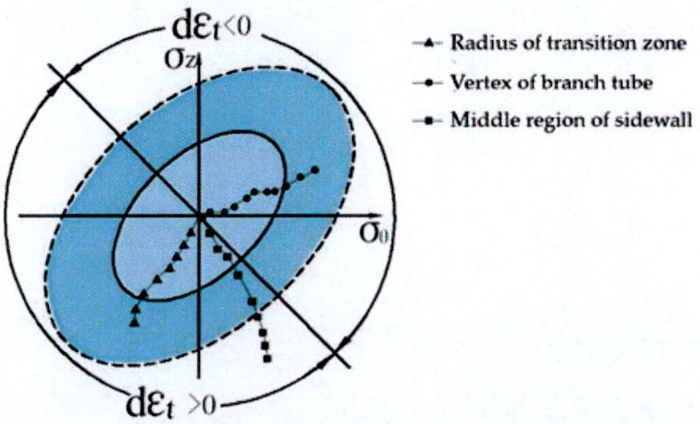

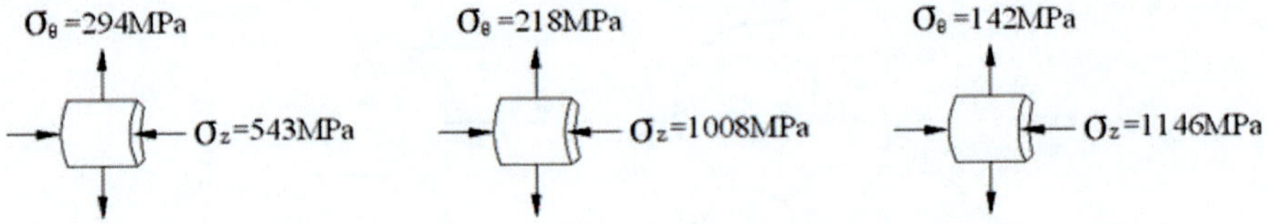

Fig. 5.17 The stress states at middle region of main tube as wrinkling occurs

thickening. The stress state of the middle region of sidewall of the main tube (point 3) is hoop tension and axial compression while the magnitude of axial compressive stress is greater than that of the hoop tensile stress, which results in thickening in this zone as well. In comparison, the vertex of branch tube (point 2 in Fig. 5.14) always experiences biaxial tensile stress state with a greater hoop stress than the axial stress, resulting in thinning.

When a loading path with a constant inner pressure is used, wrinkling defect occurs near the middle of main tube sidewall (point 3 in Fig. 5.14). Figure 5.17 gives a particular case of the stress states of point 3 during forming. As can be seen in this figure, the stress state is always one-tension and one-compression. Figure 5.17 show the stress-states under different axial feeding stages of left punch, 20, 55 and 86 mm. The inner pressure is 36 MPa for the three conditions. The hoop tensile stress of this point decreases with the proceeding of deformation, while the magnitude of axial compressive stress gradually increases. Such an exaggerated axial compressive stress may induce axial instability resulting in wrinkling defect.

When utilizing a loading path with a gradually increased inner pressure, splitting defect would shift to the vertex of branch tube (point 2). Figure 5.18 shows the stress states at different axial feeding stages, i.e. 20, 55 and 86 mm. The inner pressures are 62, 140 and 188 MPa correspondingly. As can be seen in this figure, the vertex of branch tube always sustains a bi-axial tensile stress state. With a further deformation, the magnitudes of tensile stresses in biaxial directions dramatically increase resulting in the ultimate splitting failure.

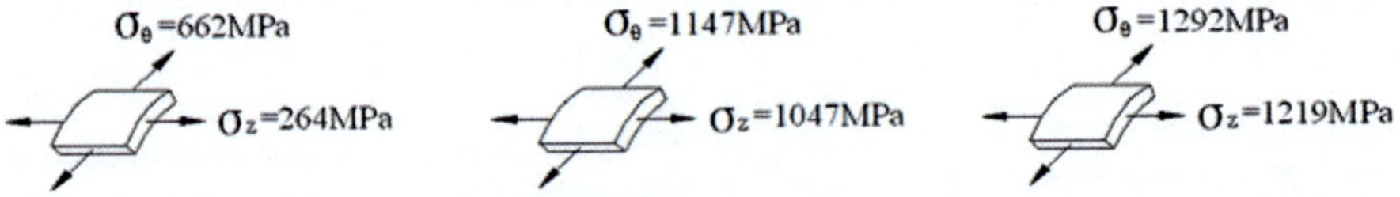

Fig. 5.18 Stress state of the vertex of branch tube

5.4 Stress-State of Corner Zone and Splitting Mechanism

5.4.1 *Stress Analysis of Corner Zone*

Filling of the corner radius dominants the later stage of hydroforming irregular cross-sectional component. Based on the above analysis, the forming pressure is inversely proportional to the value of corner radius. The successful forming of radius requires a very high forming pressure (relative radius $r/t = 3$). The stress state of corner region is a three-dimensional stress state resulting in a very complex deformation process. Through the numerical simulation of solid element, the stress state regarding the forming of corner region, occurrence and development of plastic deformation are analyzed. In addition, through which, the stress state and stress locus on the yield cylinder of typical points can be given, and visualized representations of three-dimensional normal stress graph are established for typical points.

Figure 5.19 shows the distribution of effective stress and plastic region in the corner zone of a simulated rectangular cross-section contour. The effective stress within the cross-section increases from 101 to 215 MPa gradually when the position varies from the middle point of straight side to the transition point between straight side and corner radius. Due to the friction effect, the effective stress peaks at the outside surface at the transition point between straight edge and radius zone. Such a transition zone always undergoes the maximum effective stress during forming, which is believed to be the most vulnerable region for occurring defect.

With the increase of inner pressure, the plastic region on this rectangular cross-section continuously expands, while the characteristics of expansion of straight edge zone and corner zone are different. In the die-clamping stage, the stress is dominated by bending stress. The effective stress of straight edge of rectangular cross-section

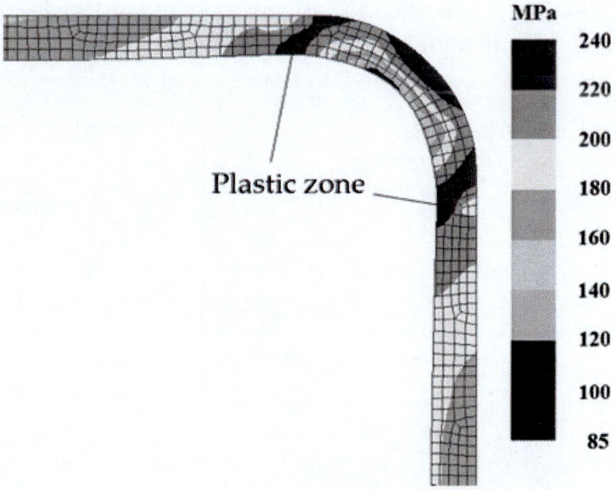

Fig. 5.19 The distribution of effective stress and plastic region of corner zone

firstly exceeds yield stress to occur plastic deformation. With the proceeding of die-clamping, the middle of radius zone subsequently enters the plastic deformation stage. Once this operation is accomplished, the in and out-side of straight edge and around a quarter of in & out-side of radius region occur plastic deformation, while the two ends of corner region still belong to elastic deformation. The effective stress of the middle of corner region becomes greater than that of straight edge, and this region has the maximum effective stress subsequently.

At the initial stage of hydroforming, the effective stress decreases when pressurizing starts. When the target pressure is reached, plastic deformation continues to occur for both the inside of straight edge zone and outside of corner zone, and the entire region of radius region follows to occur plastic deformation. With the increase of inner pressure, the plastic region quickly expands, and the rest regions all enter plastic region, except for a quarter of middle region outside the radius region, and approximately a quarter of the inside two ends regions approximately. Finally, the effective stress of rectangular cross-section exceeds the yield stress and the whole cross-section entirely enters plastic region.

5.4.2 *Mechanism of Splitting at the Corner Zone*

Figure 5.20 gives the predictions of tube outside profile evolutions during the corner filling stage obtained by the FE simulations. The free-bulging at the die radius can be always treated as a 90 ° circular arc approximately. At the beginning stage of corner filling, the variation of radius is quick. With the increase of tube material length that is in contact with to the die profile, the radius variation becomes slow when the inner pressure is greater than 50 MPa in the late stage of corner filling.

To measure the correlation between the radius of part and inner pressure during filling, a displacement sensor is mounted at the die radius of tooling, as shown in Fig. 5.21. Through measuring the displacement variations of the component radius,

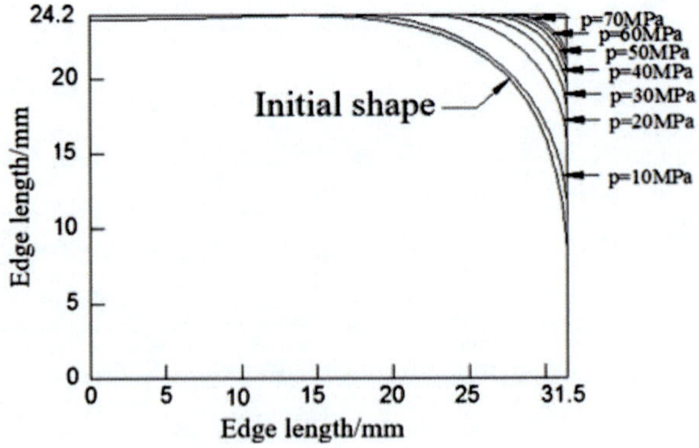

Fig. 5.20 The profile evolution in the corner filling stage

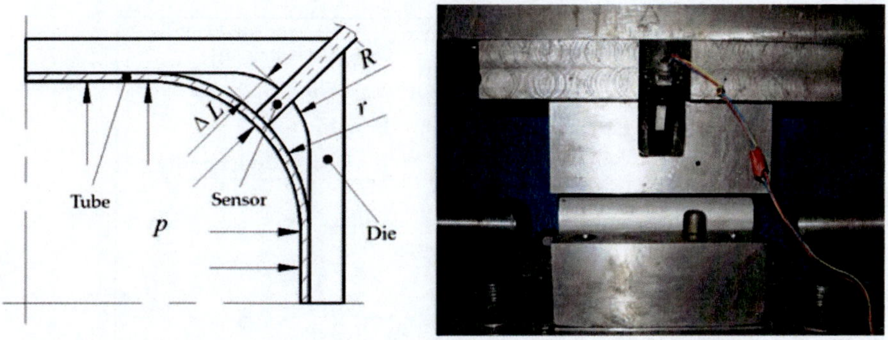

Fig. 5.21 Measurement device for corner radius

the radius under different inner pressures can be calculated using the function correlation between radius and displacement. Then the quantitative relationship between radius and inner pressure during filling can be obtained to determine the limit radius magnitude.

Figure 5.22 shows the experimentally determined corner radius and inner pressure. During the corner filling, the corner radius gradually decreases with increasing the inner pressure. The interfacial friction influences the filling stage of radius and limit corner radius significantly. The reason is believed to be that, during the late stage of filling, the interfacial friction shear stress increases with increasing the inner pressure. In addition, with the increase of contact length between the part and die, the friction effect becomes more significant results in a gradually decreased corner radius variation. If the inner pressure is continuously increased, the corner radius is unvaried anymore which results in the splitting at the transition point of radius. In terms of practical cases, when the lubricant is not used for hydroforming 5A02 aluminium alloy, splitting occurs when the relative corner radius is 7.5, then the subsequent limit relative corner radius is 7.5. In comparison, the relative corner radius is 6.5 using oil lubrication. Solid lubricant MoS2 enables to significantly improve the lubrication enabling a limit relative corner radius 4.5 to be achieved; The best lubrication performance is achieved by the polyethylene film with a limit relative corner radius approaching to 2.

When the polyethylene film is used, the variation of relative corner radius is significant when the radius value is greater than 5, while the variation becomes gentle when the value is smaller than 5. To achieve the final forming of small radius with a full contact to the die profile, the inner pressure needs to be significantly increased at the late stage of radius filling. In terms of a given transition radius, the inner pressure required for forming small radius can be reduced through improving lubrication and designing preform properly.

In order to analyze the mechanics of splitting occurred at the transition point between straight edge and corner radius. Selecting a unit length along the axis of the tube, the deformation of tube can be assumed as plane strain considering the magnitude of tube length is much greater than that of diameter.

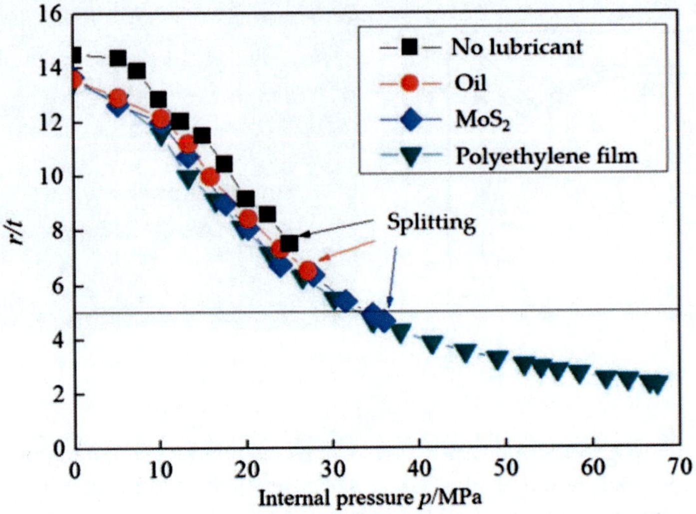

Fig. 5.22 Correlations between corner radius and inner pressure under different lubrication conditions

Figure 5.23 gives the force analysis of the corner filling. For a particular moment during the corner radius filling stage, the inner pressure is p, the instant corner radius of tube blank is r. Point O is assumed as the transition point between the straight edge and corner region, and point B is the middle point of straight edge. Then, the force equilibrium equation in the x direction at a random point A within the straight edge is established as:

$$F_A(x_A) = F_0 - \tau b \tag{5.3}$$

where $F_A(x_A)$ represents the inner force of point A along the horizontal direction (N),$F_0 = pr$ is the inner force of point O along the horizontal direction (N),$\tau = \mu p$ is the friction shear stress between the tube and die (MPa), μ is the friction coefficient, b is the horizontal distance between point A and point O.

The stress of point A in the x direction, σ_x, is

$$\sigma_x = \frac{F_A b}{t} \tag{5.4}$$

The stress of point A in the y direction, σ_y, is

$$\sigma_y = -p \tag{5.5}$$

In terms of the principal stress sequence, the first principal stress $\sigma_1 = \sigma_x$, the third principal stress $\sigma_3 = \sigma_y$. Regarding the plane strain problem, the second principal stress $\sigma_2 = \sigma_Z = (\sigma_x + \sigma_y)/2$. The principal strains are: $\varepsilon_1 = \varepsilon_x$, $\varepsilon_3 = \varepsilon_y$,$\varepsilon_2 =$

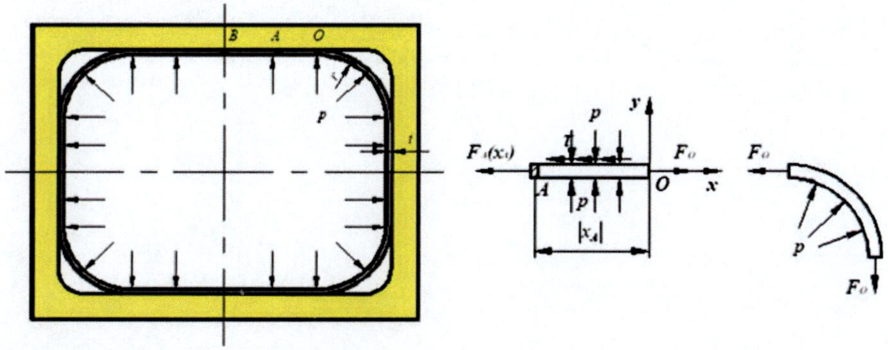

Fig. 5.23 The force analysis of corner filling stage

$\varepsilon_Z = 0$. According to the unvarying volume principle, the strain state $\varepsilon_1 = \varepsilon_x > 0, \varepsilon_3 = \varepsilon_y < 0$ can be deduced.

The effective stress σ_i of point A is

$$\sigma_i = \frac{\sqrt{3}}{2}(\sigma_1 - \sigma_3) = \frac{\sqrt{3}p}{2}\left(\frac{r - \mu b}{t} + 1\right) \tag{5.6}$$

Figure 5.24 shows the distribution of effective stress of straight edge determined from Eq. (5.6). During the filling stage of rectangular cross-sectioned corner, due to the friction effect, the effective stress gradually increases from the middle point B of straight edge to the transition point O between straight edge and corner, peaking at the transition point. Therefore, the region near the transition point is likely to experience splitting or overthinning of thickness, and a limit value exists for the corner region of the rectangular cross-section.

Fig. 5.24 The distribution of effective stress of the straight edge zone

Chapter 6
Equipment and Tooling of Hydroforming

6.1 Construction and Functions of Hydroforming Equipment

6.1.1 Construction of Hydroforming Equipment

The hydroforming equipment is mainly constructed by two systems, which are the die-clamping press and high pressure system. The die-clamping press can be either frame-typed or four column-typed. Compared to conventional presses used in stamping or forging processes, the die-clamping press is not used to form the part during the downwards stroke of ram, while only provides a force that is known as clamping force. Such a force is exerted on the dies at the end of stroke to fully close the gap between the upper and lower dies. The high pressure system is composed by five dominant sub-units including the high pressure source, horizontal hydraulic cylinder, hydraulic system, water pressure system and control system, as schematically shown in Fig. 6.1.

In general, the operation principle of hydroforming equipment is described as follows: (1) closing the gap between dies, (2) exerting a clamping force vertically, (3) filling liquid medium in the tube; (4) sealing the tube ends; (5) loading processing paths (inner pressure and axial feeding); (6) hydropiercing; (7) discharging internal liquid pressure and unloading the die-clamping force; (8) withdrawing horizontal punches; (9) opening the dies. Considering above operations, the main difference compared to conventional hydraulic presses is the capability of executing the pre-designed loading path which is a proper combination of internal pressure, axial displacement and die-clamping force. Such a unique capability is also the core of hydroforming equipment.

S. Yuan, *Modern Hydroforming Technology*,
https://doi.org/10.1007/978-981-19-5775-8_6

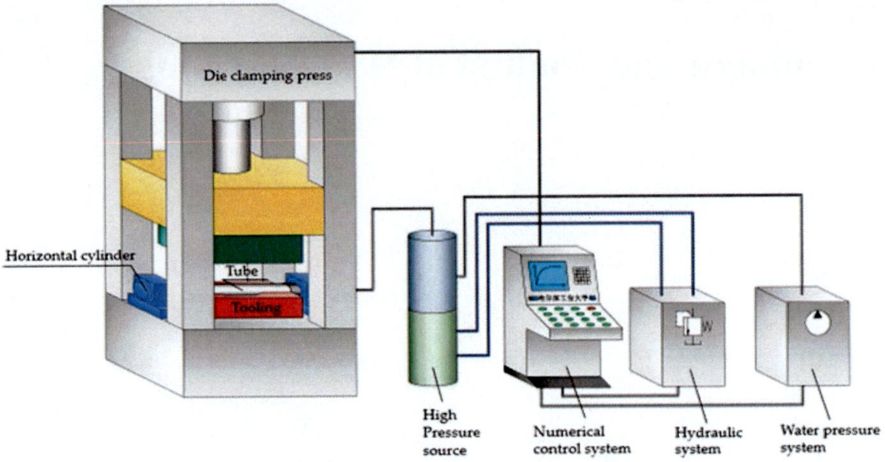

Fig. 6.1 Schematic of hydroforming equipment

6.1.2 Sub-Systems of Hydroforming Equipment

- **Die-clamping press**

The firm clamping of forming dies is the priority of hydroforming, which enables to eliminate obvious die gap at the tube ends that probably results in flash defect and unsuccessful sealing. The die-closing force is exerted by the hydraulic clamping press. Considering the press enables to exert the maximum force at any intermediate position within the total ram stroke, and the force is easy to be adjusted and maintained. Currently, the hydraulic press is extensively to be used to clamp the dies.

During hydroforming, the liquid pressure inside the tube is built up gradually. The pressure is controlled to increase following a specific loading path that is pre-designed based on process requirements. Therefore, to avoid the energy waste arising from the traditional scheme of running the press at full capacity, modern hydroforming requires the clamping force exerted on the dies to be able to vary continuously with the given liquid pressure at an instant moment during hydroforming. Figure 6.2 schematically shows the loading profiles of dominant parameters of hydroforming, i.e. die-clamping force, axial displacement and internal pressure, with the proceeding of processing time.

- **High pressure source**

To form the tubes of metallic materials, the liquid pressure used in hydroforming is normally around 300–400 MPa, or even higher. However, the maximum capability of conventional hydraulic pump is only 31.5 MPa approximately, which is unable to deform the tubes as it is significantly lower than the required high pressure. To compensate such a pressure difference and provide a high pressure for hydroforming tubes, the pressure intensifier is normally used to construct the high-pressure source.

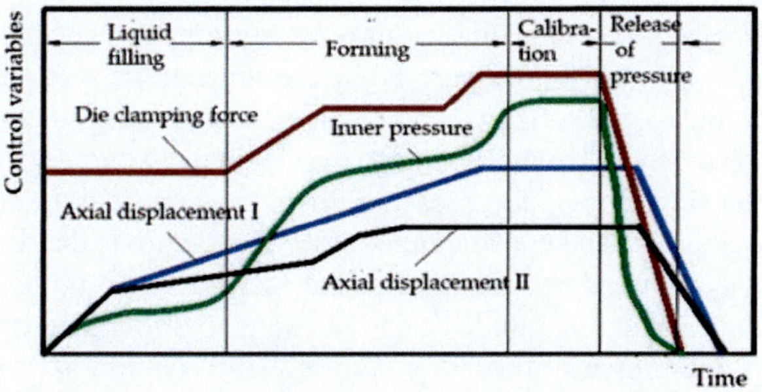

Fig. 6.2 Process parameters and loading paths of hydroforming

Figure 6.3 schematically shows the working principle of the pressure intensifier. Hydraulic oil with a lower pressure is injected into the cavity below the large piston to counter-push the piston to move upwards. According to the force equilibrium condition on the piston, $p_1 A_1 = p_2 A_2$, then the amplified pressure inside the high pressure cavity becomes $p_2 = p_1 A_1 / A_2$. A pressurizing ratio is defined as A_1/A_2 representing the ratio between the project areas of the low- and high- pressure cavities, respectively. Considering the uniform geometry of intensifier cylinder, the ratio also indicates the ratio of pressure between the two cavities of this intensifier. In terms of a practical case with a 25 MPa output from the low-pressure end, a 200 MPa pressure is obtained in the high-pressure cavity at a given pressurizing ratio of 8. Accordingly, the magnitude of amplified pressure can be further increased to 400 MPa with increasing the pressurizing ratio to 16.

Based on the pressurizing mechanism, the type of pressure intensifier can be classified as single-operation and double operation. With regard to the single-operation pressure intensifier, it owns both a high-pressure and a lower pressure cavity. The

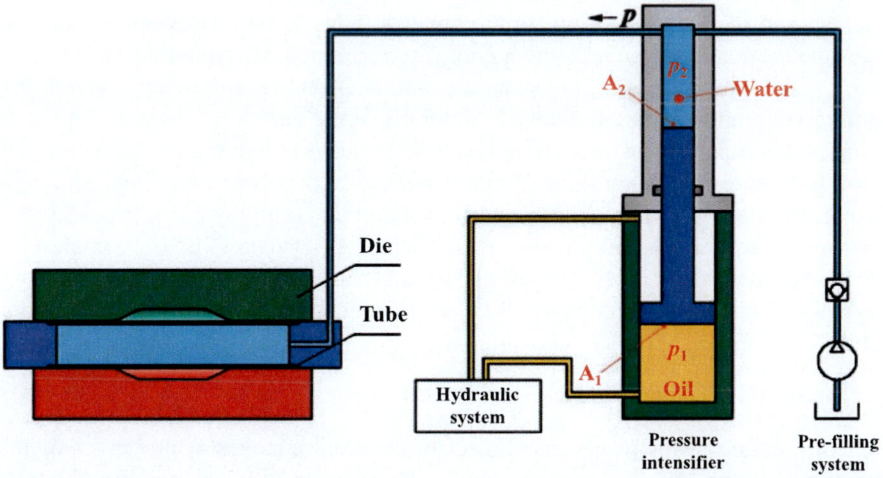

Fig. 6.3 The working principle of pressure intensifier

stroke of piston is divided into pressurizing stroke and reset stroke. High pressure is provided by the pressure intensifier during the pressurizing stroke only, while next round of pressurizing can be only occurred unless the piston accomplishes the stroke of current cycle and returns to the original position. Therefore, such a principle makes the single-operation pressure intensifier normally to be used in the one-off forming scenarios. The convenient control of pressure is the dominant advantage as the one-way valve is not required at the high-pressure outlet. When the pressure exceeds the critical magnitude, it can be reduced through controlling the withdraw of piston in the pressure intensifier. Consequently, a disadvantage of requiring high volume pressure intensifier is involved. However, practically, the volume of high pressure cavity is limited, which constrains the single-operation pressure intensifier only to be suitable for small-sized parts.

For the double-operation pressure intensifier, two high-pressure and one low-pressure cavities are designed. High pressure liquid can be continuously provided by the means of piston motion in the either left or right direction. Compared to the single-operation type, theoretically, an infinite volume of high-pressure liquid can be output using a pressure intensifier with a relatively smaller volume. This characteristic promotes double-operation pressure intensifier to be extensively applied in hydroforming systems for producing tubular parts with a wide range of volumes. Nevertheless, the pressure is unable to be reduced by adjusting the position motion when the pressure exceeds the critical magnitude, due to the existence of one-way valves at both ends of high-pressure outlets. The unloading valve or shut-off valve has to be designed in the hydraulic circulate which brings greater complexity to the pressure control system.

Hydraulic oil is normally used as medium in the low-pressure cavity, while emulsion water can be alternatively used. The advantages of using hydraulic oil are high viscosity, good sealing capability and corrosion-free of equipment and formed parts, while obvious disadvantages, such as high compression volume, high cost, difficult to clean and the contamination of components, have to be addressed. Figure 6.4 gives the correlations between the medium pressure and the compression ratio for both oil and water medium. For example, when a 400 MPa pressure is given, the compression ratio is approximately 17% for oil, while only 8% for water. The high compression capability of oil and resulted low efficiency in volume accumulation can cause huge energy consumption. While using water as pressure medium, although the compression percentage is small, corrosion problem is highly likely to be induced for both the equipment and formed parts. Therefore, emulsion that is composed by 5–10% emulsive oil and rest of water is normally used as the pressurizing medium for hydroforming, which contributes to addressing the disadvantages of hydraulic oil and avoiding rust problem cooperatively.

- **Horizontal servo hydraulic cylinder**

Reliable sealing needs to be guaranteed by the axial punches at proper moments during hydroforming. Accompanying with the variation of inner pressure, the tube ends are pushed into the die cavity by the axial punches. Normally, axial punches

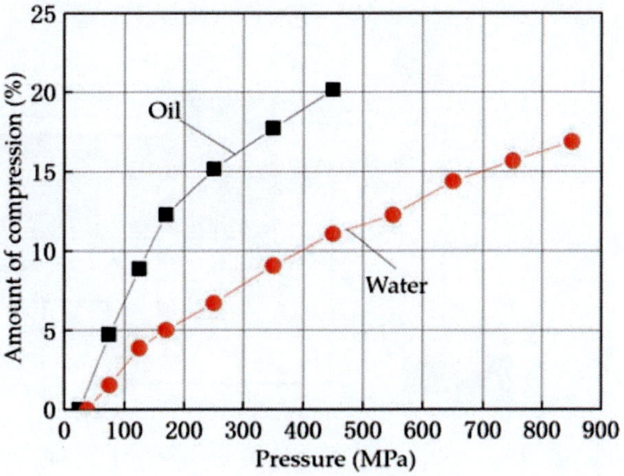

Fig. 6.4 Correlations between pressure and compression ratio of oil and water medium

are mounted on the piston rods of horizontal cylinders. Then, the cylinders drive the piston to accomplish the motion of punches horizontally. For the hydroforming of multi-branch tubular parts, e.g. Y-shaped or T-shaped tubular parts, an additional punch which is named as counter punch is used to exert a back force on the branch tube, in addition to the axial sealing at the tube ends using horizontal punches. Therefore, more horizontal cylinders are required to be configurated to achieve the independent displacement control of each punch.

For the applications requiring axial displacement of punch owning high precision, the servo hydraulic cylinder is commonly used enabling displacement to be instantly adjusted by the feedback of sensors. The piston displacement in oil cylinder is precisely controlled by the servo-control system with a variety of servo-valves. As illustrated in Fig. 6.5, the horizontal cylinder is mounted on the press bed or the base plate of die set using the holding plate of the oil cylinder. This fix type enables to compensate the counter force generated by the punch movement. Alternatively, extra pulling rods can be assembled between horizontal cylinders to construct horizontal press.

- **Hydraulic system**

The driving forces of both the low-pressure cavity in the pressure intensifier and the horizontal oil cylinder are provided by the hydraulic pump station. The pump station should be able to provide sufficient oil flux to guarantee the rapid feeding of oil cylinder and rapid pressurizing of the pressure intensifier. To avoid the high power of hydraulic system, energy accumulation device can be used to provide sufficient flux enabling the rapid pressurizing using a decreased pump power subsequently. For hydraulic servo system, high precision filtration needs to be configurated to the pump station. In addition, extra cooling is used to reduce oil temperature to address the danger of severe temperature increase of hydraulic system in mass production.

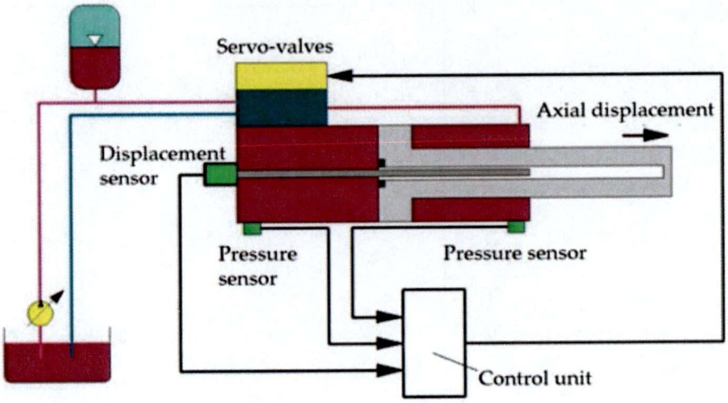

Fig. 6.5 The working principle of horizontal servo hydraulic cylinder

- **Water pressure system**

For hydroforming equipment, water pressurizing equipment is used to achieve functions of rapid filling of the emulsion, recycle and filtration. Rapid filling refers to a series of operations before building up the inner pressure, including filling emulsion, discharging gas and filling of the high pressure cavity of the intensifier. In order to increase the efficiency, water pressure equipment is ought to own a relatively high flux and a certain pressure. The emulsion should be able to be successfully recycled and filtered by this water pressure equipment after hydroforming, in preparation for next production cycle.

- **Computer control system**

The above presented five elements are connected as an integrated system through a computer-centric control system. Aiding by the computer neuro system, the automation and required production rhythm can be achieved following reasonable process, production sequence and predesigned loading path.

For industrial applications, Industrial Personal Computer (IPC) or PLC is the core of computer control system which utilizes other control elements including data acquisition card, pressure and displacement sensors and signal amplifier. Specific software is required for the control system enabling to emit signals to the control elements, such as servo valve and electromagnetic valve, according to the predesigned loading path. Subsequently, the executive elements, such as intensifier and horizontal cylinders, are driven after receiving signals. In the meantime, the data feedback regarding the inner pressure and axial displacement variations can be obtained from the pressure and displacement sensor, respectively, and interactively returned to the computer system, enabling control parameters to be output to operate the individual element. Then, axial displacement and inner pressure are simultaneously controlled by the loading path to accomplish the full automated hydroforming production.

6.2 Parameters of Hydroforming Equipment

6.2.1 Definitions of Main Parameters

(1) Die-clamping press

The main parameters of the hydraulic press include nominal die-clamping force, dimensions of press bed, maximum stroke, crosshead height and velocity of the slide.

The nominal die-closing force (F_n) is the critical parameter of the clamping press, which determines the tonnage of equipment. The force represents the maximum compressive force that the press can be exerted nominally. Theoretically, its magnitude equals to the liquid pressure multiplied by the total working area of the piston theoretically. F_n is determined by the maximum clamping force required in the calibration stage of hydroforming, with a certain allowance considering the variation of component standard additionally.

The dimensions of press bed ($L \times B$) represents the length and width of the bed within the press columns. As can be seen in the typical clamping press with three frames and four columns shown in Fig. 6.1, the bed refers to the horizontally projected working space of press. In general, for hydroformed parts requiring the horizontal cylinders to be distributed on both sides of press, the width of bed (BB) is determined by the width of both the part and die, while the length (L) is determined by the length of part, length of punch, die base plate and holding plate of horizontal cylinder; For parts requiring horizontal cylinder mounted on one side of press (e.g. U-shaped auxiliary frame), the whole profile of die and horizontal cylinders, and the center of dies needs to be overall considered to determine the bed dimensions, which is beneficial for avoiding waste of platform space and an unbalanced loading of press. In the meantime, sufficient operation space needs to be reserved for the convenience of maintenance and exchange of dies, punches and horizontal cylinders, as well as extra operations, such as lubrication and process monitor. The dimensions of bed directly affect both the dimensions and weight of the designed three frames. In addition, in-service properties including the rigidity, strength and precision of equipment are also dependent on the dimensions of bed significantly.

The maximum stroke (S) represents the maximum movable distance of the slide crosshead, which is determined by the maximum working distance of forming a particular part. The magnitude directly affects the length of main working cylinder, length of returning cylinder, piston length and the total height of hydraulic press frame. For parts with a large height gradient between the tube ends, increased maximum stroke is recommended to satisfy the assembling and disassembling of parts; while for parts with a small gradient in the height direction, proper spacing insert is used to reduce the slide crosshead stroke, that is beneficial for reducing the opening and closing time of forming dies.

Operation height (H) represents the distance between the top surface of press bed and bottom surface of slide crosshead when the slide crosshead motive frame is positioned at the upper limit position. Such a height determines the achievable operation space of clamping press in the vertical direction, which depends on the rig height, magnitude of working stroke, and the space used for mounting tube. The total height, stability of hydraulic press and the height of plant are greatly affected by this operation height. Therefore, the height should be designed at a minimal value under the premise of satisfying process requirements.

The velocity of slide crosshead includes both the clamping (rapid downwards movement) and reset speed (return to original position). Considering the clamping force only works once the dies are firmly closed, before that, rapid motion of the slide crosshead is recommended enabling a higher productivity. In addition, a switch of altering velocity between quick and slow mode is needed by the die-clamping press. The slow mode is switched on to avoid the severe crush of forming dies and impact on the hydraulic system, when the die-closing is approached to finish. The calculation of liquid supply of pump station is directly affected by this slide crosshead velocity.

(2) High-pressure source

The high-pressure source in hydroforming is normally generated using the single-operation pressure intensifier, which is used for guaranteeing a sufficiently high medium pressure to deform the tube. The main parameters of high-pressure source are the maximum pressure and volume of high-pressure cavity. The maximum pressure is determined by the required liquid medium pressure used in the calibration stage of hydroforming. A further amplitude adjustment by multiplying a certain amplified coefficient depending on specific part and material.

The volume of high-pressure cavity mainly refers to the volume of high pressure medium discharged by the single-operation pressure intensifier during one stroke. In comparison, the medium volume discharged by the double-operation pressure intensifier is huge due to the reciprocate movement of position. Then, theoretically, there are no particular requirements needed for the volume of high-pressure cavity. While in terms of the single-operation pressure intensifier, the pressure is increased ultimately up to the calibration pressure following the loading path within one stroke, therefore, the volume of high-pressure cavity needs to be calculated and designed considering the volume variation of part during hydroforming, as well as the compression percentage of liquid under high pressure.

(3) Horizontal hydraulic servo cylinder

Piston cylinder is normally used as the horizontal cylinder with main parameters including maximum force, stroke and maximum velocity. While in terms of servo cylinders, the precision of stroke control is also an important parameter. The maximum thrust refers to the pushing force delivered by the horizontal cylinder when the pump station runs at the maximum working pressure. The magnitude of thrust is determined by the required axial feeding force of horizontal punch movement in hydroforming.

The stroke of horizontal cylinders refers to the maximum achievable displacement of cylinder, which should satisfy the space requirements of punch feeding and withdrawing in hydroforming. However, the larger the stroke of oil cylinder, the greater the length of cylinder. Correspondingly, the dimensions of bed and cost of the die-closing force are increased.

(4) Hydraulic system

The dominant parameters of hydraulic pump station are the maximum pressure and flux of oil pump. The flux of driving pump in the intensifier determines the piston velocity, which affects the pressurizing rate of the intensifier accordingly. The maximum pressure is determined by the maximum force at the low-pressure end, and subsequently the maximum achievable inner pressure of the pressure intensifier. To determine the flux of driving pump, the velocity of piston in the pressure intensifier is calculated first using the volume of high-pressure liquid high enough to hydroforming a tubular part. The magnitude of flux is able to be used for calculating the power of pump further considering the maximum force. To increase the pressurizing rate of hydroforming system, the power of oil pump needs to be controlled below the critical value. The objectives of decreasing power and increasing efficiency of hydroforming system are normally achieved by the cooperative contributions of hydraulic pump and energy accumulation device. For the horizontal cylinder, the flux of driving pump determines the velocities of the mounted piston and punch; the maximum compressive force determines the maximum push force enabling to be exerted by the cylinder. Therefore, the flux and power of horizontal driving pump should be calculated using the punch displacement, cross-section area of piston and the working pressure of pump station. When a same driving pump is used for the pressure intensifier and horizontal pump, flux and pressure need to be simultaneously considered for selecting a proper driving pump.

(5) Water pressure system

The main units of water pressure system include rapid pre-filling pump, filling pump for intensifier and recycling pump. The flux of pump for rapid pre-filling is determined by the volume of tubular part and extent of discharging air, which is calculated according to the maximum dimension of tube and lead time for filling. The flux of pump for liquid compensation is calculated according to the volume of high-pressure cavity and in-service frequency of the pressure intensifier. The flux of pump for recycling is calculated based on the cycle time and recycled emulsion volume.

Therefore, the flux of pump for rapid filling and pump for liquid compensation have great influence on the production efficiency, and the flux of recycling pump is of virtually importance for the recycling of emulsion and protection of plant environment.

6.2.2 Determination of Main Parameters

Among the variety of technical parameters of hydroforming equipment, nominal die-clamping force, maximum pressure of pressure intensifier and maximum thrust of horizontal cylinder are the three key parameters affecting the equipment scale. In this section, the determination of the three main parameters is introduced.

(1) Nominal die-clamping force

Hydraulic clamping press is used in hydroforming processes for a series of tubular products with various length and diameters. Considering the required forming pressure is different, the required die-clamping force for hydroforming a specific part needs to be calculated according to part length, tube diameter and forming pressure. For the convenience of engineering reference, nominal die-clamping force can be selected the standard with one series relatively higher than needed magnitudes.

(2) Maximum pressure

The required forming pressure is calculated based on the thickness, minimal corner radius and yield stress of material for a specific part.

(3) Maximum force of horizontal cylinder

The axial feeding force is calculated using the tube diameter, thickness, yield stress of tube material and required calibration pressure of the target component. Then, according to the standard series of horizontal cylinder with relatively higher maximum values. The commonly used maximum thrust of horizontal cylinders in industrial production include 1000, 1500 and 2000 kN.

6.2.3 Recommendations of Standard and Parameters of Hydroforming Equipment

From the view of economical cost-effectiveness and equipment versatility, proper selection of hydroforming standard enables to satisfy industrial requirements of production efficiency, part exchange and modification, and save the investment of equipment.

Taken some industrial cases as examples, the hydraulic press with a 10,000 kN nominal die-clamping force is suitable for producing small-scale parts with a 1 m equivalent length. In this book, the equivalent length is defined as the obtainable part length under this nominal die-clamping force for a 100 mm tube diameter and 100 MPa forming pressure. While if the tube diameter is 50 mm, the equivalent length becomes 2 m correspondingly. Similarly, 20,000 kN and 50,000 kN tonnage hydroforming press are suitable to produce parts with an equivalent length 2 m and 5 m, respectively.

6.3 Typical Structures and Characteristics of the Die-Clamping Press

Using the stroke of main oil cylinder, clamping press used in hydroforming can be classified as long-stroke and short-stroke category. When the die operations, i.e. die-clamping, opening and locking, are all accomplished by the main oil cylinder, the working stroke of main cylinder is required to be relatively long, and press with such a cylinder is defined as the long-stroke. While regarding most of stroke of die clamping and opening is additionally compensated by the lift cylinder (assist cylinder), and only the locking operation is accomplished by the main cylinder, the category of such hydroforming press becomes short-stroke.

From the view of press body structure, alternatively, the hydroforming press can be also divided into four-columns type, frame-type and open-type (C-type). In comparison, two-columns and wedge types are seldomly seen nowadays.

To date, the leading companies of producing hydroforming systems are mainly located within Europe, such as Schuler Germany, SPS Germany, AP&T Sweden, Schafer etc. Among the products delivered by these companies, the maximum high pressure source can reach up to 400 MPa. For passenger automobile parts, the capacities of hydroforming presses are mostly within the range of 5000–6000 tonnage. While for the truck parts, the capacity can go up to 12,000 tonnage, and the bed dimension is 6 m × 2.5 m.

6.3.1 The Long-Stroke Die-Clamping Press

The long-stroke die-clamping press can select traditional hydraulic press using the main cylinder to accomplish the lifting and clamping of forming dies, and exert clamping force during hydroforming. Figure 6.6 shows the schematic of typical structure of such a press. The stroke of main cylinder is normally greater than 400 mm.

Using traditional hydraulic press to construct long-stroke hydroforming equipment exhibits a series of advantages, which are given as follows:

- Characteristics of good universality, structured equipment and serialization of components enable a convenient maintenance and product upgrade to be achieved.
- The hydraulic clamping press can be alternatively converted to traditional hydraulic press to be used for stamping production when the horizontal cylinders and tooling are removed.

However, disadvantages of using long-stroke clamping press in hydroforming system are still inevitable, including:

- The capacity of main cylinder is huge, and a high flux pump is required;
- Great compression percentage of hydraulic oil induces a long lead time for building up of the clamping force resulting in a severe energy consumption.

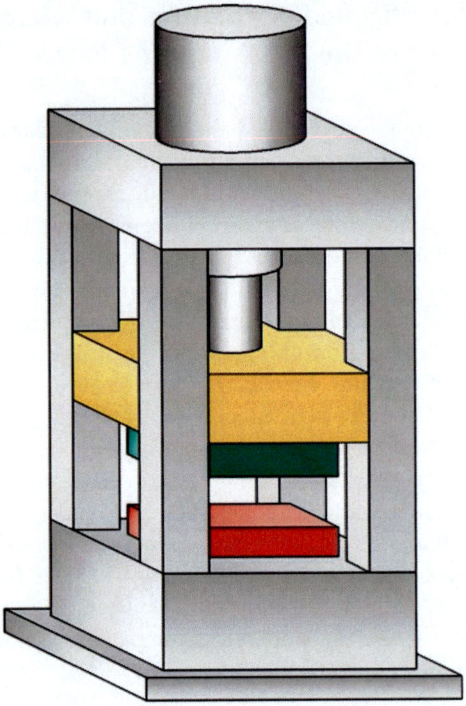

Fig. 6.6 The schematic of long-stroke die clamping press

- Main cylinder is constantly used to perform the lifting and closing of the forming dies. Consequently, the operation time is relatively long and productivity is low.

Long-stroke hydroforming equipment is normally constructed based on the column-typed press, such as the four-columns structure with three crossheads and four columns. In terms of the manufacture requiring relatively small capability, two-columns press can be also adopted.

Four-columns clamping press has a closed body structure which is composed by upper crosshead, slide crosshead, lower bed and four supporting columns. The main cylinder is mounted inside the upper crosshead with a piston rigidly connected to the slide crosshead. The columns are locked to the upper crosshead and bed using locknuts. The slide crosshead is driven by the main cylinder to slide vertically according to the guidance of columns. Upper and lower forming dies are mounted on the upper crosshead and bed, respectively. During hydroforming, the clamping force is sustained by the machine frame composed by upper crosshead, bed and columns.

The most significant advantage of four columns type press is the large operation space and flexibility, which contributes to arranging the layout of horizontal cylinders and forming dies, especially for the hydroformed parts with U or S-shaped curved axis and multi-branch tubular parts. The horizontal cylinders and forming dies can be easily distributed in arbitrary direction of press bed for the convenience of processing. In addition, due to the individual manufacture of crosshead and columns, compared to the welded frame-type press, the process difficulty and the complexity in processing sequence are significantly reduced, resulting in a lower total equipment cost.

However, there are still a few disadvantages existing for the four-columns die-closing press, which are:

- The cross-section area of column is small resulting in a poor bending-resistance, low rigidity of whole press and large elastic deformation.
- The clearance between crosshead and column is difficult to be adjusted, which results in a relatively lower precision of crosshead motion compared to the frame-type press. In addition, such a clearance reduces the capability of sustaining unbalanced loading.

For the 30,000 kN hydroforming press produced by the Schuler Germany, the machine is four-columns type with a nominal die-clamping force of 30,000 kN. A maximum inner pressure, 400 MPa, can be achieved by the pressurizing system, and the volume of high-pressure cavity in the pressure intensifier is 4 L. The bed dimension is 3150 mm $\times$ 1400 mm. Such a big platform enables productions using one die set for two parts to be achievable in terms of forming parts requiring relatively small die-clamping force. In addition, hydropiercing device is configurated to make the hole features before releasing the medium pressure. Full automation is achieved in the system achieving a significantly increased productivity. The mass production volume can reach 750,000 parts per year calculated on three working periods per day.

6.3.2 The Short-Stroke Die-Clamping Press

The short-stroke press refers to the machine with stroke of main cylinder smaller than 50 mm. To achieve a quick operation speed, the die operations, i.e. opening and closing, are accomplished by the assisted lifting cylinder with a small capacity. The short stroke characteristic results in a small volume of main cylinder, enabling to build up the clamping force quickly and increase productivity.

Figure 6.7 shows the schematic of the short-stroke press. The opening and clamping operations of dies are performed by the small tonnage long-stroke cylinder that is mounted on the upper crosshead. Figure 6.7a shows the state of die-opening, two rigid positioning inserts are fixed on the two horizontal spacer cylinders. These two inserts are pushed into the gap between the top of machine body and slide crosshead from both sides in the die-closing state, as shown in Fig. 6.7b. Then, the clamping force is alternatively exerted by the short-stroke cylinder implemented in the bottom of machine body. Using such a structure, the bottom short-stroke cylinder is only required to compensate the elastic deformation of equipment frame and dies, as well as eliminate the initial vertical gap of the positioning inserts. Hence, the stroke can be very small.

However, the short-stroke press is only used for providing maximum force within a very short stroke, which limits its application in hydroforming production. The

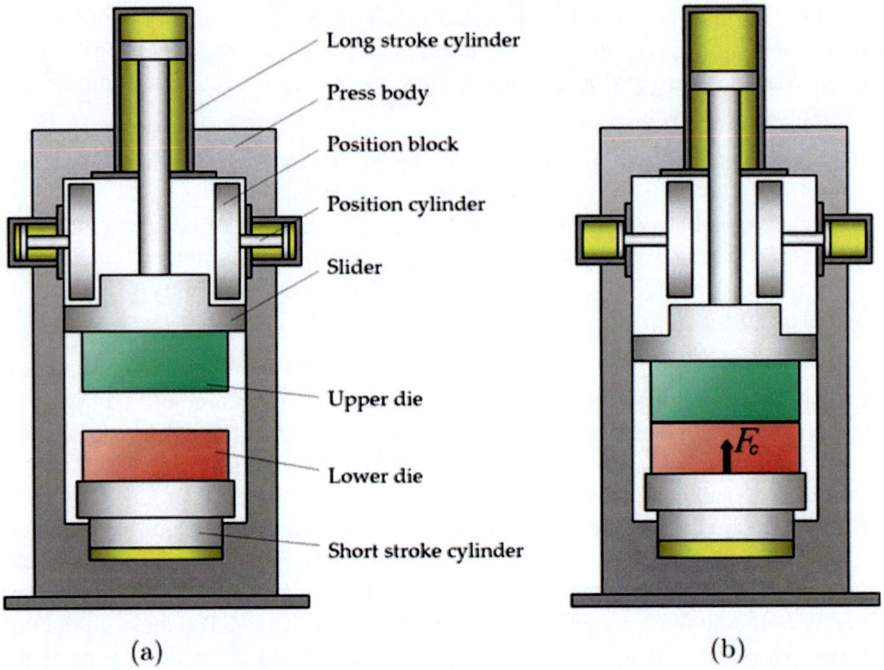

Fig. 6.7 Schematic of short-stroke hydroforming press. **a** Structure and **b** die-clamping state

versatility of such equipment is very poor. Moreover, two horizontal cylinders and a long-stroke cylinder are additionally constructed in the press resulting in a more complex mechanical structure and higher requirement of equipment rigidity.

Frame-type machine body is normally adopted by the short-stroke hydroforming press. The frame-type body prefers to utilize hollow box structure with front and back sides open and left and right sides closed. The columns can select either rectangular or π shaped cross-section. Two pairs of guide rails are used inside which can be adjusted by blots to guarantee the precision of slide crosshead motion.

The main advantages of frame-type die-closing press are good bending-resistance and high precision of stroke. In addition, the welded frame-type press body reduces the total weight significantly, and is easy to be transported and assembled. While the smaller operation space under the same bed dimension is an obvious drawback compared to the four columns-typed structure. Furthermore, the whole frame requires annealing treatment and subsequent machining, which adds the total production cost on. Due to the good rigidity, the frame-type machine body is more suitable for hydroforming presses requiring larger tonnage.

Schaefer Sweden company designed a new short-stroke clamping press with merits of closely-packed structure and easy assembly. The total height is 5.1 m. The press can be directly installed on the concrete ground without the huge foundation. Schuler and AP&T also developed mechanical die-closing devices for hydroforming press, such a device successfully achieves the built-up of die-clamping force without using complex hydraulic system, which contributes to significantly shortening the production cycle 25%. Schuler Germany and Stuttgart university collaboratively developed a frame-type short-stroke hydroforming press with a nominal die-closing force of 35,000 kN. The stroke of main cylinder is only 50 mm and the operation pressure is 60 MPa. The bed dimension is 2500 mm × 900 mm. The achievable maximum forming pressure is 420 MPa.

Four-columns type machine body can be also adopted to construct the short-stroke hydroforming press. AP&T Sweden developed a short-stroke hydroforming press HF50000 with die-locking cylinders mounted within the bed. Regarding the key parameters, the nominal die-clamping force is 50,000 kN, while the stroke is only 20 mm. Two cylinders installed on the upper crosshead enables to accomplish the motions of lifting and closing the dies. The pushing force is 2000 kN and the maximum stroke is 1100 mm. The usable dimension of bed is 3000 mm × 2500 mm. A maximum 400 MPa pressure and 4L volumed high-pressure cavity of pressure intensifier are configurated to the hydroforming system satisfying requirements of forming.

Asian countries also contributed to the development of hydroforming presses. Institute of water pressure industry in Japan developed a mechanical die-locking device to utilize small tonnage press to produce large-scale hydroforming parts; Kawasaki Oil Industry in Japan developed a hydroforming press with a nominal die-clamping force of 50,000 kN, which was upgraded from the conventional hydraulic press.

6.4 Large-Scale Hydroforming Equipment

6.4.1 Servo-Control of Ultra-High Pressure and Multi-axial Displacement

To form qualified parts without wrinkling and splitting defects, firstly, a simultaneous and precise combination between inner pressure and axial displacement is a key capability of hydroforming equipment. Then, the process parameters need to follow the predesigned loading paths, and to be loaded digitally. Servo-control is the core approach to accomplish above requirements, and the biggest difficulty of applying it in hydroforming is the precise control and decoupling of the mutual disturbance between the ultra-high pressure of liquid medium and axial displacement. This is the key difference compared to commonly seen servo-controls of rigid displacement.

The precision of control is determined by various factors considering process characteristics of hydroforming, those are:

Under the condition of ultra-high pressure, the compression percentage of liquid medium is relatively large, which affects the resolution of pressure control. The axial feeding of horizontal cylinder results in the decrease of tube volume, which subsequently induces the rapid increase of tube pressure and loss of control accuracy of inner pressure.

The inner pressure also acts on the horizontal cylinder, which requires the displacement control system to be capable of responding quickly.

The velocities of horizontal cylinders during rapid feeding and returning should be sufficiently high, while high accuracy of displacement control is required to guarantee reliable feeding.

6.4.2 Computer Numerical Control System and Control Software

Normally, the computer numerical control (CNC) press utilizes multi-axial motion control cards, such as Siemens 840D system, to achieve multi-axial servo control. However, these motion control cards are unsuitable for the control strategy of hydroforming as the programming rules are unable to fulfill the special requirements of hydroforming. To overcome this limitation, PLC-based control system and software are developed. The main characteristics of this system are summarized as follows:

High precision of loading path of displacement and inner pressure. The control system selects high-order spline to perform interpolation, and extracts the eigenvalues with big variations as second-order characteristic points to control, enabling the contradiction between the loading rate and accuracy of processing paths of forming to be solved. Forefront technologies of forming time and control accuracy can be achieved globally.

High production efficiency. The instant monitor of forming status results in the full automation in the hydroforming production line.

High safety and reliability of the system. A series of overloading protection approach and safety designs are used to guarantee the safety of equipment and operators. The status of all key elements in operation are instantly monitored through feedback design. Different alarming approach is classified and controlled based on the safety level, which guarantees both the production safety, and normal working of equipment.

Figure 6.8 shows a typical GUI of hydroforming control system developed by Harbin Institute of Technology (HIT) in China. Using this user interface, process parameters, such as inner pressure, axial displacement and die-closing force, can be instantly observed and monitored. Then, aspects of fault diagnosis, module monitor, system maintainability, are significantly enhanced, enabling the engineers to solve most of faults with shortened maintenance time.

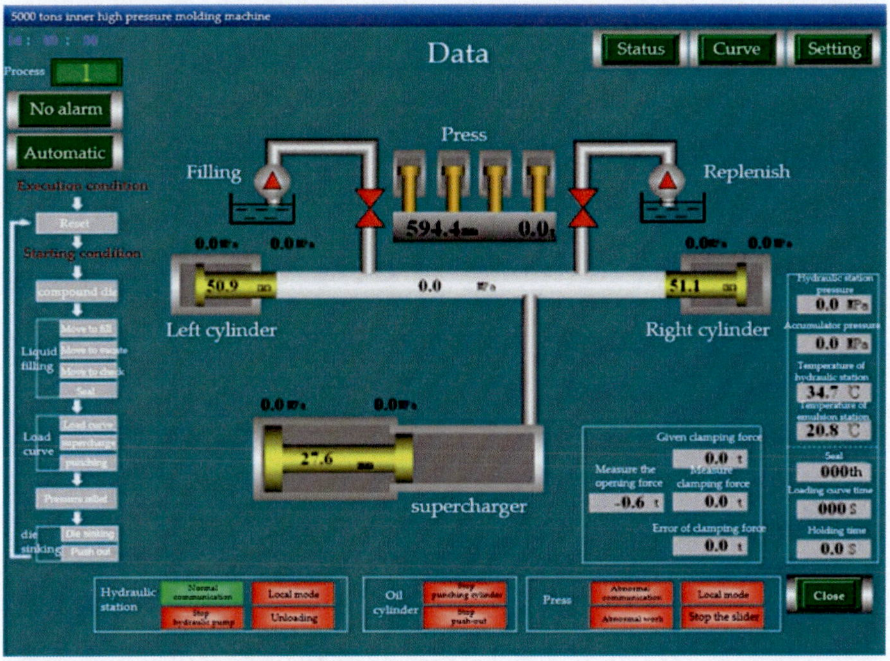

Fig. 6.8 User interface of control system

6.4.3 Series of Hydroforming Equipment Developed by HIT

Figure 6.9 shows the layout of hydroforming system. Seven sub-system standard modules are incorporated from the views of manufacture, assembly, debugging and system integration (SI), which are: the die-clamping press, horizontal cylinder, forming dies, hydraulic system, water-pressure system, high pressure source and control system. To improve the SI level, hydraulic system, water-pressure system and high pressure source are integrated as a whole system. Then, the hardware of hydroforming system is only composed by four dominant elements: die-clamping press, horizontal cylinder and die, high pressure station and control system.

The representative characteristics can be identified for the hydroforming system:

(1) The control systems of die-clamping press and hydroforming system are independent, which can be online connected to compose the integrated hydroforming system, enabling either full-automation production or individual service to be delivered.
(2) The die-closing force can be simultaneously adjusted in accompany with the inner pressure. The synchronous control during hydroforming contributes to reducing die dimension, external loading and deformation significantly. The dimensional accuracy of component is increased.

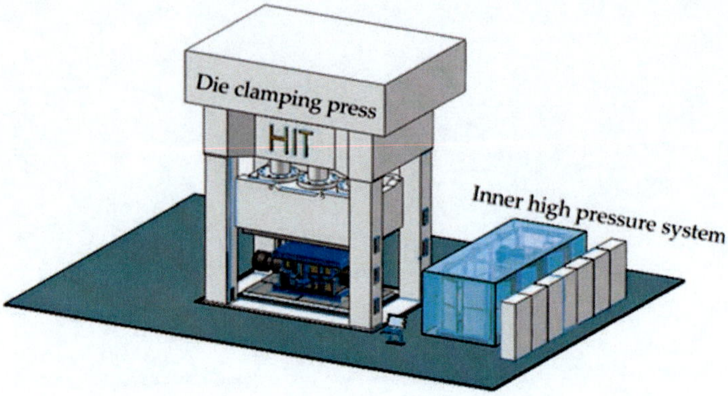

Fig. 6.9 The layout of CNC hydroforming equipment (Clamping press, inner high pressure system)

(3) Precise closed-loop control of process parameters results in a digitalized loading. Under the condition of ultra-high pressure, the synergy between inner pressure and axial displacement is achieved by the cooperation with hardware and software according to the forming process requirements. Digitalized loading is achieved based on the pre-designed path.
(4) High precision and efficient CNC system and control software. For example, the strategy of using the combination between robust and PID control proposed by HIT is a leading methodology. The administration method is leveled according to the parameters, then a user-friendly interface is obtained enabling easier operation and management, which can be used to diagnose the abnormal variations of key parameters and protect the normal service of equipment.
(5) Highly reliable and low-cost high pressure source. High pressure source is the core of hydroforming system that needs to satisfy mass production requirement (200–300 thousand parts per year).
(6) Energy efficiency design. The hydraulic system adopts the approach of using oil pump and energy accumulator simultaneously, which enables the power of hydroforming system to be reduced significantly. Fig. 6.10 shows the large-scale hydroforming equipment for industrial production developed by HIT, China.

6.4.4 The Layout of Hydroforming Production Line

The main operation sequence of hydroforming production includes pre-bending of tube blank, lubrication, preforming, hydroforming calibration and hydropiercing, examination of positioning holes, laser-cutting of tube ends and post-cleaning. Correspondingly, Fig. 6.11 shows the layout of a hydroforming line for producing automobile auxiliary frame part. In this line, a tube bending machine, preforming press, hydroforming system, laser cutting machine, and necessary lubrication and cleaning equipment function together to construct the complete production line. Firstly, the

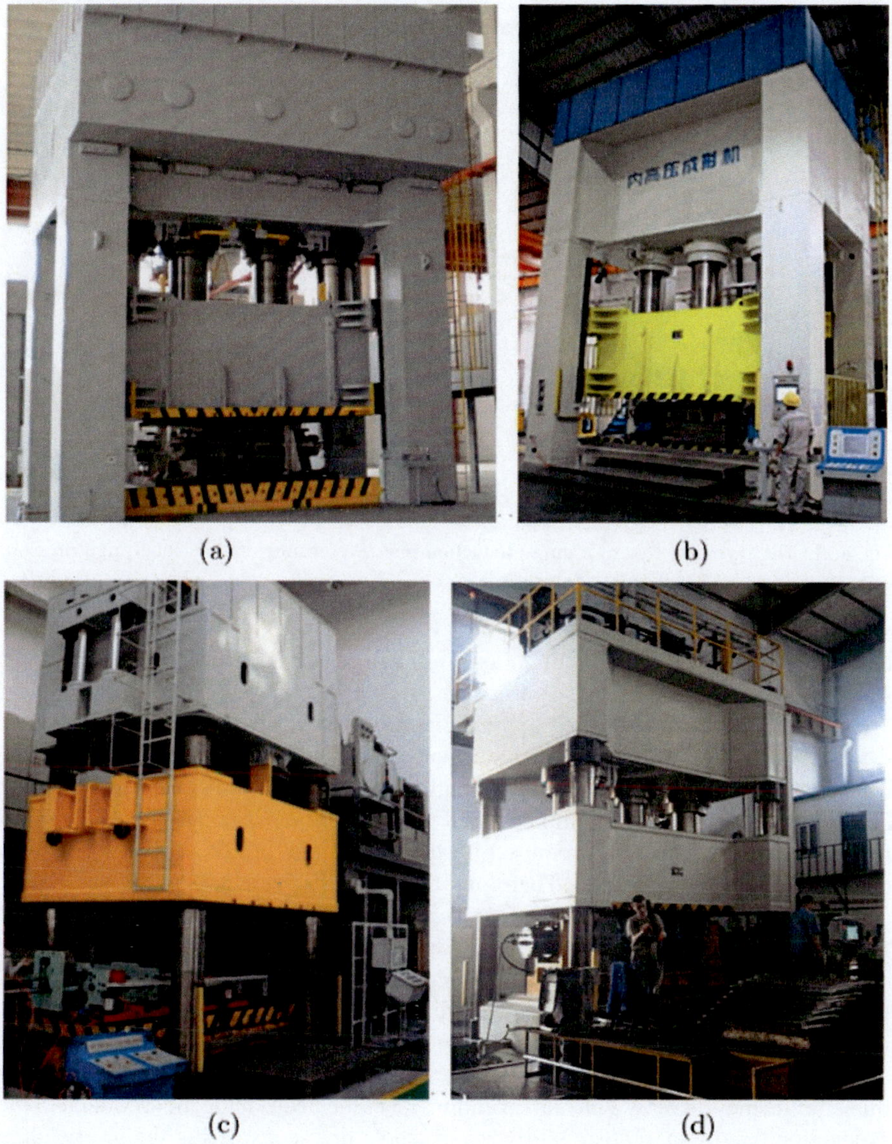

Fig. 6.10 The series of CNC hydroforming systems for industrial production. **a** 5000t three axis servo control, **b** 3000t three axis servo control; **c** 2000t five axis servo control, **d** 2000t three axis servo control

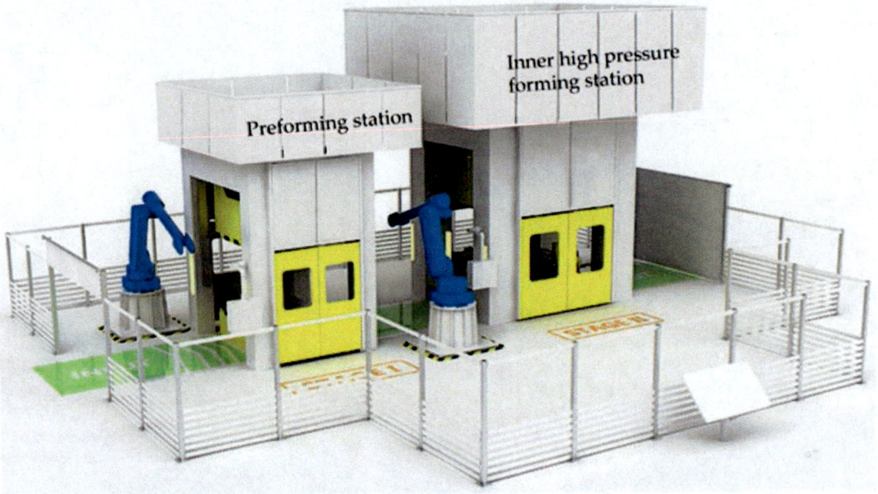

Fig. 6.11 The layout of hydroforming production line (Preforming station, inner high pressure forming station)

tube is bent to a shape with a similar axis of the final part and lubricated on the outer surface; Secondly, the tube is preformed using either press or other approaches; Then, the preform is transferred to the clamping press for hydroforming; after that, the positioning accuracy of pierced holes on the part is examined, after which, part ends are trimmed using laser cutting. Finally, the part needs to be cleaned and properly stored in the plant. It should be noted that, for some particular parts difficult to form, annealing treatment is required for the preform before final hydroforming to eliminate the work-hardening. Therefore, the lubricant needs to be removed before annealing and re-sprayed after, in order to avoid the contamination of heat treatment facility and plant environment due to the dust arising from the hot lubricant.

The transfer of part between each operation can be either done by the human laboring manipulation or robotics, depending on the production rhythm, equipment investment and production cost. The production rhythm varies between 30 s per part to 2 min per part according to the automatic degree. With regard to products requiring higher efficiency, at least 2 NC tube bending machine needs to be incorporated in the production line to guarantee the rhythm of whole line, considering the NC bending occupies the majority of production cycle.

To date, the hydroforming production line is still mainly used for the production of single part. In the future, the production line is believed to be developed towards owning functions of rapid die exchange, rapid part update, high suitability of equipment. High versatility and flexible design will be achieved by the future hydroforming production line to reduce the idle time and increase the productivity.

6.5 Tooling of Hydroforming

6.5.1 The Structure and Material of Forming Dies

Figure 6.12 shows a typical tooling of hydroforming die sets. The whole die set is mainly composed by upper die 1, lower die 2, left punch 3 and right punch 4. Similar to the rigs of other forming processes, additional upper plate 5 and bottom plate 6 are individually mounted on the slide crosshead 7 and press bed 8.

The die cavity is designed using a methodology of die insert design based on different working conditions and machining characteristics. Additional guiding and locking devices are necessary considering the position of forming die shifts horizontally during the hydroforming of a part with complex curved axis. The horizontal punches are particular elements of the die set, which are used to seal the tube and axially feed the material. The sealing at the punch front is very critical determining the success of hydroforming and productivity. The diameter and length of the punch vary with the diameter and length of the tube. Dovetail grooves used for sustaining loading can be selectively machined on the two sides of upper and lower plates, for the convenience to locate the flange of horizontal cylinder during die-clamping and establish a force equilibrium to balance the axial punching force on the punches.

The selection of die material should follow the below basic criteria:

(1) The die material needs to meet the service requirements of hydroforming, owning good abrasive resistance, high toughness and good fatigue crack resistance. Therefore, different material and specific heat treatment is used based on the working conditions of different segments of the die-set. For example, sealing is achieved by the compressive deformation of the tube ends through axial punches, then the tube is compressed to achieve axial feeding. Therefore, severe friction exists at the sealing regions of tube ends resulting in materials with good abrasive resistance preferably used. The die structure at this region is recommended to design individual insert, on one hand, the tool life is prolonged,

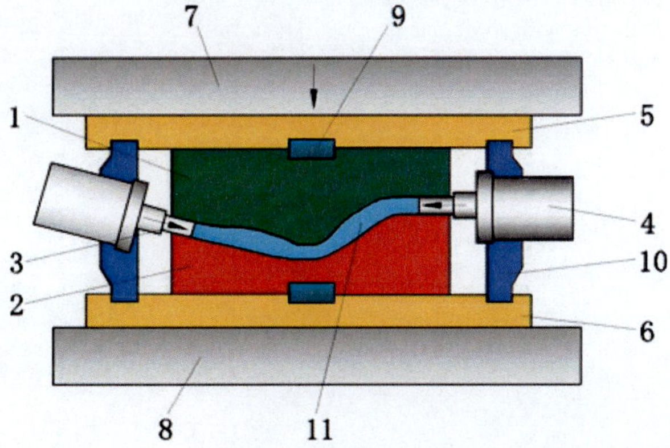

Fig. 6.12 Schematic of a typical die set of hydroforming. 1-Upper die; 2-lower die; 3-left punch; 4-right punch; 5-upper plate; 6-bottom plate; 7-slide crosshead; 8-press bed; 9-pin; 10-support of horizontal cylinder; 11-tubular part

and on the other hand, the insert is easy to replace after damage. In terms of die cavity, die material with good fatigue crack resistance is preferable to be used for extending the tool life, considering the local stress at the inner radius of cavity is normally very high under the ultra-high inner pressure.

(2) The die material needs to meet the process requirement of machining, which are malleability, hardenability and machinability. The hydroforming dies are normally complex-shaped requiring high dimensional accuracy and surface roughness, which brings great difficulties in machining. Therefore, materials with good machinability are recommended to satisfy the technical requirements of the die.

(3) The die material needs to meet the economical requirement. Low-cost die materials are reasonable candidate considering production volume, product properties and process parameters. In addition, the tooling structure can be further optimized according to the cost. With regard to small volume products with low forming pressures, good carbon structural steel, such as #45, can be used to manufacture tooling, while for large volume products with high forming pressures, alloyed tool steel is recommended.

6.5.2 *Effects of Thickness of Die Sidewall and Bottom Plate*

Figure 6.13 shows the force state of the die with a circular cavity during hydroforming to illustrate the effects of die cavity dimensions. The key geometrical dimensions of hydroforming are inner diameter of die cavity (d), width of sidewall (B) and height of bottom plate (H). The tube effect can be ignored as the tube thickness is small. Correspondingly, the pressure exerted on the die cavity is equal to the inner pressure within the tube, $p_i\, p_i$. For a specific part, the sustainable die-clamping force and inner pressure of the die and its whole profile are determined by B and H, as the inner diameter of cavity, d, is a constant.

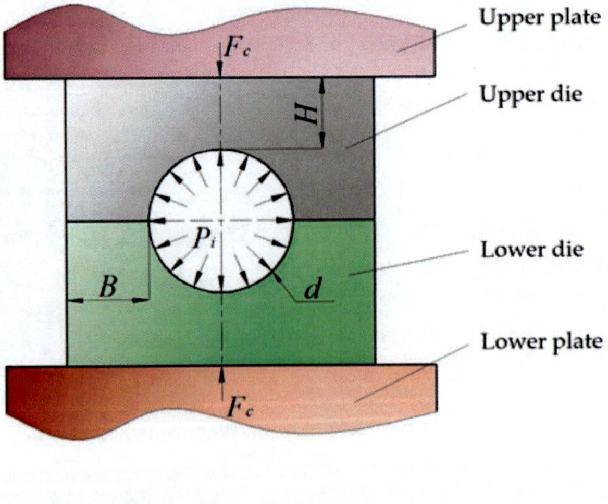

Fig. 6.13 The force analysis of the die with a circular profiled die cavity during hydroforming

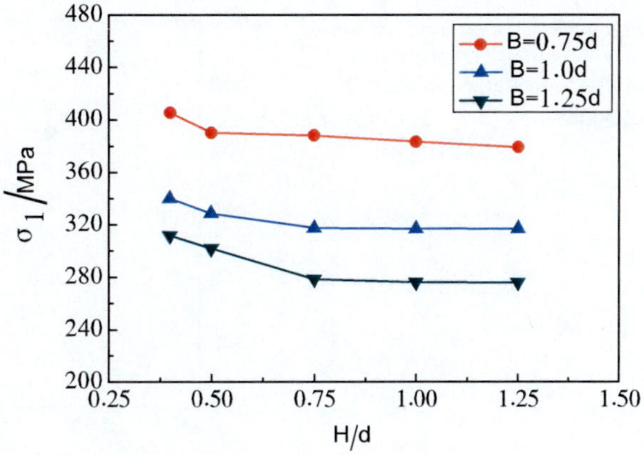

Fig. 6.14 The variation of maximum hoop tensile stress of the die with the increase of height

Taken a circular-shaped die cavity as an example to numerically analyze the above dimension effect on the stress and deformation of the die. Figure 6.14 gives the variations of maximum hoop tensile stress with various height, H, under a given combination between inner pressure and die-clamping force. With the increase of H, the maximum hoop tensile stress within the die gradually decreases. However, when the relative height H/d is greater than 0.75, effect on the maximum hoop tensile stress is very limited with further increasing the bottom height. The reason is believed to be that, the die height mainly affects the magnitude of horizontal stress of the die bottom, that is dependent on the inner pressure rather than the die-clamping force. Therefore, the further the distance to the die cavity of material along the bottom, the less effect on the stress states of die cavity. From the view of engineering application, $H/d = 0.75$ can be used as the reference value to determine the minimal height of bottom in the die design.

Figure 6.15 gives the variations of maximum hoop tensile stress with the width of the die at a relatively height $H/d = 0.75H/d = 0.75$. With increasing the sidewall thickness, the maximum hoop tensile stress is significantly reduced. When the relative width B/d is greater than 1.0, the tensile stress in the die gradually decreases. Similarly, the die deformation is monotonously varied with the stress. It should be noted that, the above conclusions are only the preliminary design criteria based on the proposed simplified model, modifications are still required according to the die material, formed parts and cross-section shape of die cavity in practical applications.

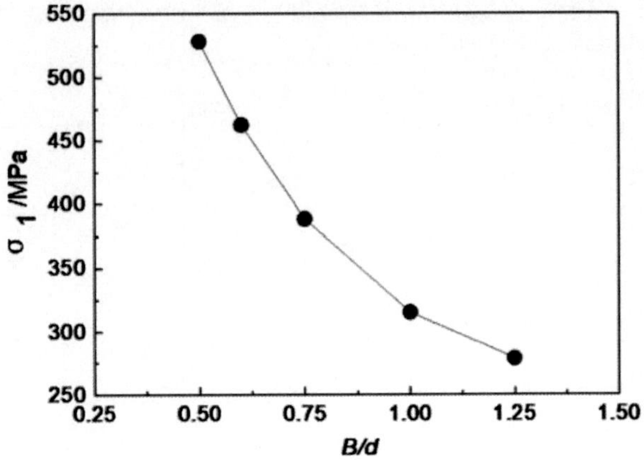

Fig. 6.15 The variation of maximum hoop tensile stress of the die with the increase of width

Chapter 7
Sheet Hydroforming Process

7.1 Procedure and Characteristics of Process

7.1.1 Forming Procedure

The principle of sheet hydroforming is utilizing liquid as the forming medium to transfer loading, by which a sheet is deformed according to the die profile under the pressure of fluid medium to become desired parts. It is also call hydro mechanical deep drawing, hydro-deep drawing and hydro-drawing. The procedure of sheet hydroforming process can be divided into four steps, as shown in Fig. 7.1. In step one, the hydraulic pump is activated, liquid medium is filled into the cavity until reaching the top surface of female die (Fig. 7.1a). The liquid is used to provide counter pressure on the deformed sheet metal, therefore, the liquid chamber can be also named as counter pressure pot. In step two, the blank is mounted on the female die followed by giving a blankholding force F_Q on the flange (Fig. 7.1b) to clamp it; Then in step three, punch (male die) travels downwards into the cavity of female die. A counter pressure is built-up using natural pressurizing or hydraulic system for the liquid medium inside the cavity, which forces the deformed sheet to be firmly attached on the punch (Fig. 7.1c). During the deep drawing, liquid continuously flows outwards underneath the flange region due to the compression of punch movement, enabling a fluid lubrication to be achieved until the end of forming (Fig. 7.1d).

Figure 7.2 illustrates the fluid lubrication and beneficial friction, which are the unique advantages of sheet hydroforming. A beneficial friction force, F_f, is generated at the interface between sheet and punch, resulting from the counter pressure of liquid chamber. Normally, the greater the counter pressure, the larger the resulted friction force. When a critical magnitude of the counter pressure is reached, the flange of part is separated with the corner radius of female die due to the effect of liquid pressure, resulting in the absence of friction. Regarding the unsealed condition of flange outer edge (Fig. 7.2a), liquid inside the female die cavity is forced to flow outwards from the interface between part flange and female die. Hence, a beneficial liquid lubrication can be achieved for the whole flange zone, which effectively

S. Yuan, *Modern Hydroforming Technology*,
https://doi.org/10.1007/978-981-19-5775-8_7

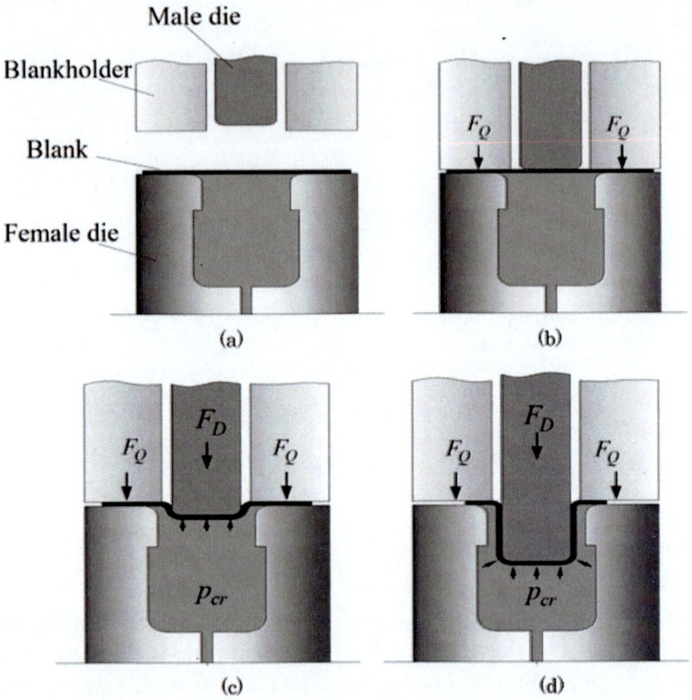

Fig. 7.1 The procedure of sheet hydroforming process. **a** Filling liquid; **b** exerting blankholding force; **c** forming part; **d** end of forming

decreases the interfacial friction. However, an unavoidable disadvantage is the difficulty in precisely controlling the pressure of liquid chamber. While regarding the sealed condition shown in Fig. 7.2b, the liquid medium is unable to flow outwards underneath the flange. Although liquid lubrication cannot be achieved, the pressure of liquid chamber can be adjusted using the unloading valve. Using such a tooling configuration, initially the liquid pressure is built up using the natural pressurizing approach achieved by the downwards compression of punch. Therefore, the liquid pressure at the early stage of forming is normally insufficient to compensate the tensile stress of the sheet at the corner radius of female die, which could result in the splitting defect. To address such a limitation, the hydraulic pump is required to be activated once the blankholding force is exerted on the flange, which enables the pressure to be quickly increased by filled liquid. After the pressure reaches a certain magnitude, punch is subsequent to travel downwards into the cavity of female die to conduct the sheet hydroforming.

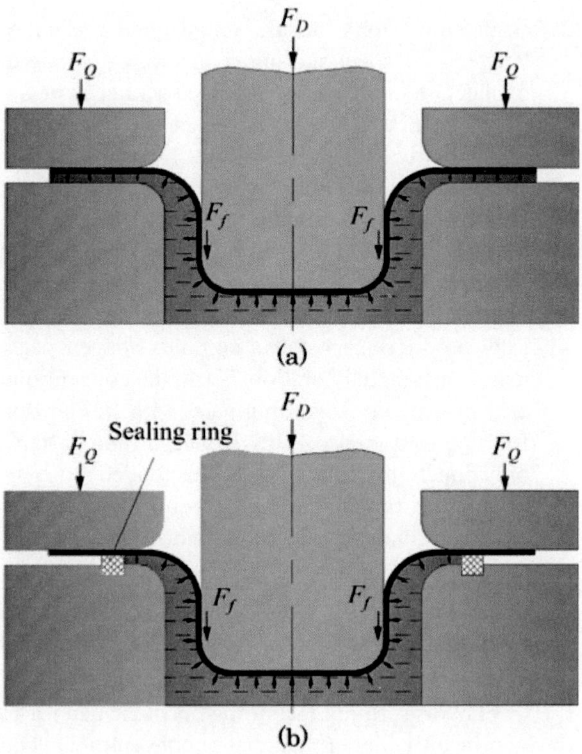

Fig. 7.2 Illustrations of the liquid lubrication and beneficial friction of sheet hydroforming process. **a** No sealing, and **b** sealing

7.1.2 Characteristics of Sheet Hydroforming

Compared to the conventional deep drawing, there are obvious advantages of sheet hydroforming process summarised as follows:

(1) High forming limit. Due to the influence of counter pressure of liquid, the deformed part is firmly compressed to the punch resulting in a beneficial friction; Liquid lubrication is generated at the zones of corner radius and flange of the female die to reduce the unbeneficial friction. Moreover, the load capacities of punch corner radius and force transition zone are increased as well. The above effects enable to increase the forming limit of part subsequently, and reduce the number of deep drawing operations. For instance, regarding a 0Cr18Ni9 stainless steel deep cylinder with a 0.7 mm thickness and 50 mm diameter, a qualified part can be formed in single operation using sheet hydroforming at a liquid chamber pressure of 86.5 MPa. A highest draw ratio 3.36 is successfully obtained. It should be noted that, when the pressure of liquid chamber is 0, the sheet hydroforming process becomes the conventional deep drawing, and the achievable drawing ratio is only 2.31. For a SPCC cylinder part with a great drawn-in and relatively small radius, the achievable drawing ratio is around 2.4, while the ratio is only 1.3 using conventional deep drawing.

(2) High dimensional accuracy and good surface quality. The liquid lubrication between the sheet and upper surface of female die contributes to the drawn-in of sheet and reducing the surface scratches of formed part. The surface quality of initial sheet can be maintained still for the outer appearance of formed part. Such an advantage makes sheet hydroforming especially suitable for forming sheet parts with high requirements of surface quality.
(3) Less forming operations. Resulting from the high forming limit, normally only a single operation of deep drawing is needed, which is beneficial for reducing the number of energy-consuming processes, such as intermediate forming operations and annealing treatments.
(4) Low production cost. As complex-shaped parts can be successfully formed using one single operation. Then, the conventional die sets for multi-operations are not needed; Regarding parts with similar dimensions or thickness, a set of forming tools can be utilised; Regarding parts with very complex geometries, only punch (male die) with corresponded dimensions of the part, the unnecessity of complex female die cavity requiring to be fitted with the punch results in the significantly decreased production cost.

7.1.3 Application Fields of Sheet Hydroforming

The sheet hydroforming technology is dedicated for parts with a degree of deformation significantly higher than that of conventional deep drawing, such as cylindrical-shaped, conical-shaped, parabolic cross-section shaped, and box-shaped parts. In addition, sheet parts with negative stamping angle, conventional deep drawn ones requiring complex die cavity profile. Parts with materials those are difficult to form, can be also formed utilizing sheet hydroforming. These parts are mainly applied in the aircraft industry for satisfying pneumatics properties, e.g. cowling and helmet, as well as the panel structures in automotive industry, e.g. the engine cover.

The suitable materials of sheet hydroforming include mild steels, high strength steels and stainless steels etc. Utilizing the sheet hydroforming, cylinders with extremely large depth can be obtained using single operation deep drawing, enabling the process to be especially suitable for low ductile high strength steel or aircraft industry preferred aluminium alloys, such as 5A06 or 2024.

In terms of the thickness requirements of raw sheets, sheet hydroforming process is mainly suitable for thin sheets with a thickness lower than 3 mm. Due to the characteristics of high forming limit and relatively low efficiency, this process is normally applied in the relatively low-volume of production of panel components.

7.2 The Calculation of Main Process Parameters

The main process parameters of sheet hydroforming process include the critical pressure of liquid cavity, p_{cr}, the forming force, F_Q, the blankholding force, F_Q, etc., as illustrated in Fig. 7.1.

7.2.1 The Critical Pressure of Liquid Cavity

The critical pressure of liquid cavity refers to the minimal counter pressure of liquid enabling to bulge the sheet to be separated from corner radius of female die during sheet hydroforming. Under this critical pressure, besides enhancing the advantage of the beneficial friction at the tooling interface, the connection between sheet and corner radius of female can be avoided and the unbeneficial friction at the interface between sheet and female die is eliminated, which contributes to increasing the forming limit.

Regarding the sheet hydroforming with a fixed clearance for circular cylindrical parts, when the sheet is separated from the corner radius of female die (see Fig. 7.3), the counter force of the liquid requiring to separate the sheet from the female die, is equal to the vertical deep drawing force of the sheet at the entrance of female die, the above force equilibrium equation is established in the following equation.

$$2\pi R_p \cdot t \cdot \sigma_z = \pi[(R_d + r_d)^2 - R_d^2] \cdot p_{cr} \tag{7.1}$$

After the substitution, the Eq. (7.1) can be modified as:

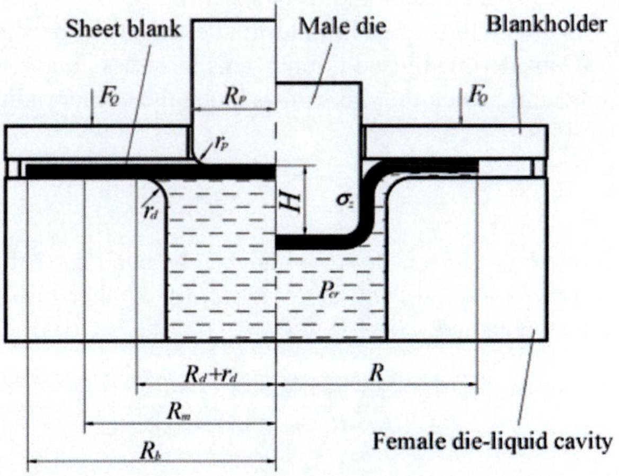

Fig. 7.3 The calculation scheme of critical pressure of liquid cavity

$$\sigma_z = \frac{r_d(2R_d + r_d) \cdot p_{cr}}{2R_p t} \tag{7.2}$$

where, p_{cr}—the pressure of liquid chamber (MPa);
R_p—the corner radius of punch (mm);
σ_z—the tensile stress in the vertical direction (MPa);
R_d—the radius of female die (mm);
r_d—the corner radius of female die (mm);
t—the sheet thickness.

When the fixed clearance is designed for the sheet hydroforming process, the friction between sheet and blankholder can be neglected. Based on the satisfied differential equilibrium equation and yield criterion of the deformed flange zone during hydroforming a cylindrical part, there are

$$\frac{d\sigma_r}{dr} + \frac{\sigma_r - \sigma_\theta}{r} = 0 \tag{7.3}$$

$$\sigma_r - \sigma_\theta = \sigma_s \tag{7.4}$$

where σ_r—the radial stress (MPa);
σ_θ—the circumferential stress (MPa);
σ_s—the flow stress of sheet material (MPa).

After the integration of Eq. (7.3) using the boundary conditions: $r = R$, $\sigma_r = 0$, the radial stress can be obtained as:

$$\sigma_r = \sigma_s \cdot \ln(\frac{R}{r}) \tag{7.5}$$

where R—the instant radius of sheet (mm);
r—the radius of an infinitesimal element on the sheet (mm).

Considering the additional tensile stress that is caused by the bending and unbending when the sheet flows along the corner radius of female die,

$$\sigma_w = \frac{\sigma_s \cdot t}{4(r_d + \frac{t}{2})} = \frac{\sigma_s \cdot t}{4r_d + 2t} \tag{7.6}$$

where σ_w—the tensile stress induced by bending (MPa).

Combining Eqs. (7.2), (7.5) and (7.6) together, the following equilibrium equation can be obtained:

$$\sigma_s \cdot \ln(\frac{R}{R_d + r_d}) + \frac{\sigma_s \cdot t}{4r_d + 2t} = \frac{r_d(2R_d + r_d) \cdot p_{cr}}{2R_p t} \tag{7.7}$$

The critical pressure of liquid chamber can be obtained after arranging Eq. (7.7):

$$p_{cr} = \frac{2R_p t\ [\sigma_s \cdot \ln(\frac{R}{R_d+r_d}) + \frac{\sigma_s \cdot t}{4r_d+2t}]}{r_d(2R_d + r_d)} \tag{7.8}$$

Equation (7.8) represents the mathematical expression of critical liquid chamber pressure when situations of the sheet is separated from the corner radius of female, unfunctional sealing and overflow happen. As can be seen from the equation, the magnitude of critical pressure is affected by not only the corner radius of female die, but also the sheet thickness, corner radii of punch and the inner female die.

Considering the effect of material hardening on the flow stress during drawn-in, normally, the drawing force maximizes when the diameter of flange zone is deformed to 85% of initial sheet size. Consequently, the required critical pressure of liquid chamber is obtained with the mathematical expression given as:

$$p_{cr} = \frac{2R_p t\ [\sigma_s \cdot \ln(\frac{0.85R_b}{R_d+r_d}) + \frac{\sigma_s \cdot t}{4r_d+2t}]}{r_d(2R_d + r_d)} \tag{7.9}$$

where R_b—the initial radius of workpiece.

For the work-hardened materials, the flow stress σ_s in Eqs. (7.8) and (7.9) needs to be obtained using the strain-hardening equation. As an approximate approach, the mean value between yield and ultimate tensile strength of material can be used.

For the sheet hydroforming of box-shaped parts, the deformation type of straight wall zone is bending, while that of the corner radius is deep drawing. Normally, for the deformed zone, the tensile stress caused by the bending of straight wall is smaller than the tensile stress of corner radius caused by the deep drawing. Hence, the pressure of liquid chamber for lifting the straight wall zone is smaller than that at the corner radius zone. Then, the critical liquid pressure is determined by the required critical pressure of lifting the straight wall zone of box-shaped part. According to the deformation characteristics, a quarter of the part is selected for the analysis. To simply the calculation, the metal deformation of zones those are neighbouring the straight wall zone is ignored. As shown in Fig. 7.4, selecting a ribbon element with a width of Δy within the straight flange of sheet, the force due to the exerted pressure of liquid chamber on the corner radius of narrow this ribbon is:

$$F' = p_{cr} \cdot \Delta y \cdot r_d \tag{7.10}$$

The force that is required to lift the sheet at the entrance of the female die is:

$$F'' = \Delta y \cdot t \cdot \sigma_w \tag{7.11}$$

where σ_w—the tensile stress caused by bending and unbending.

Similar to Eq. (7.6), according to the force equilibrium equation $F = F'$, the critical pressure of liquid chamber for sheet hydroforming of box-shaped part can be obtained as:

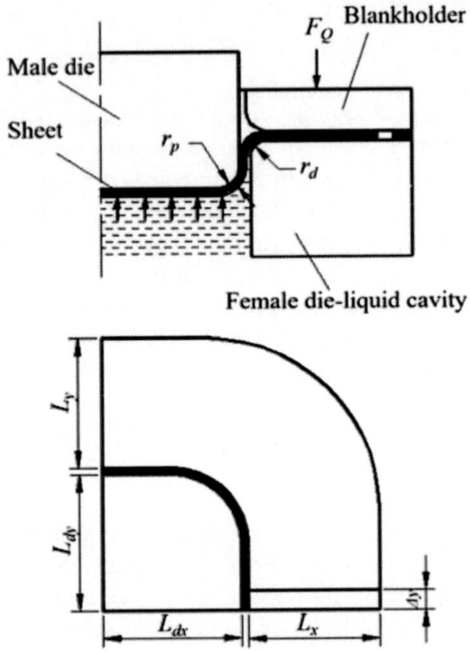

Fig. 7.4 The schematic of sheet hydroforming of box-shaped part

$$p_{cr} = \frac{\sigma_s t^2}{r_d(4r_d + 2t)} \tag{7.12}$$

7.2.2 The Forming Force

The forming force, F_D, of sheet hydroforming is composed by two parts, $F_D = F_1 + F_2$, where F_1 represents the drawing force in conventional deep drawing (kN), F_2 is the counter force arising from the pressure of liquid chamber (kN).

The conventional drawing force, F_1, can be calculated using the empirical equation as:

$$F_1 = \pi d_p t \sigma_b K_d \tag{7.13}$$

where t—the sheet thickness (mm);

d_p—the diameter of deep drawn part (mm);

K_d—a coefficient related to the drawing ratio and relative thickness, $K_d = 0.2 \sim 1.1$.

The counter force due to the pressure of liquid chamber, F_2, can be calculated using the below equation:

$$F_2 = \frac{\pi d_p^2}{4} p_{cr} \tag{7.14}$$

Then, the forming force F_D of sheet hydroforming is:

$$F_D = \pi d_p t \sigma_b K_d + \frac{\pi d_p^2}{4} p_{cr} \tag{7.15}$$

With regard to the highly strained deep drawn part or the high strength sheet parts, the pressure of liquid chamber p_{cr} is normally relatively large, resulting in a dramatically increased equipment tonnage, compared to the conventional deep drawing. For instance, for the cylindrical part with a diameter of 20 mm, when the pressure of liquid chamber is 50 MPa, $F_2 = 157$ ton. In order to reduce the tonnage of equipment or energy consumption, columns for reducing force can be considered to be incorporated into the tooling design. Through the mutual effects of ejecting cylinder and columns for compensating pressure, the pressure of liquid chamber is avoided to be directly exerted on the bottom of deep drawn part, which contributes to reducing the forming force effectively.

7.2.3 Blankholding Force

After the occurrence of overflow during sheet hydroforming, the fluid pressure distribution underneath the flange zone of sheet varies with three typical zones, which are: (1) the liquid pressure near the corner radius of inner female die is approximately same with the liquid chamber pressure, p_{cr}, (2) the pressure underneath the flange zone decreases gradually from the inner to the outer of the flange, and (3) the pressure at the outer periphery of sheet is zero. Ideally, the blankholding force of sheet hydroforming should be able to balance the force on the flange exerted by the liquid, which varies with the proceeding of deep drawing. Therefore, the calculation is very complicated. For the purpose of selecting equipment parameters, the following equation can be used to calculate blankholding force,

$$F_Q = \frac{p_{cr}}{2} S_f \tag{7.16}$$

where S_f is the area of the sheet flange. It should be noted that, the blankholding force needs to be adjusted according to the specific condition of parts.

7.3 The Limit Drawing Ratio and Type of Defects

7.3.1 The Limit Drawing Ratio

Due to the effect of liquid chamber pressure, the resulted beneficial friction and liquid lubrication during sheet hydroforming enables to significantly increase the forming limit. Drawing ratio, that is defined as the ratio of initial sheet dimension to diameter or side length of formed part, is an important parameter to evaluate the forming limit. Table 7.1 summarizes the limit drawing ratios of typical mild steel and aluminium alloy parts. Using the common process conditions of sheet hydroforming, for cylindrical parts, the limit ratio of 08AL mild steel can be increased to 2.55 compared to that of conventional deep drawing, 2.2, and the ratio for A1100 aluminium alloys is increased to 2.61 compared to 2.1 using conventional approach; For the box-shaped parts, the limit drawing ratio of A1100 aluminium alloy is able to be increased to 2.9 from conventional deep drawing, and an increasement of drawing ratio from 2.1 to 2.6 is found for mild steel. With regard to parts with complex curved surfaces, such as conical-shaped, parabolic cross-section-shaped, the sheet hydroforming process can also accomplish the forming using a single operation instead of 3 or 4 operations of conventional deep drawing. Considering the complexity and deformation characteristics of the curved structure of conical-shaped parts, the maximum degree of deformation is lower than that of cylindrical parts with a flat nose.

The limit drawing ratio of sheet hydroformed parts is determined by pressure of liquid chamber, blankholding force, pressurizing approach, corner radius of die etc. The beneficial friction can be sufficiently enhanced when the pressure of liquid chamber is higher than the critical magnitude, hence, the forming limit increases with increasing the pressure within a reasonable range. When the blankholding force is too small, the pressure of the liquid chamber is unable to be built up or too low, which results in the lowered forming limit; When the blankholding force is too high, the liquid lubrication is difficult to achieve resulting in an increased unbeneficial friction at the flange, and the forming limit is decreased subsequently. The pressurizing approach includes natural pressurizing and forced pressurizing. The natural pressurizing utilizes the compression of liquid achieved by the punch downwards movement. Normally, the built up pressure is small at the initial stage of forming,

Table 7.1 The limit drawing ratios of typical mild steel and aluminium alloy parts

Material	Part	Sheet hydroforming	Conventional deep drawing
08Al mild steel	Cylindrical	2.55	2.2
	Box-shaped	2.6	2.1
A1100 aluminum alloy	Cylindrical	2.61	2.1
	Box-shaped	2.9	2.65

which decreases the forming limit. In comparison, the approach of forced pressurizing using the hydraulic pump or pressure intensifier to fill liquid into the liquid chamber, enables to increase the pressure at initial stage of forming. Then, the splitting defect at the early stage of forming is avoided and the forming limit is increased subsequently. The effect of corner radius of inner female die on the forming limit is that, with increasing the value of radius, the forming limit increases. Once the increasement reaches a critical value, e.g. 13 times higher than the sheet thickness, the forming limit is maximized without further increasing; The forming limit increases with the increase of punch corner radius, and maximizes after reaching a critical value without further increasing, e.g. 5 times of the corner radius of punch. Then, the forming limit is determined by the splitting at the corner radius of female die, rather than that at the corner radius of punch.

Similar to the effect of tool corner radius, the forming limit is also significantly affected by the type of forming process. For instances, regarding the cylindrical parts of A1100 aluminium alloy, the sheet hydroforming with radial pressure can increase the critical drawing ratio to 3.31, while to 4.92 using sheet reverse hydroforming. For the SUS304 stainless steel cylindrical part that is hydroformed at a temperature of 90 °C, the critical drawing ratio is increased to 3.3, compared to that of 2.0 using conventional deep drawing at room temperature. For A1050 cylindrical part of aluminium alloy, the critical drawing ratio can reach 3.4 using a non-isothermal sheet hydroforming at 200 °C.

7.3.2 The Type of Defects

Splitting and wrinkling are the main defects of sheet hydroforming. The splitting defect of conventional deep drawing normally occurs at the corner radius of punch at the initial stage of deep drawing. While for sheet hydroforming, due to the introduce of liquid chamber pressure effect etc., splitting defects may occur at different locations and stages during forming, as summarized in Fig. 7.5.

The splitting of corner radius of punch mainly occurs at the early stage of forming (Fig. 7.5a), which is normally caused by two reasons: (1) when the pressure of liquid chamber is too low, beneficial friction and liquid lubrication cannot be established effectively; (2) the large tensile stress greater than material strength resulting from the relatively large fluctuation of liquid chamber pressure. In addition, a too small corner radius of punch is also a possible reason inducing splitting. Accordingly, the approaches of addressing the splitting of punch corner radius are, (1) increasing the pressure of liquid chamber via either hydraulic pump or pressure intensifier to avoid a too low pressure; (2) using the overflow valve to instantly control the pressure of liquid chamber to avoid the large fluctuation; and (3) reasonably increasing the punch corner radius to reduce the tensile stress.

The splitting near the corner radius of inner female die mainly occurs at intermediate or late stage of forming (Fig. 7.5b), the reason is mainly caused by the severe thinning due to the reverse bulging of sheet when the pressure of liquid chamber is

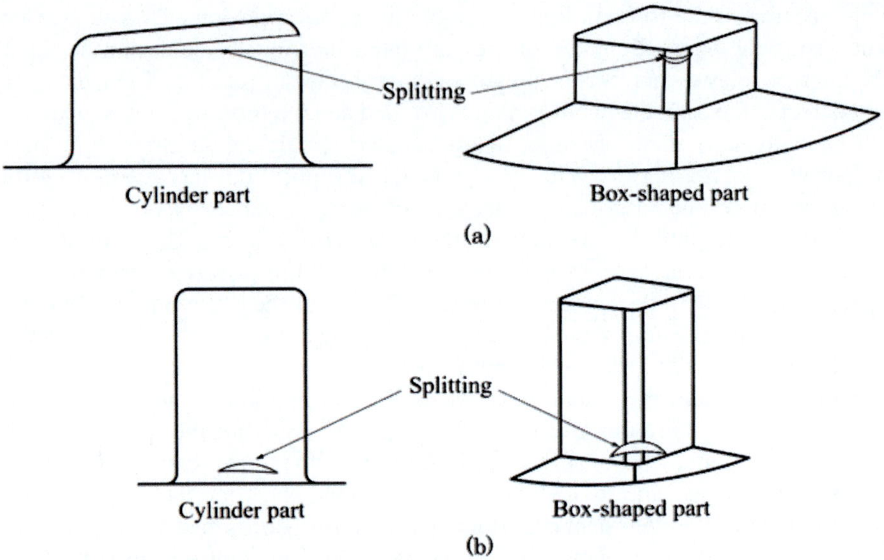

Fig. 7.5 The splitting defects in sheet hydroforming

too high; In addition, a too small corner radius of female die is also easy to induce splitting at the late stage of forming. Accordingly, the approaches to address the splitting of sheet at the corner radius of female die are proposed as follows, (1) reduce the pressure of liquid chamber reasonably, (2) decrease the thinning extent resulted from the reverse bulging of sheet near the corner radius of punch, and (3) increase the corner radius of female die properly.

In terms of sheet parts with different draw ratios, to overcome the above two types of splitting, there exists a reasonable range of liquid chamber pressure. Figure 7.6 shows the correlation between pressure of liquid chamber and the drawing ratio. With increasing the drawing ratio, the variation range of liquid chamber pressure becomes narrower. There are upper and lower limit magnitudes of the liquid chamber pressure. Pressure either below the lower limit or above the upper limit is likely to induce splitting. The main reason is that, with the increase of drawing ratio, to avoid splitting of punch corner radius, the required liquid chamber pressure is higher, while such a higher pressure simultaneously increases the unbeneficial friction between blankholder and sheet flange. In the meantime, the thinning of reverse bulging is easy to generate under a high liquid chamber pressure. Therefore, splitting near the corner radius of female die is induced. The simultaneous requirements of increasing pressure to overcome splitting at the punch corner radius, and decreasing pressure to overcome splitting near the corner radius of female die, narrow down the range of reasonable liquid chamber pressure window. Theoretically, with increasing the drawing ratio, there is a unique critical liquid chamber pressure for the limit drawing ratio.

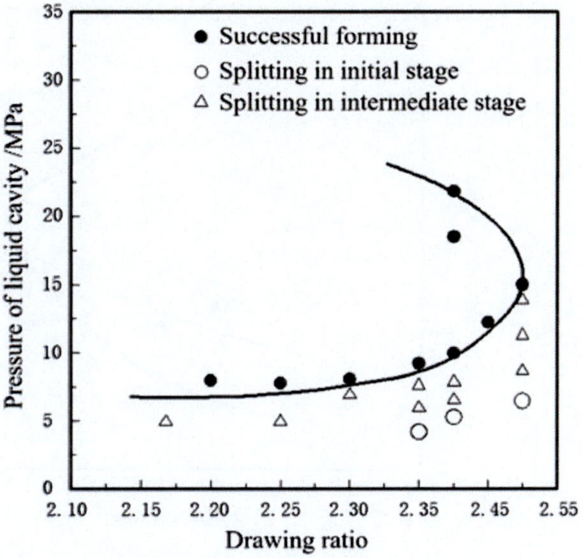

Fig. 7.6 The correlation between drawing ratio and pressure

The effect of blankholding force on the above splitting defects is also very complicated. For sheet hydroforming, the blankholding force is not only used to clamp sheet conventionally, but also affects the built-up of liquid cavity pressure significantly. When the blankholding force is too small, splitting can occur. The reason is that, a sufficiently large liquid cavity pressure cannot be built up under the conditions of natural pressurizing or a too large corner radius of female die. Subsequently, the beneficial friction is insufficient to induce splitting. On the contrary, when the blankholding force is too large, a resulted high liquid cavity pressure may induce the reverse splitting of sheet at the corner radius of female die, that is also a commonly seen defect. Therefore, the blankholding force also has an upper critical magnitude. The correlation between blankholding force and drawing ratio is shown in Fig. 7.7. In practical processes, the blankholding force needs to be adjusted based on the specific requirements of components, including adjustments of constant blankholding force and deep drawing stroke-based blankholding force.

Figure 7.8 shows the wrinkling defects occurred in sheet hydroforming. The wrinkling of flange zone shown in Fig. 7.8c is a common defect of deep drawing, which mainly occurs when the blankholding force is lower than the lower limit shown in Fig. 7.7. Such a defect can be avoided through increasing the blankholding force.

Figure 7.8b shows the wrinkling of unsupported zone of part with a curved surface. The wrinkling locates above the middle of deformed curved surface, which is mainly caused by the tangential compressive stress arising from the absence of die support on the sheet and a relatively small pressure of liquid chamber. Regarding the box-shaped part shown in Fig. 7.8c, wrinkling at the corner of straight walls also exists. Such a defect is induced by the experienced bending deformation of straight edge and partial deep drawn deformation at the corner. The part of straight wall is easy

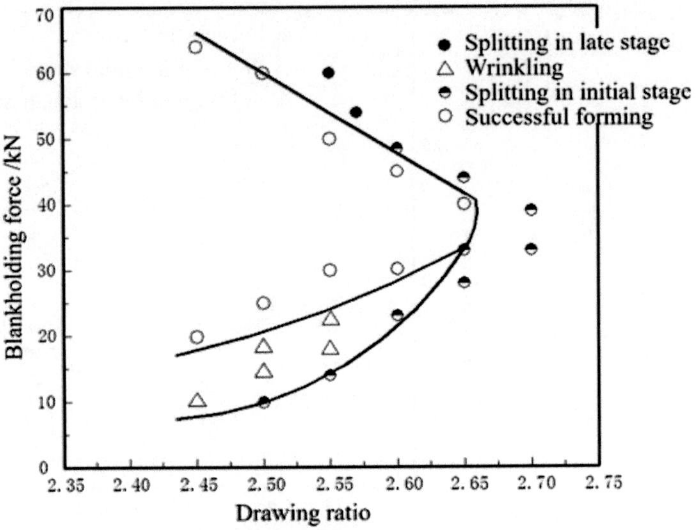

Fig. 7.7 The correlation between blankholding force and drawing ratio

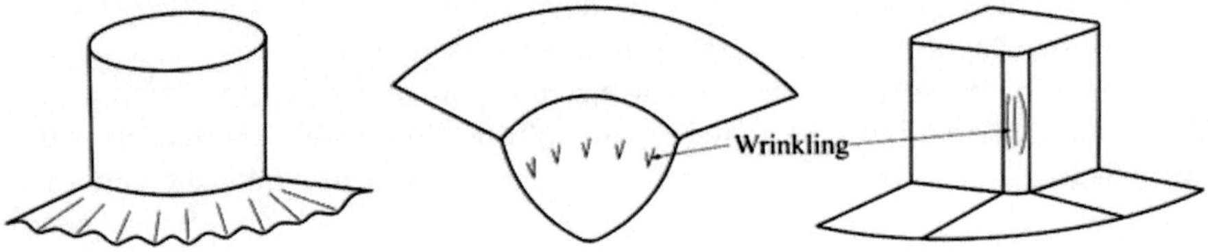

Fig. 7.8 Typical wrinkling defects in sheet hydroforming

to attach on the die under the influence of liquid chamber pressure, and subsequent friction effect. The flow of extra material from the corner of straight edges to the straight wall is affected, resulting in wrinkling occurred at the edge. This defect can be addressed through decreasing the liquid chamber pressure and optimization of sheet dimension (e.g. using initial sheet with trimmed corners).

7.3.3 The Stresses During Sheet Hydroforming Defects

During the sheet hydroforming of parts with curved contours, wrinkling defect mainly occurs at the unsupported zone of curved surface, while splitting mainly occurs near the corner radius of punch. To obtain a thorough understanding and reveal the occurrence mechanism of wrinkling and splitting defects, stress states of typical zones on the part were computed and analyzed using finite element simulations. Three typical locations were selected along a cross-section of part, where point 1 locates in the flange zone, point 2 locates in the unsupported zone of entrance of female die cavity and point 3 locates at the bottom of curved surface. The computed radial tress (σ_r)and circumferential stress (σ_θ) at different stages of forming are shown in Fig. 7.9.

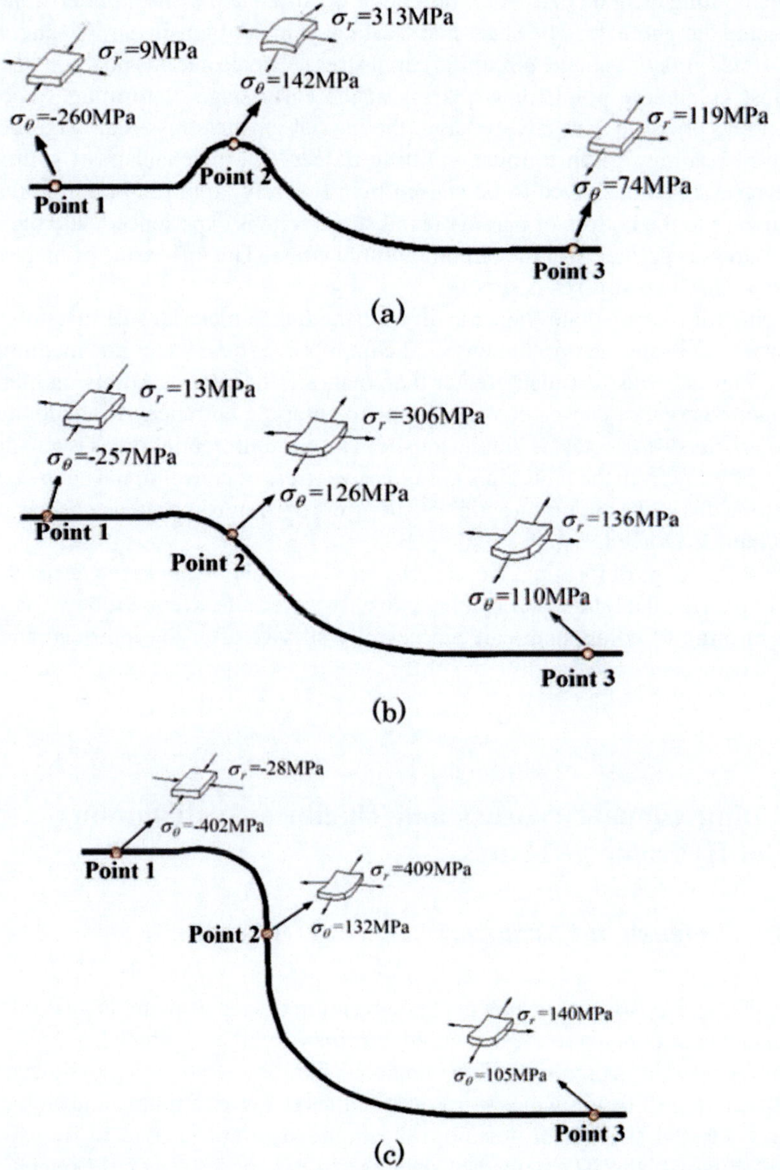

Fig. 7.9 Stress states of typical locations during the different stages of sheet hydroforming

During the early stage of forming, the flange zone material experiences tensile stress in the radial direction and compressive stress in the circumferential direction. The magnitude of circumferential stress is significantly greater than that of radial stress, which results in a tendency of wrinkling. The wrinkling occurred in the flange can be effectively controlled, via using either reasonable blankholding force or blankholding clearance; Reverse bulging is occurred at the unsupported zone near the female die entrance. The sheet here sustains a biaxial tensile stress state with a radial stress greater than the circumferential stress. Therefore, this position becomes the most vulnerable point for wrinkling at the early stage of forming. When the pre-bulging pressure is relatively large, the biaxial tensile stresses are significantly increased resulting in an ultimate splitting defect. The mechanism of splitting at the reverse zone is believed to be caused by a severely large pre-bulging pressure. With regard to the bottom of part, a biaxial stress state is experienced, and the radial tensile stress is greater than the circumferential stress. The difference of magnitudes between these two stresses is small.

During the intermediate stage of forming, the flange material still sustains a one-dimensional tensile and one-dimensional compressive stress state. The magnitude of circumferential stress is much greater than that of radial stress. At this moment, the unsupported zone of curved surface begins to enter the entrance of female die, and the experienced stress state is biaxial tensile; The circumferential stress is tensile, and totally different from the latitudinal compressive stress of conventional deep drawing. Therefore, due to the circumferential tensile stress of the unsupported zone, wrinkling defect can be avoided.

At the late stage of forming, the stress states of typical positions are consistent with those of intermediate stage of forming, with only varied stress magnitudes. Therefore, if the forming of component can successfully survive after the intermediate stage without occurrence of wrinkling or splitting defect. A sound part can be successfully formed.

7.4 Dimensional Accuracy and Thickness Distribution of Hydroformed Parts

7.4.1 Dimensional Accuracy

The accuracy of part dimensions using conventional deep drawing is guaranteed by the precision of punch, female die and their proper clearance, which brings stringent requirements of high precision of forming dies. For the deep drawing of thin sheets, a small clearance of forming dies brings difficulties in the assembling and adjustment. While using the sheet hydroforming, the liquid medium is used to transfer load instead of female die. The deformed sheet is firmly compressed on the punch due to the liquid pressure. Then, parts with high dimensional accuracy can be also obtained

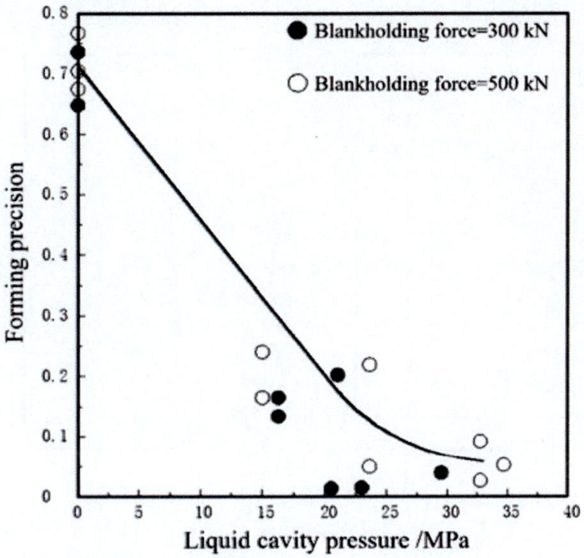

Fig. 7.10 Correlation between precision and liquid cavity pressure

without giving a small die clearance. The process is especially suitable for forming parts with a curved surface, such as the conical-shaped part.

During the sheet hydroforming, the shape precision of part sidewall is affected by the liquid pressure. Figure 7.10 shows the correlation between percentage of deformed part attached on the die representing the forming precision, and pressure of liquid chamber. The higher the pressure, the greater the additional tensile stress on the sidewall arising from the reverse bulging of liquid pressure. The higher the capability of maintaining formed shape of sidewall, the higher the percentage of part attached on the punch. Even using a die clearance with 2 times of sheet thickness, a part with better forming precision can be still obtained, compared to using conventional deep drawing.

7.4.2 The Thickness Distribution

For the sheet hydroforming of circular cylindrical part, the thickness variation is thickening of the upper zone and thinning of the lower zone of the part. The unvaried thickness line locates within the thinning zone with a distance of 1/3 part height to the top. The thinning zone is larger than the thickening zone, and the severest thinning occurs near the corner radius of punch. Figure 7.11 shows the thickness variation of cylindrical part. The maximum variation is −24% in an AA1100 aluminium alloy part with a drawing ratio of 2.4. Under a same deformation extent, due to the effect of liquid chamber pressure, the beneficial friction between sheet and punch enables to minimize the thinning of thickness, which contributes to obtaining a

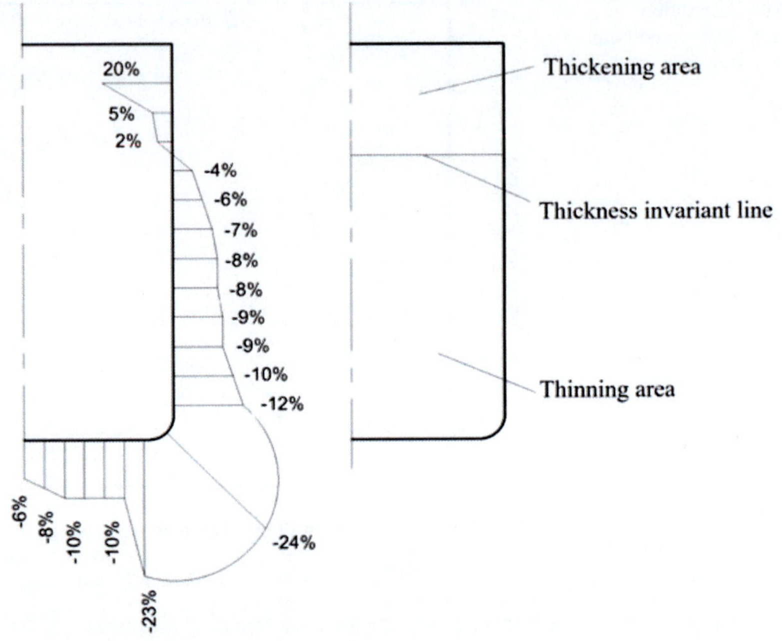

Fig. 7.11 The thickness variation of the cylindrical part

better uniformity of thickness using sheet hydroforming compared to conventional processes. Normally, the greater the liquid pressure the less the thinning. In the meantime, the beneficial friction can also effectively reduce the tensile stress of part near the punch corner radius, which allows a further thinning of part without splitting occurred.

Figure 7.12 shows the thickness variation percentages of typical cross-sections, which are sectioned along the central line of straight wall and middle of side vertically. For the straight wall, the maximum thickness variation percentage is smaller than that of corner of edge, and there is no significant thickening here, which is believed to be that the specific deformation type. Essentially, the deformation type of straight wall of box-shaped part is bending, while that of corner of edge is deep drawing. During the forming of straight wall, an unsupported zone without the die contact exists. Reverse bulging and bending occur simultaneously due to the liquid chamber pressure, resulting in thinning of this unsupported region of straight wall. In comparison, the corner of edges occurs deep drawing deformation. The flange material flows into the female die and becomes thickening. Simultaneously, thinning occurs due to the reverse bulging of liquid pressure. The overall trend resulting from the mutual effect of above types of deformation is the gradual thickening for the corner of straight edges.

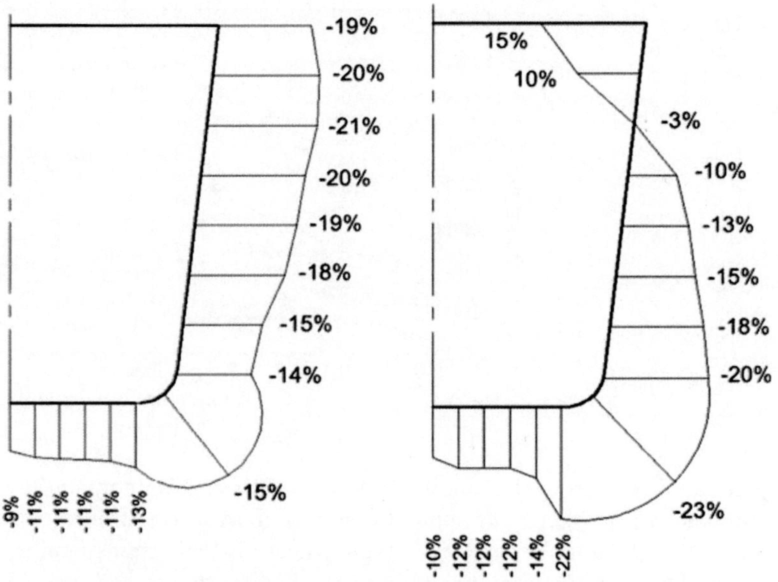

Fig. 7.12 The thickness variation of box-shaped part

The dominant factors affecting the thickness variation percentage are draw ratio, blankholding force, pressure of liquid chamber, pressurizing path and the mechanical properties of materials (*n* and *r* value). Under a same condition, the larger the draw ratio, the greater the maximum thickness variation percentage. Using a same draw ratio, the higher the blankholding force, the greater the maximum thickness variation percentage; the higher the pressure of liquid chamber, the smaller the maximum thickness variation percentage. When the liquid chamber pressure is consistent, pressurizing path with a certain initial pressure is able to decrease the variation percentage. The greater the *n* and *r* value, the better the formability and the smaller the maximum thickness variation percentage. In addition, such a variation percentage is also affected by the part shape. Using a same sheet, forming a part without a conical feature brings a smaller thinning percentage compared to forming a conical-shaped part. The smaller the tapered angle, the smaller the thinning percentage.

7.4.3 Springback Variation

The sheet hydroforming enables to effectively reduce the springback extent and subsequently increase the dimensional accuracy of part. In order to obtain a quantitative correlation between springback amount and the pressure, five sets of sheet hydroforming experiments using different typical material were performed. The

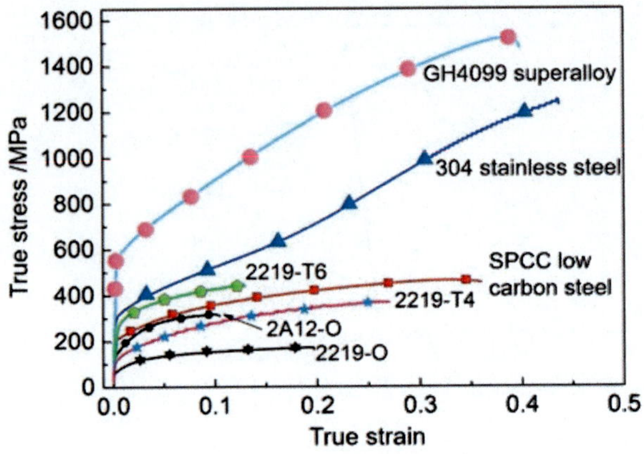

Fig. 7.13 The true stress–strain curves of typical materials

tested materials were mild steel, stainless steel, superalloy, 2A12 aluminium alloy and 2219 aluminium alloy. Figure 7.13 shows the stress–strain curves of these materials. Among them, 2219 aluminium alloy is a typical heat-treatable strengthening alloy, which exhibits different formability varying with the heat treatment temper. Therefore, the springback variations of annealing (O), natural ageing (T4) and artificial ageing (T6) tempered, 2219 aluminium alloy component were investigated.

Three liquid chamber pressures, 10, 20 and 30 MPa, were used. The measurement location of springback locates in the middle of component. The measurement directions were 0°, 45° and 90°. Each direction was measured three times and the mean value was selected. In order to increase the measurement accuracy, the part was sectioned to measure the thickness considering the practical thicknesses of the part were different. The magnitude of springback is equal to the difference between outer diameter and the sum of punch diameter plus twice of thickness. The relative springback amount is defined as the ratio of springback magnitude to the diameter of cylindrical part (punch diameter).

Table 7.2 and Fig. 7.14 gives the experimentally determined relative springback amount of different cylindrical parts of different materials. As can be seen from the results: the higher the pressure of liquid chamber, the smaller the relative springback amount. However, for different materials, the specific correlation between springback amount and pressure varies. With regard to materials with relatively lower yield strength, e.g. mild steel and aluminium alloys, the relative springback amount drops dramatically with increasing the liquid chamber pressure. For example, once the pressure reaches 30 MPa, the reduction of relative springback amount is less than 1%. In comparison, the reduction is only 0.18% for the Ni-based superalloy. Obviously, it can be concluded that the variation trend is determined by the material mechanical properties. Considering yield strength of both Ni-based superalloy and stainless steel are relatively high, the resulted extent of springback is greater. Hence, for mild steel and aluminium alloys, using a relatively low liquid chamber pressure can limit the relative springback amount below 1.0%, while for stainless steel and superalloy, a greater liquid chamber pressure is required to obtain a similar springback extent.

Table 7.2 The relative springback amount of different materials (%)

Liquid cavity pressure/MPa	Mild steel	2A12 Aluminium alloy	Stainless steel	Ni-based Superalloy	2219 aluminium alloy		
					O temper	T4 temper	T6 temper
0	2.52	2.64	3.08	3.07	1.88	1.89	1.78
10	2.10	2.17	2.84	2.96	1.43	1.71	1.69
20	0.68	1.11	2.68	2.96	1.20	1.56	1.67
30	0.58	0.92	2.44	2.89	0.92	1.14	1.41

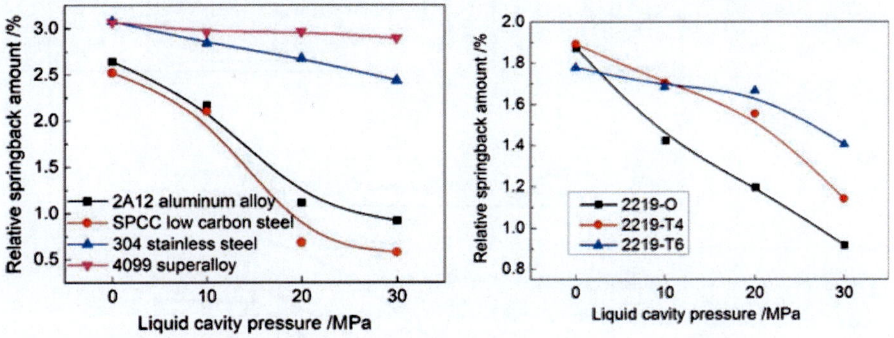

(a) (b)

Fig. 7.14 Relative springback amounts of typical materials. **a** Different materials; **b** 2219 aluminium alloys at different tempers

With regard to 2219 aluminium alloys with different heat treatment conditions, the yield strength of O-tempered alloy is lowest and subsequent reduction of springback is largest. When the pressure of liquid chamber reaches 30 MPa, the extent of springback is lower than 1.0%; In comparison, for the artificially aged alloy, the magnitude of springback varies insignificantly with the change of liquid chamber pressure. The relatively springback amount is only decreased 0.37% when the pressure is increased to 30 MPa, which is believed to be caused by the high yield strength of age hardened aluminium alloys.

7.5 The Equipment and Tooling of Sheet Hydroforming

7.5.1 The Construction of Sheet Hydroforming Equipment

The equipment of sheet hydroforming is mainly composed by the main press and sheet hydroforming system, enabling the instant control of process parameters, including liquid chamber pressure, displacement of deep drawing, blankholding

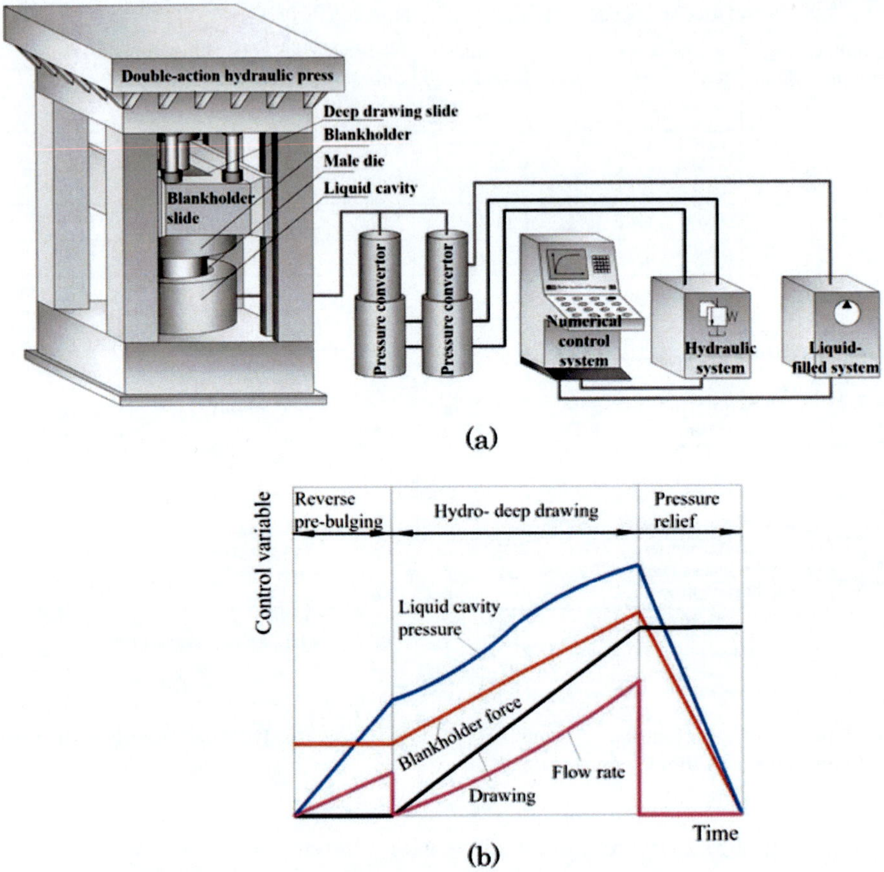

Fig. 7.15 The construction of sheet hydroforming equipment. **a** Construction of equipment; **b** process parameters

force and liquid volume, to be achieved. Figure 7.15 shows the system construction and process parameter variations. The main equipment is normally the double-action hydraulic press with functions of providing required drawing force and blankholding force during forming; The sheet hydroforming device includes the pressure convertor, hydraulic system and numerical control system.

7.5.2 The Characteristics and Main Parameters of Sheet Hydroforming Equipment

The main parameters of sheet hydroforming equipment include deep drawing, blankholding force, bed dimensions and pressure of liquid chamber. Normally,

Fig. 7.16 The 1300 tonnage sheet hydroforming equipment

the drawing force and blankholding force are determined by the part dimension and required liquid chamber pressure. Then, the dimension of bed is determined according to the dimensions of forming die set. Finally, forming equipment can be selected considering the above strategies. Figure 7.16 shows a 1300 t tonnage sheet hydroforming equipment. The series of advanced equipment own the following characteristics:

(1) The integrated control scheme a with good system implementation and high flexibility is adopted. The software and hardware systems function independently, which enable to be used as both the communication equipment for supporting the hydraulic press, and specific equipment of sheet hydroforming.
(2) The capability of simultaneous multi-directional loading. According to the geometry and material of formed parts. The normal, reverse and radial pressures can be either individually or mutually exerted to change the stress states, enabling parts with a large ratio of height to diameter and complex curved contour to be successfully formed.
(3) The closed-loop control of liquid pressure and high precision. Displacement, velocity and blankholding force of hydraulic press, and pressure and flux of high-pressure source driven by the multi-loop hydraulic system, can be controlled

using a closed-loop scheme and loaded under a specific path, as well as their synchronism. A process control with high precision can be achieved.

(4) The equipment has a large volume of high pressure liquid, which is suitable for the forming of large-scale parts. A parallel synchronism control with the large volume high pressure source, contributes to achieving the flux control of high pressure liquid with a large volume. The maximum volume can reach 5 m^3 satisfying the requirements of forming large panel parts.

(5) The utilization of water medium is environmentally friendly to guarantee a clean production plant. The working medium of sheet hydroforming is free of environment contamination. The clean part surface is easy to be stored, transferred, post heat-treated and surface treated.

7.5.3 The Structure and Materials of Die-Sets

The structure of sheet hydroforming die-sets is simple, especially that female die is not necessary to be machined a consistent cavity profile with the punch. The machining work is significantly reduced as the machining of female die is more difficult than that of punch in conventional deep drawing. The key difference between sheet hydroforming and conventional deep drawing is the design of liquid chamber, besides punch, female die and blankholder. The liquid chamber is required to be connected to the hydraulic system or pressure adjustment devices (e.g. overflow valve) for the purpose of establishing a closed pressure cavity during deep drawing that can be hydraulic controlled. Figure 7.17a shows the structure of liquid chamber. A structure of connecting multiple segments is used for both the female die (also named as draw ring) and liquid chamber, which is convenient for exchanging elements of the forming dies with similar dimensions resulting in a reduction of tooling cost. With regard to the complex-shaped but shallow parts, the depth of liquid chamber can be small resulting in the selection of integrated structure composed by designing the liquid chamber and female die as a whole body, as shown in Fig. 7.22b, as the multiple-segment structure adds the machining cost on. In some particular scenarios, the increase of wall thickness of cavity is incapable of satisfying the strength requirement, additional pre-tighten structure is used outside the liquid chamber to ensure the in-service safety.

The tooling parameters of sheet hydroforming dies mainly include corner radius of punch and female die, corner radius of blankholder and the clearance of dies. The corner radius of punch is designed according to the part dimensions. Components with relatively large depths can be also deep drawn even punch radius is equal to zero. Regarding the deep drawn parts with relatively large degree of deformation, the corner radius of female die can be given 6–10 times of sheet thickness, which is slightly larger than that of conventional deep drawing. The corner radius of blankholder is approximately 5 times of sheet thickness to avoid the splitting at the late stage of forming arising from the reverse bulging of liquid pressure. The requirement of the clearance between punch and female die is not vigorous, which is around 1.2 times

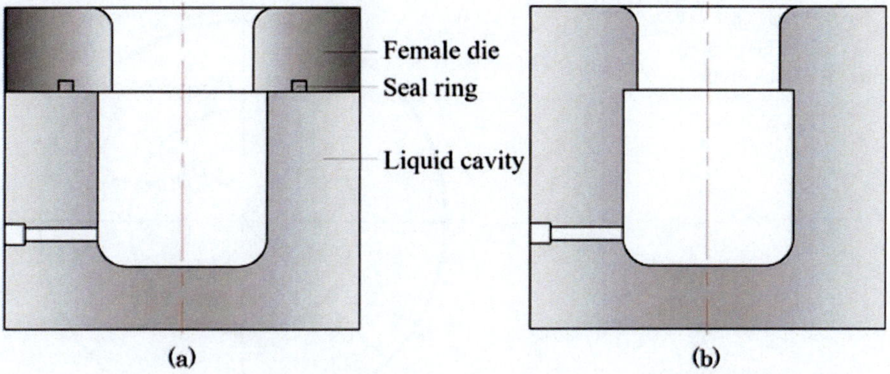

Fig. 7.17 The schematic of liquid cavity structure. **a** Segmented structure; **b** integrated structure

of sheet thickness. The dimensional accuracy of part is guaranteed by the mutual effects of punch and pressure of liquid chamber.

With regard to the materials of forming dies, considering the liquid chamber is subjected to an internal high pressure, high strength and toughness materials are the ideal candidates for the liquid chamber. When the pressure of liquid chamber is relatively low, cast steel can be adopted, while wrought steels have to be used for cases of liquid chamber sustaining high pressures. In terms of large complex parts, the punch material could be made using resin or cement with metal fibers, which contributes to further decreasing the tooling cost.

The element of sheet hydroforming die set subjected to liquid pressure is the cavity for filling liquid. The sustained pressure can reach up to 100 MPa. Therefore, the materials of liquid pressure cavity must own sufficiently high rigidity and strength to avoid failure of sealing that could be caused by the too large elastic deformation under high pressures, or the fracture of forming dies. According to the longitudinal and transverse dimensions of deep drawn parts, the liquid chamber can be designed as thick-walled cylinders with either a circular or rectangular cross-section. For the cavity with a circular cross-section, a stress state of radial compressive stress and circumferential tensile stress is sustained by the liquid chamber and corresponding distribution is shown in Fig. 7.18.

The point that is subject to failure locates on the inner wall of liquid chamber, with mathematical expressions of radial compressive stress and circumferential tensile stress given as:

$$\sigma_r = \frac{p_{cr} r_a^2}{r_b^2 - r_a^2}\left(1 - \frac{r_b^2}{r_a^2}\right) \tag{7.17}$$

$$\sigma_\theta = \frac{p_{cr} r_a^2}{r_b^2 - r_a^2}\left(1 + \frac{r_b^2}{r_a^2}\right) \tag{7.18}$$

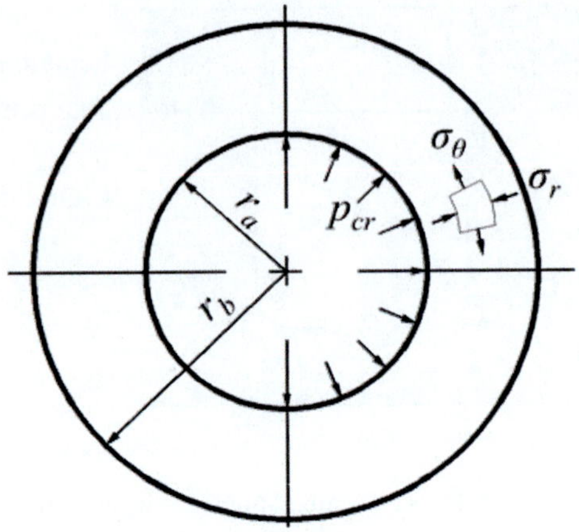

Fig. 7.18 The true stress–strain curves of typical sheet materials

According to above equations, for the convenience of discussion, a ratio of outer diameter, r_a, and inner diameter, r_a, is defined and expressed as m. For a constant r_a, the ratio m increases with increasing the r_a, then the effective stress at the inner wall decreases. However, when $m > 4$, the stress variation becomes stable without an obvious decrease. As can be seen, when $m \leq 4$, the effective stress of the inner wall can be reduced by increasing outer diameter or thickness to increase the strength of liquid chamber; While when $m > 4$, a further increase of outer diameter or thickness is unable to decrease the sustained stress of liquid chamber and improve the strength of structure. Multi-laminated pre-stressed structure could be considered.

7.6 The Sheet Hydroforming with Radial Pressure

7.6.1 Process Procedure

Unlike the conventional deep drawing, the positive (forward) sheet hydroforming with radial pressure is to exert a dependent controllable radial pressure on the outer periphery of sheet. The given radial pressure promotes the material flow at the flange and aids the deep drawing of punch to deform the sheet. Figure 7.19 shows the process procedure. Resulting from the positive radial pressure, the radial tensile stress of the deformation zone of part can be effectively decreased, suppressing the overthinning of dangerous zone and increasing the forming limit. In the meantime, at both the upper and lower surfaces of the sheet contacting with blankholder and female die, respectively, liquid lubrication is generated. Then, the friction constraining material flow is reduced enabling a further decrease of radial tensile stress and an increase of the loading capability of material to further increase the forming limit.

To achieve the operation of positive pressurizing in the radial direction, two pressure intensifiers and a pressurizing control system are required to be configurated into

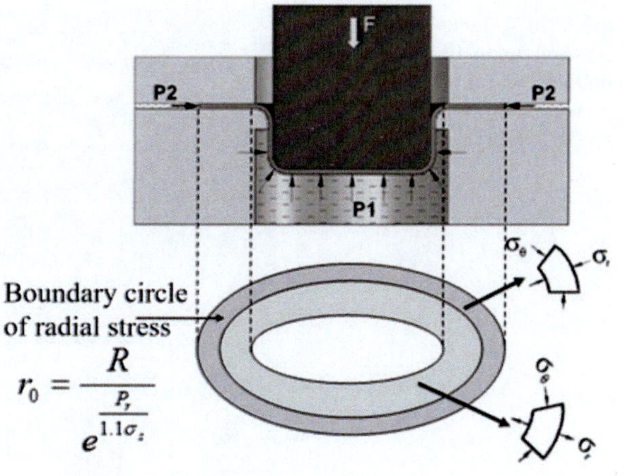

Fig. 7.19 Procedure of sheet hydroforming with positive radial pressure

the sheet hydroforming equipment. The radial pressure is independent on the pressure of female die cavity. The control of such pressure can be optimized according to deformed material and forming limit to increase the controllability of process. Using the sheet hydroforming with radial pressure, an obvious boundary circle of radial stress distinguishing the sign of radial stress is generated within the flange zone. With increasing the radial pressure, the position of this circle moves towards the entrance of female die. The tensile stresses at dangerous positions are gradually decreased. Then, the forming limit can be increased and the thinning of thickness can be suppressed.

7.6.2 Effect of Radial Pressure on the Forming Limit

As discussed in the above section, altering the radial pressure enables to form cylindrical parts with a flat bottom of aluminium alloy with different drawing ratios. Figure 7.20 shows the deep drawn cylinders and corresponding limit drawing ratio.

The limit drawing ratio without using radial pressure is 2.4. After giving a radial pressure, the limit drawing ratio increases gradually with the increase of radial pressure. When the radial pressure is 15 MPa, the limit drawing ratio is 2.66, and such ratio is further increased to 2.8 using a 35 MPa radial pressure, which is a 16.7% improvement compared to conventional deep drawing. The reason is that, the greater the exerted radial pressure, the smaller the radial tensile stress sustained by the flange material, which contributes to suppressing splitting.

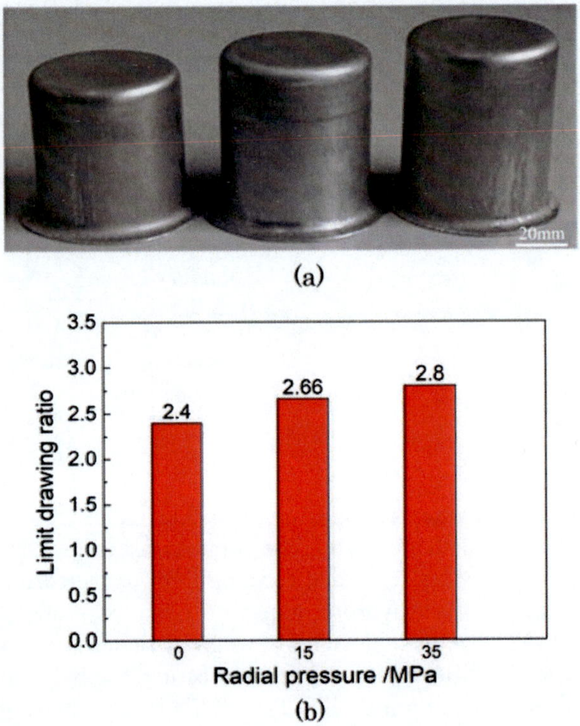

Fig. 7.20 Effect of radial pressure on the forming limit. **a** Typical parts; **b** limit drawing ratios at different radial pressures

7.6.3 Effect of Radial Pressure on the Thickness Distribution

The parts formed using different radial pressures were sectioned along the axis of formed cylinder to measure thickness. Figure 7.21 shows the measured thickness distributions. As can be seen in this figure, the thickness of the bottom zone is nearly unchanged. Thinning occurs at the corner radius of punch and part of the cylinder sidewall approaching the bottom, while thickening occurs at the top mouth of cylinder. The locations are all identified at the corner radius of punch. When the radial pressure is 15 MPa, the maximum thinning percentage is 18%. Such a maximum thinning percentage gradually decreases with the increase of radial pressure. The magnitude is decreased to 10% when the radial pressure is increased to 25 MPa. Furthermore, the position of unvaried thickness on the sidewall also shifts downwards with increasing the radial pressure, which indicates a relatively more uniform thickness distribution.

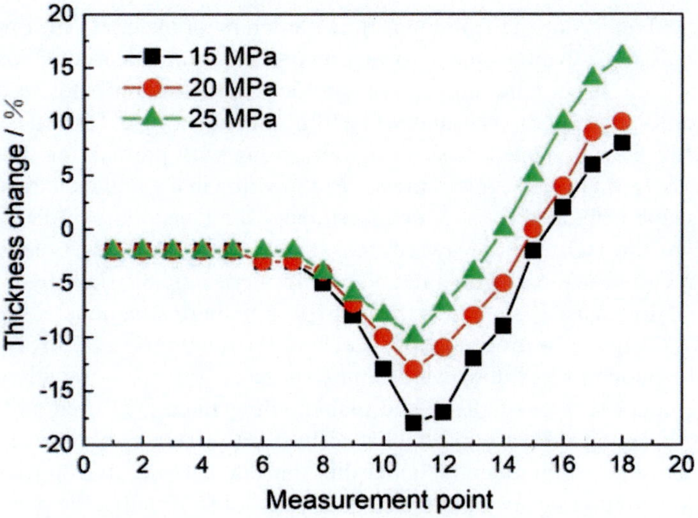

Fig. 7.21 The effect of radial pressure on the thickness distribution

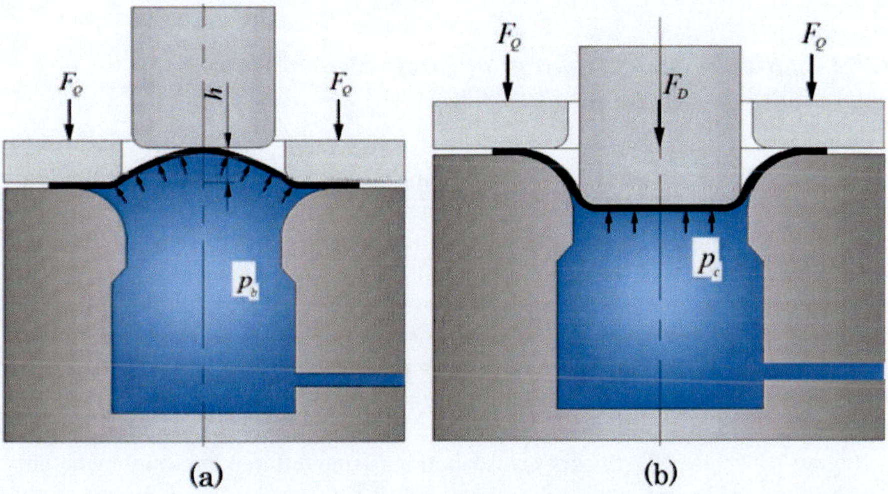

Fig. 7.22 The schematic of sheet hydroforming with pre-bulging. **a** Pre-bulging; **b** deep drawing

7.7 The Sheet Hydroforming with Pre-bulging

7.7.1 Process Procedure

With the growing applications of high strength steels, such as DP steel and TRIP steel, the conventional deep drawing exists non-uniform and insufficient deformation drawbacks, which is due to the limited strain hardening of material. The objective of using pre-bulging is to adjust the deformation of the bottom material of component, and increase the strain hardening extent of sheet, which contributes to obtaining sufficient rigidity, strength, resistance of bending and buckling of formed parts. Such a modified process is named as sheet hydroforming with pre-bulging. Essentially, the sheet is bulged under the initial pressure of medium in the liquid chamber. A pre-designed shape of bulged sheet is obtained under the constraints of die. Then, the sheet is deep drawn with the downwards movement of the punch under a certain liquid pressure. A cylindrical part with a flat bottom formed using the sheet hydroforming with pre-bulging is used as a case to further illustrate in this section.

Figure 7.22 shows the procedure of sheet hydroforming process with pre-bulging. Initially, the punch moves downwards and terminates at a position with a distance of h to the upper surface of sheet, and blankholding force is exerted on the sheet. Then, a pre-bulging pressure is established by pressurizing the liquid chamber to pre-bulge the sheet. Subsequently, under this constant pressure, the punch moves to the position with a same level parallel to the blankholder. Finally, the punch further moves downwards to accomplish the deep drawing of part at an instantly adjusted pressure of liquid chamber.

7.7.2 Thickness Distribution of Sheet Hydroforming with Pre-bulging

During sheet hydroforming with pre-bulging process, the degree of pre-bulging deformation directly determines the subsequent deformation process. When the pre-bulging deformation is small, effect of strain hardening is relatively small. While if the pre-bulging deformation is too severe, material accumulates heavily at the gap between punch and blankholder, which subsequently induces a folding defect. For the sheet hydroforming with pre-bulging process of DP590 dual phase steel, folding defect can be generated during ironing when the relative pre-bugling height is 30%. A sound cylindrical part with flat bottom can be formed when the relative pre-bulging height is smaller than 25%.

Figure 7.23 shows the thickness distributions using different pre-bugling heights. As can be seen in this figure, the variations of thickness distribution are similar for the drawn parts with different pre-bulging heights. The thickness distributions are relatively uniform at the bottom of parts. Within the bottom area, the region approaching to the corner radius experiences the compressive deformation during the ironing, which results in a relatively increased thickness. Similar to the thickness distribution of conventional deep drawing, the maximum thinning zone still locates at the corner radius of punch. For the straight wall, the thickness gradually increases from the bottom to the top, and maximizes at the region approaching the flange zone. When the relative pre-bulging height is 25%, the thickness at the corner radius of flat bottom of cylinder is minimal, with a thinning percentage of 14.5%; The

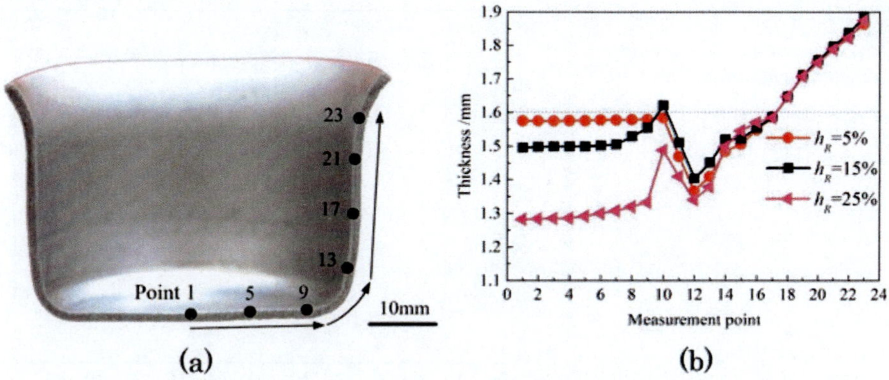

Fig. 7.23 The part formed by sheet hydroforming with pre-bulging and its thickness distribution

difference of thickness within the bottom is small, and the thinning percentage is 11.5%; The thickening percentage at a location on the bottom near corner radius (3/4 of radius to the center) is 3%; The thickness at the bottom of straight wall maximizes with the thickening percentage reaches up to 17%. Compared to the conventional deep drawing of flat bottom cylinders, the pre-bulging approach enables the thinning percentage at the bottom to be increased 11%, while thickness of the straight wall and corner radius of part vary insignificantly. With increasing the relative pre-bulging height, the thickness at the bottom of part gradually decreases, while the thickness distributions of rest regions are similar indicating the thinning of the flat bottom of cylinder is mainly resulted from the thinning during pre-bulging.

7.7.3 The Strengthening of Sheet Hydroforming with Pre-bulging

Using the method of sheet hydroforming with pre-bulging can increase the straining degree of the bottom zone of cylindrical part, and the hardness subsequently due to work hardening. When the relative pre-bugling height is 20%, the hardness distribution of the hydroformed flat cylinder with pre-bulging is shown in Fig. 7.24. As can be seen from this figure, the Vickers hardness at the bottom of part is minimal, with a magnitude of 260 HV. Vickers hardness of the thickening zone at the bottom is higher than the neighboring zones. The hardness of the straight wall gradually increases from the top to bottom, with a maximum magnitude of 338 HV at the bottom. The global hardness difference of the cylinder with a flat bottom is 78 HV. Compared to the conventional deep drawn cylinder, a 24% increase of hardness is found for the bottom, while insignificant variation of hardness for the straight wall. The global hardness difference decreases 38% indicating a more uniform deformation and strength distribution can be obtained using the pre-bulging approach.

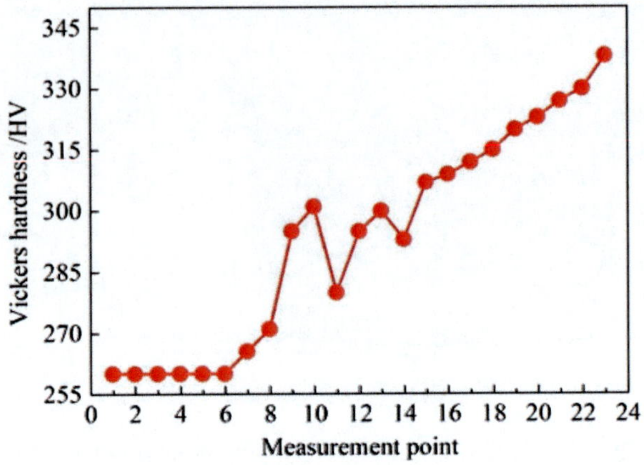

Fig. 7.24 The hardness distribution of the cylindrical part with a flat bottom

The essence of such improvement is the enhanced strain hardening due to the pre-bulging of cylinder bottom. Considering the straight wall is mainly achieved by the deep drawn of flange zone, the pre-bulging results limited effect on this area, therefore, the magnitudes of hardness and its distribution is nearly same with those of conventional deep drawing.

Figure 7.25 shows the hardness variations of parts hydroformed with different pre-bulging heights. The hardness at the center of bottom increases with the increase of pre-bulging height, while it is nearly unchanged for the bottom end of straight wall. When the relative pre-bulging height is 5%, the hardness difference is 114HV; when the relative pre-bulging height is 25%, the Vickers hardness at the center of bottom reaches 281 HV, and the global hardness difference within the part is reduced to 52 HV, achieving a 54% reduction. Compared to the conventional deep drawing, the hardness at the center of bottom obtains a 31% increasement to 66HV. The global hardness difference decreases to 65 HV obtaining a 56% reduction. The above results have shown that, regarding the cylindrical part with a flat bottom, using the sheet hydroforming with pre-bulging, the increase of pre-bulging height can increase the hardness at the bottom to subsequently reduce the global hardness difference and obtain a more uniform distribution of hardness. As there exists a quantitative relationship between hardness and strength, the above hardness variation can indirectly illustrate that the strength of part becomes more uniform using the approach of pre-bulging.

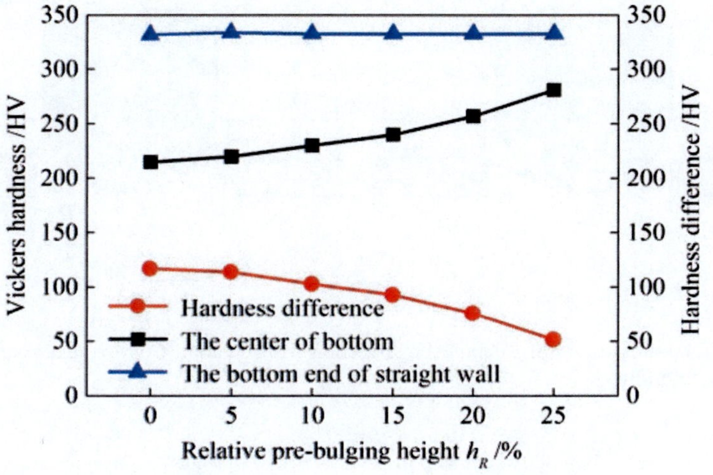

Fig. 7.25 Effect of pre-bulging height on the hardness of flat bottom cylinder part

7.8 The Sheet Hydroforming of Typical Parts

7.8.1 Parabolic Cross-Section Shaped Parts

Unlike the cylindrical part with a flat bottom, the parabolic cross-section-shaped part has a relatively sharp bottom, and the cross-section geometry is a parabolic, belonging to a complex-shaped curved surface feature, which is difficult to form using conventional deep drawing, requiring at least 5 operations of deep drawing with intermediate annealing to accomplish the forming normally. The production efficiency and percentage of qualification are relatively low. For the parabolic cross-section shaped component, how to overcome the wrinkling and splitting defects is the key determining the success of forming, as there exists an unsupported zone between the punch and female die. To address this challenge, using sheet hydroforming enables to improve the stress states of the part during deep drawing. Figure 7.26 shows the comparisons of stress states between conventional deep drawing and sheet hydroforming. As shown in this figure, under the influence of pressure of liquid chamber, a soft draw bed was generated resulting in a decreased area of unsupported zone. The conventional circumferential compressive stress is changed to tensile stress, resulting in the reduced tendency of wrinkling. However, the pressure of liquid chamber should be strictly controlled to avoid the splitting defect of material at this zone due to overthinning.

The diameter of entrance of female die was 120 mm, the corner radius was 8 mm. The initial sheet used for forming had a diameter of 250 mm. The raw material was aluminium alloy A1060, and the thickness was 1 mm. The mechanical properties are summarized in Table 7.3.

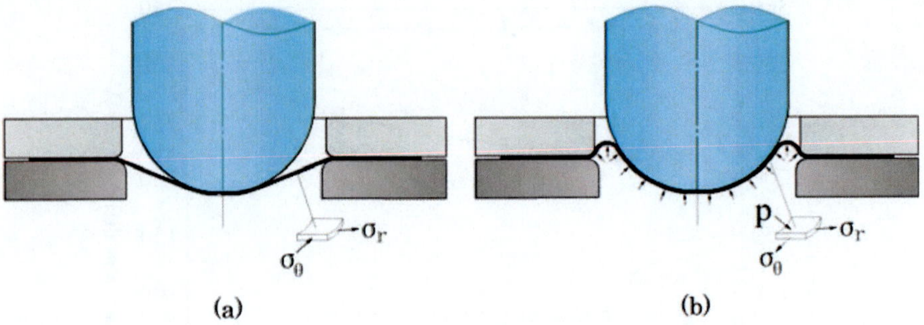

Fig. 7.26 The stress states of parabolic cross-section-shaped part. **a** Conventional deep drawing; **b** sheet hydroforming

Table 7.3 Mechanical property summary of A1060 for hydromechanical deep drawing

Yield strength σ_s	Ultimate tensile strength σ_b	Elongation δ	Strain hardening component n	Thickness anisotropic component r
28 MPa	69 MPa	43%	0.17	1.08

During the sheet hydroforming of parabolic cross-section shaped part, the pressure of liquid chamber must gradually increase according to a specific path with the proceeding of deep drawing stroke, to guarantee the continuous existence of soft draw bead used for suppressing wrinkling. Figure 7.27 shows the reasonable loading window of liquid chamber pressure, the shaded zone represents the forming window, while the unshaded zones above and below the forming zone is splitting and wrinkling zone, respectively.

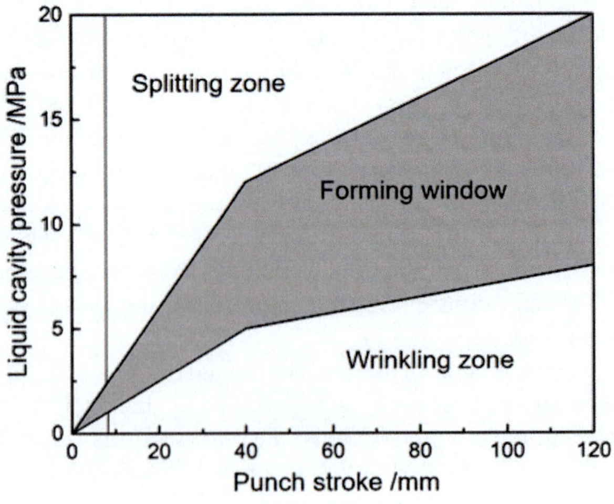

Fig. 7.27 The loading window for a parabolic-shaped part

Fig. 7.28 A typical parabolic cross-section shaped part using sheet hydroforming

The critical boundary with varied slopes of avoiding wrinkling is divided into two stages. At the initial stage of deep drawing, the width of unsupported zone is relatively large which is prior to induce wrinkling. A rapid pressurizing rate is required to avoid wrinkling. When the stroke reaches a certain magnitude, the area of unsupported zone decreases, the strain hardening of material within this region increases, which can also decrease the tendency of wrinkling. Therefore, the liquid chamber pressure is only required to be increased a certain amount. The critical boundary with varied slopes of avoiding splitting can be also divided into two stages. At the initial stage of deep drawing, the strain hardening rate of material is higher, and the strength of dangerous cross-section increases quickly as well. Then, the pressure of liquid chamber can be also increased quickly. With the proceeding of deep drawing, the hardening rate of material decreases and becomes gentle, the increasing of strength of the dangerous cross-section becomes gentle as well. Then, the pressure of liquid chamber is required to increase slowly to avoid splitting. Figure 7.28 shows a typical parabolic cross-section-shaped part formed within the proper loading window of liquid chamber pressure.

7.8.2 Box-Shaped Panel with a Single Curvature

With regard to the forming of box-shaped shell with a single curvature, the unsupported area is large, and the deformation degree in longitudinal and latitudinal directions are different. A relatively large compressive stress in the latitudinal direction in the unsupported zone, which is easy to induce wrinkling defect. Therefore, overcoming the wrinkling defect of unsupported zone is the key of successful forming. Figure 7.29 shows a typical box-shaped component with a single curvature.

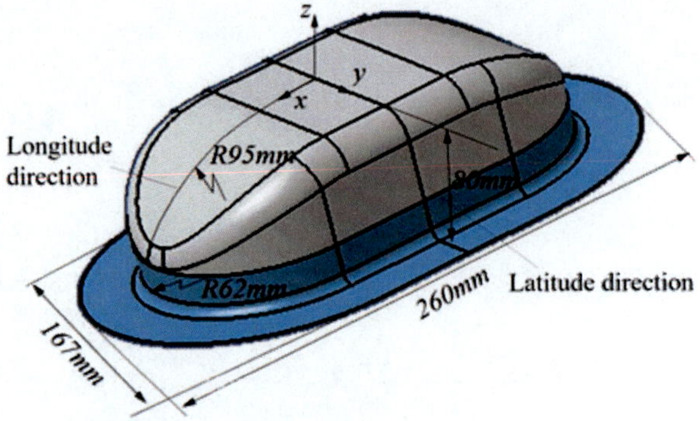

Fig. 7.29 The shape and dimensions of a box-shaped panel with single curvature

Table 7.4 Mechanical property summary of SUS304 stainless steel

Yield strength σ_s	Ultimate tensile strength σ_b	Elongation δ	Strain hardening component n	Thickness anisotropic component r
180 MPa	340 MPa	19%	0.19	0.8

The material used was 2A12 aluminum alloy sheet at annealing condition with an initial thickness of 1 mm. The mechanical properties are summarized in Table 7.4.

The enhanced strain hardening effect due to pre-bulging can effectively increase the wrinkling resistance of the unsupported zone and reduce the area without die support. More importantly, using the pre-bulging can change the stress states of material, which decreases the latitudinal compressive stress and changes the location of strain circle boundary moving towards the entrance of female die, enabling the tendency of wrinkling to be decreased.

Figure 7.30 shows the computed shape and thickness distribution of parts formed using different liquid chamber pressures. When the pressure varies between 5 and 30 MPa, the minimal thickness varies insignificantly within a range between 0.81 and 0.82 mm. The positions of minimal thickness all locate within the corner radius of punch. When the pressure is 15 MPa, the minimal thickness is 0.82 mm satisfying the thinning requirement still. The above FE results illustrate that, although the liquid chamber pressure enables to effectively control the thinning, the minimal thickness variation is insignificant with increasing the pressure of liquid chamber. In the meantime, when the liquid pressure increases from 5 to 30 MPa, although the wrinkling can be reduced, it cannot be entirely eliminated. The main reason is that the relatively large latitudinal compressive stress is unable to be eliminated through increasing pressure only, and the material at the transition zone is still unable to firmly compressed on the die and wrinkling may occur.

With further increasing the liquid chamber pressure to 40 MPa, due to the relatively high pressure, a great pre-bulging amount of the unsupported zone is obtained, when the punch stroke reaches 10 mm at the initial stage of deep drawing. The pre-bulged

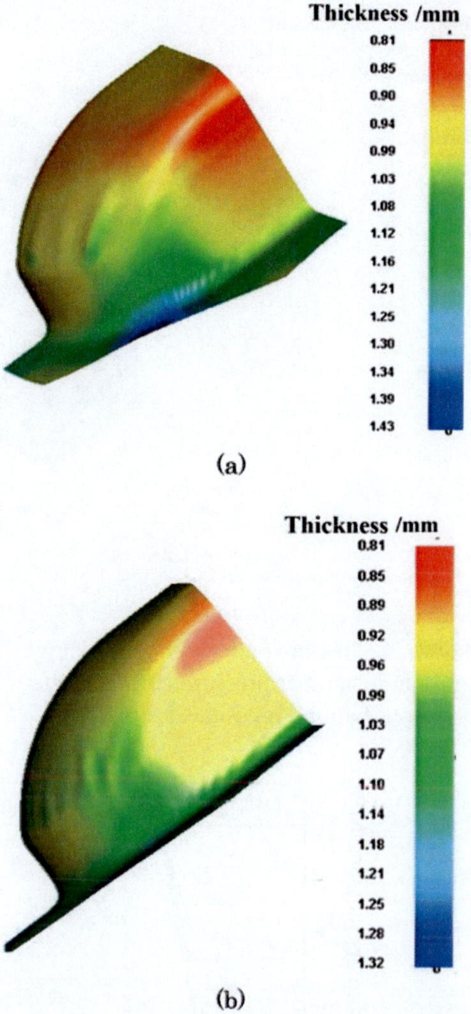

Fig. 7.30 Effect of liquid cavity pressure on the thickness distribution. **a** 5 MPa; **b** 30 MPa

material sustains relatively large biaxial tensile stresses with a maximum thinning percentage of 44.3%, that is far exceeding the maximum elongation of material. A severe splitting of deep drawing has been resulted, as shown in Fig. 7.31.

The above results have shown that, conventional deep drawing only relies on the liquid pressure to adjust the forming states of complex-shaped parts with curved surfaces, which brings limited improvements on the formability. For such a complex box-shaped part with single curvature, the approach of increasing liquid chamber pressure only is unable to simultaneously satisfy the avoidance of wrinkling and splitting. Therefore, pre-bulging approach is required to be adopted to address the above limitation. Based on the analysis of effects of liquid pressure on the wrinkling and splitting, the preliminarily reasonable liquid pressure is 15 MPa. Figure 7.32

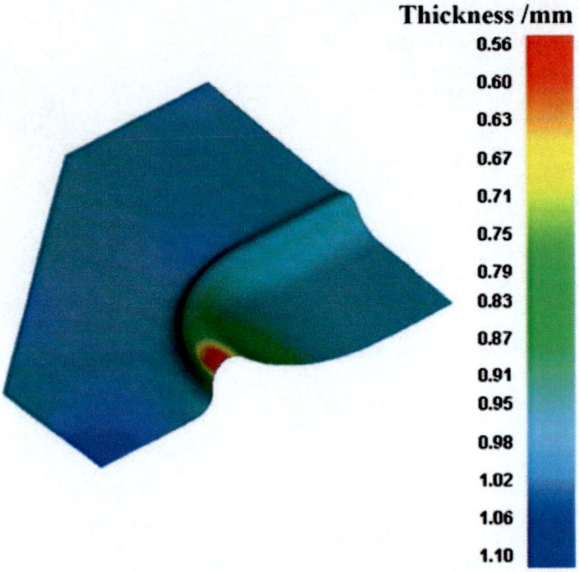

Fig. 7.31 The thickness distribution o under a liquid pressure of 40 MPa

shows the loading paths with different pre-bulging pressures at a consistent 15 MPa of liquid chamber pressure at the late stage of hydromechanical deep drawing. The used pre-bulging pressures are 1.5, 2, 3.5, 5 and 8 MPa.

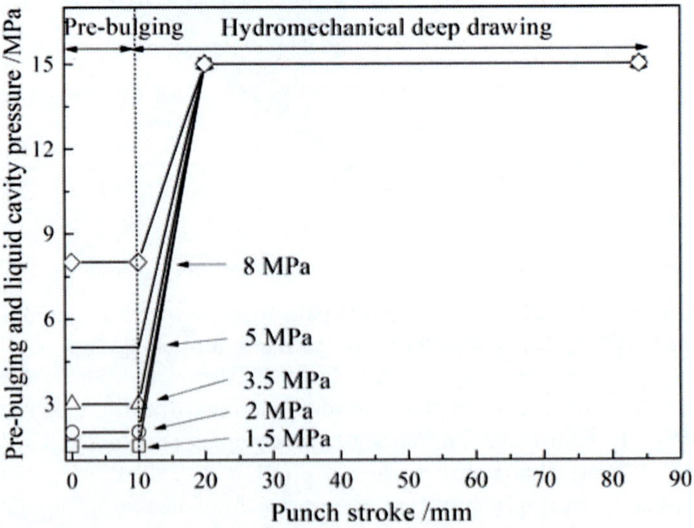

Fig. 7.32 The loading paths of pre-bulging and subsequent hydroforming of a box-shaped part with single curvature

Figure 7.33 shows the variation of the latitudinal stress of the curved surface with the proceeding of punch stroke. The pressures of pre-bulging initially and liquid chamber lately are 1.5 MPa and 15 MPa, respectively. The magnitude of compressive stress increases dramatically with increasing the deep drawing stroke. When the stroke reaches 20 mm (25% H), the latitudinal stress maximizes at the region of wrinkling with a magnitude of −170 MPa; Then, with a further increase of deep drawing stroke, although there exists small fluctuations of the compressive stress, the overall trend of such stress is decreasing. Due to the pre-bulging pressure is relatively small, at the initial stage of deep drawing, although the sheet at the curved zone is in contact with the die due to the liquid pressure, wrinkling defect can be still generated here considering the latitudinal compressive is much greater than the liquid compressive stress. While at the late stage of deep drawing, although the liquid pressure increases and latitudinal stress decreases gradually, the extra accumulated material cannot be fully unfolded, resulting in the uneliminated wrinkles on the several positions of the part.

When the pre-bulging pressures are 3.5 and 5 MPa, wrinkling defects are avoided, while the splitting defect can be observed at the transition of curved surface when the pre-bulging effect is increased to 8 MPa. Based on the above FE simulation results, the reasonable pre-bulging pressure range is between 2 and 8 MPa. In terms of using the pre-bulging approach to control the wrinkling at the unsupported zone, the deformation degree of pre-bulging is required to be controlled properly. Either a too low or too high pressure, may result in the unsuccessful suppression of wrinkles and splitting defect, respectively. Figure 7.34 shows a qualified part using a pre-bulging pressure of 3.5 MPa and liquid chamber pressure of 15 MPa. As observed from the formed part, the surface quality is good. The forming precision is high satisfying the service requirement.

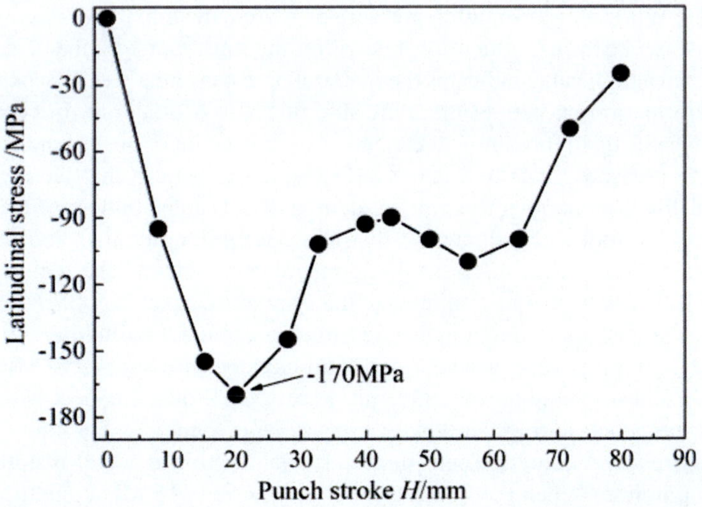

Fig. 7.33 Effect of pre-bulging pressure on the latitudinal stress variation

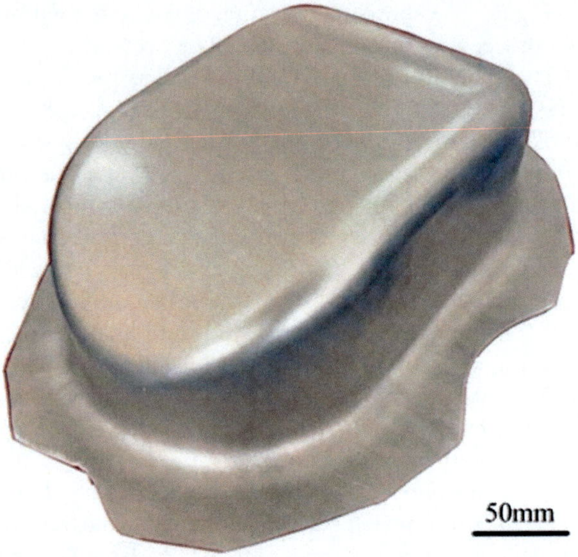

Fig. 7.34 A box-shaped component with single curvature of aluminium alloy

Figure 7.35 shows the effective strain distribution on the symmetrical plane sectioned in the longitudinal direction of part. The used pre-bulging pressures were 2 and 8 MPa. From the center to the edge along the sectioned arc, the distribution of effective strain exhibits a mountain-profiled variation. At a 2 MPa pre-bulging pressure, the maximum effective strain increases first and peaks at the corner radius of punch with a magnitude of 0.45, then the effective strain slightly decreases and increases again at the entrance of female die. The magnitude at the spinodal point is 0.22. From this position further, the effective strain increases significantly reaching 0.55 finally. While at a pre-bulging pressure of 8 MPa, the maximum effective strain degree has exceeded 0.7, which locates within the unsupported zone of this cross-section. Such a high value indicates the material here has already experienced a large plastic deformation exceeding the allowable ductility of material. In comparison, the effective plastic deformation at the center of bottom increases insignificantly for pre-bulging pressures at both 2 and 8 MPa, which illustrates that the pre-bulging pressure delivers an unobvious strain hardening effect on the bottom of the part.

Figure 7.36 shows the thickness distributions of longitudinal and latitudinal symmetrical planes under different liquid chamber pressures. The region between dash lines has a relatively large thinning of the symmetrical plane of sheet. This region belongs to the unsupported region during forming. For the longitudinal symmetrical plane, the thinning is severer when the liquid chamber pressure is 10 MPa, with a minimal thickness magnitude of 0.838 mm. The reason is that, the sheet cannot firmly attach on the punch during the downwards stroke of punch as the liquid chamber pressure is relatively small. Then, the interfacial friction is small resulting in an increased thinning; When the liquid chamber pressure is 15 MPa, due to the effect of friction maintaining, the thinning of the unsupported zone is weaken, and the

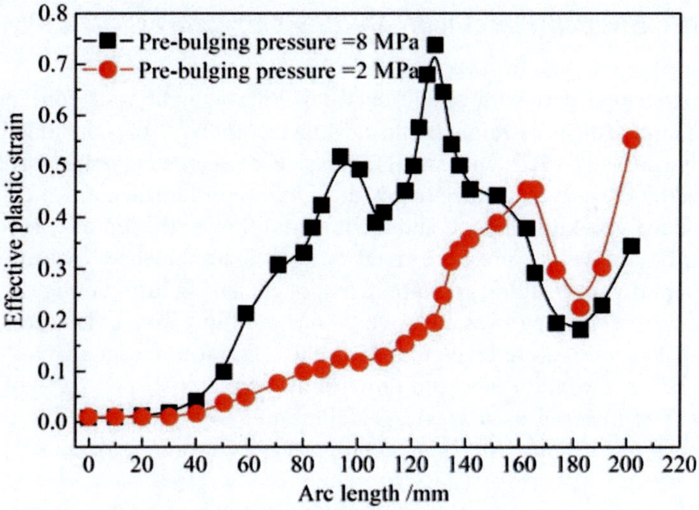

Fig. 7.35 Effect of pre-bulging pressure on the effective plastic strain

minimal thickness is 0.882 mm, indicating a uniform deformation. Regarding the latitudinal symmetrical plane, the severest thinning occurred at the corner radius of punch. The position owning minimal thickness locates at the top of corner radius of punch.

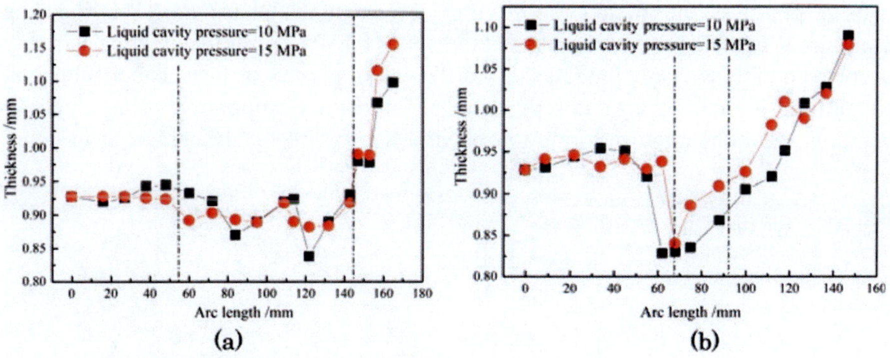

Fig. 7.36 The thickness distributions at different liquid cavity pressures. **a** Longitudinal direction; **b** latitudinal direction

7.8.3 *Box-Shaped Panel with Double Curvatures*

For the box-shaped part with single curvature, the straight wall zone below the entrance of female die is a straight surface, while that for box-shaped part with double curvature is a curved surface with small radius. Hence, a greater compressive stress is experienced by the curved surface below the entrance of female die, which is highly likely to induce wrinkling defect, and the material flow at the top of curved surface is difficult. Experiments were performed to investigate the sheet hydroforming of box-shaped part with double curvature. Components using different liquid chamber pressures and pre-bulging pressures were successfully formed. The variations of thickness distributions were given to measure the dimensional accuracy of part.

Figure 7.37 shows the shape and dimensions of a box-shaped part with double curvatures. The material used was 2A12 aluminium alloy in annealing condition. The thickness was 1.5 mm. Mechanical properties were summarized in Table 7.4 in previous section.

Figure 7.38 illustrates the effect of pre-bulging pressure on the defects occurred during the sheet hydroforming of box-shaped part with double curvature. When the pre-bulging pressure is 1 MPa, obvious wrinkles exist at the curved surface of part end. At this moment, there still exists a relatively large unsupported zone although liquid pressure functions. The liquid pressure is unable to eliminate the latitudinal compressive stress entirely resulting in wrinkles exist on the curved surface of part end which cannot be eliminated entirely.

With increasing the pre-bulging pressure, bulging deformation occurs for the supported zone sheet due to the given pre-bulging pressure. The higher the pre-bulging pressure, the greater the bugling deformation and the more obvious the thinning; When the pre-bulging pressure reaches 8 MPa, splitting defect has already occurred on the reversely bulged sheet at the end of curved surface. The mechanism of splitting is that, during the bulging of sheet at the unsupported zone, a certain bending and unbending deformation occurs. Constraint by the flange at the end,

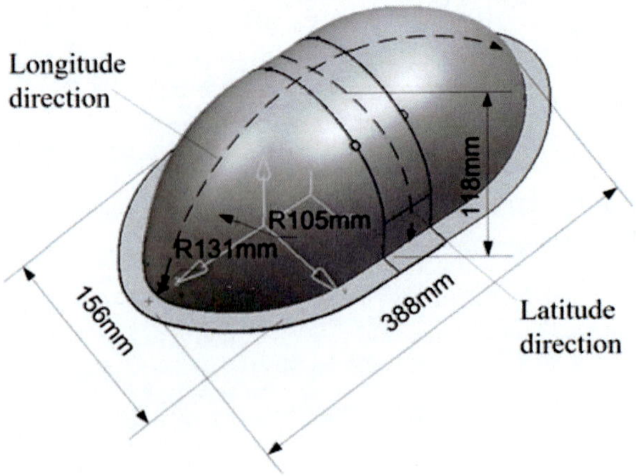

Fig. 7.37 The shape and dimensions of box-shaped component with double curvature

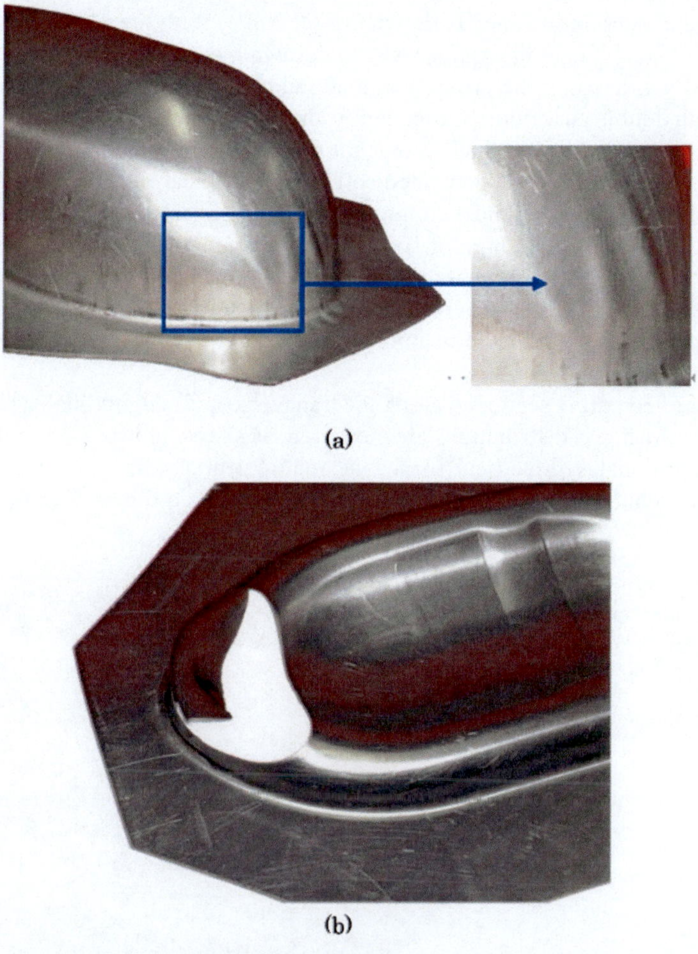

Fig. 7.38 Effect of pre-bulging pressure on the defects. **a** Wrinkling (the pre-bulging pressure 1 MPa); **b** splitting (the pre-bulging pressure 8 MPa)

dramatic increased thinning is resulted, especially for the regions with a relatively small curvature. The critical stress of bulging deformation is lower compared to that at the straight edge, resulting in the thinning of sheet and ultimate splitting.

According to the finite element simulations and theoretical analysis, using a pre-bulging pressure of 3 MPa, qualified part can be formed when the liquid chamber pressure is controlled within the range of 10–20 MPa. The experimental results have shown that, within the initial pre-bulging stage (10% of total stroke), the sheet is mounted between the punch and female die. The punch is fixed and liquid chamber pressure is increased to establish initial pre-bulging pressure. The sheet within the unsupported zone occurs reverse bulging deformation towards the direction of punch. As there is no material flow at the flange, pure bulging deformation is occurred for the sheet at the unsupported zone, as shown in Fig. 7.39a. With the further downwards motion of punch, flange material has already drawn into the die when a 75% stroke is achieved, while using the pre-bulged material can satisfy the forming of curved surface with double curvature. Subsequently, the flow at the end of flange is delayed; With further proceeding of punch stroke, the reversed sheet gradually flows into the female die, and the previous generated soft draw bead near the end curved surface with double curvatures is almost eliminated. The flange material begins to flow into the female die, then, the flange of formed curved shell is mainly achieved by the deep drawing, while the flange at the sidewall occurs bending deformation; When a full stroke (100%) is achieved and the forming is finished. The curved surface at the end of part still has no defect indicating a qualified part can be formed, as shown in Fig. 7.39b.

The qualified part is sectioned along the transverse and longitudinal symmetrical planes. The thickness distributions are measured, as shown in Fig. 7.40. With regard to the longitudinal symmetrical plane, the minimal thickness is 1.30 mm with a thinning percentage of 13.3%, indicating a relatively large thinning extent. This is

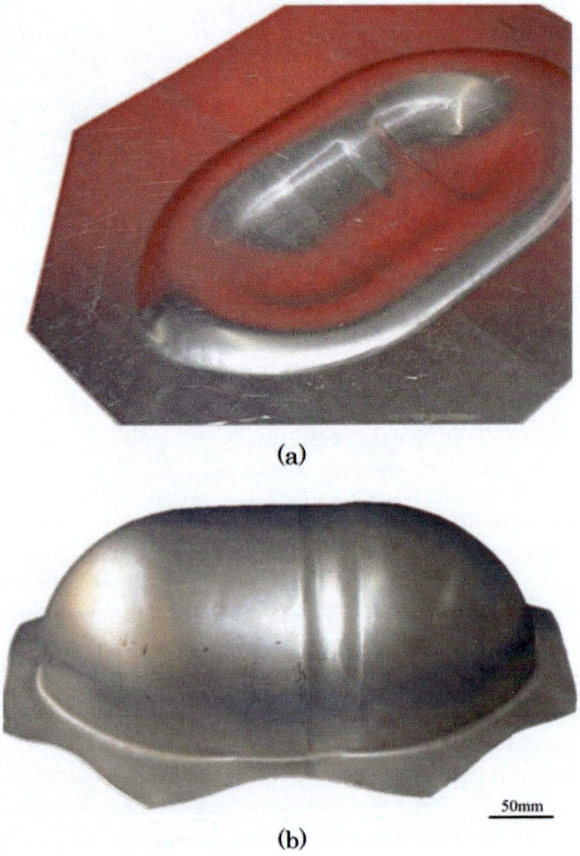

Fig. 7.39 The shape and dimensions of box-shaped part with double curvature. **a** Initial stage of fog; **b** end of forming fog

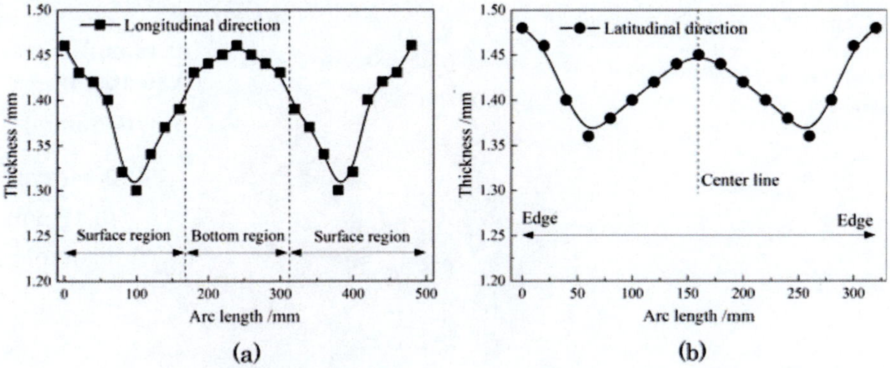

Fig. 7.40 Thickness distributions on the part. **a** Longitudinal direction; **b** latitudinal direction

mainly caused by the relatively small liquid chamber pressure during forming at this moment. Although the beneficial friction between sheet and punch can be effectively increased due to the liquid chamber pressure, and the thinning can be suppressed to a certain extent. The bulging still dominates the deformation resulting in thinning. The deformed sheet gradually attaches to the die, and the attached material is no longer experienced thinning. With regard to the latitudinal symmetrical plane, the minimal thickness is 1.36 mm, and the thinning percentage is 9.3% considering the initial thickness is 1.50 mm. The reason is that, regarding the sheet material at this measurement location during forming, the deep drawing deformation at the end of flange is later than the bending deformation of flange used to form sidewall. The bending deformation of flange is constraint resulting in a greater flow resistance of workpiece at this position. Then, the material drawn into the female die occurs bending and unbending deformation, resulting in the thinning of thickness.

The dimensional accuracy of the box-shaped part with double curvatures can be represented by the maximum clearance between the different cross-sections of part and the designed profile plate, among a series of cross-sections selected with a 10 mm interval. Figure 7.41 shows the dimensional accuracy measurement results of the box-shaped part with double curvatures. The maximum clearance is 0.22 mm locating at a cross-section within the sidewalls, while the minimal clearance is 0.1 mm locating at a cross-section within the curved surface of double curvatures, which is smaller than the 0.5 mm design requirement. Compared to the curved surface part with double curvatures, the springback at the sidewall is greater. The main reason is due to the bending and unbending. During the forming of box-shaped part with double curvatures, the initially reverse bulged material occurs a second round of building during the attaching to the punch surface. At the late stage of forming, the flange material flows into the female die becoming end curvature has a dominant type of deep drawing deformation, and the local curvature of the female die cavity is relatively small resulting in a greater degree of material deformation; After the bending and unbending process of material flowing pass the corner radius of female die, a significant work-hardening is obtained. Such a deformation type

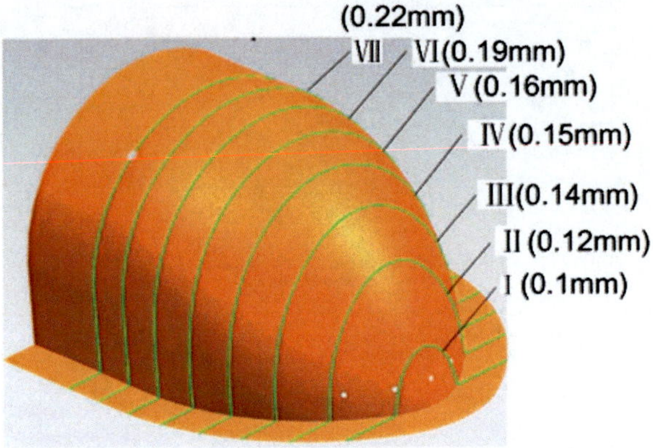

Fig. 7.41 The dimensional accuracy of the profile of box-shaped part with double curvatures

and the geometry characteristics of part enables the material at the complex profile to obtain sufficient hardening, then the springback of box-shaped part with double curvature becomes small resulting in a high dimensional accuracy.

7.8.4 The Hydroforming of 2219 Aluminium Alloy Sheet

2219 aluminium alloy is a typical heat treatment strengthening aluminium alloy with good stress corrosion resistance, toughness, weldability and relatively high strength. Such an alloy has gained extensive applications in the aerospace rocket structure, such as the low-temperature fuel tank component. Components with excellent mechanical properties can be obtained using a proper combination of sheet hydroforming and heat treatment processes. The used sheet is 2219 aluminium alloy in annealing condition with a thickness of 1.5 mm. Figure 7.42 shows the mechanical properties of 2219 aluminum alloy after a solution heat treatment and artificial ageing. As shown in this figure, both the yield and ultimate tensile strength are increased significantly after the T6 treatment. The ratio of yield strength to ultimate tensile strength is minimal for the solution heat treated temper (W-condition), indicating the sheet material has a relatively large plastic straining range enabling to be sustained a certain plastic deformation; While after the artificial ageing, both the yield and ultimate tensile strength are significantly increased compared to those of solution heat treated state. The elongations of the three tempered material are shown in Fig. 7.43. After the ageing treatment, elongation shows an obvious reduction indicating a decreased ductility. Such an artificially aged condition is unsuitable for plastic deformation with large straining requirement. With the performing of heat treatment, the hardness of material continuously increases with an average value reaching 128.1 HV, which is an improvement of 123.95% compared to that of O condition.

A 2219 aluminium alloy cylindrical part with a flat bottom was formed using the sheet hydroforming process. The drawing ratio was 1.94. Figure 7.44 shows

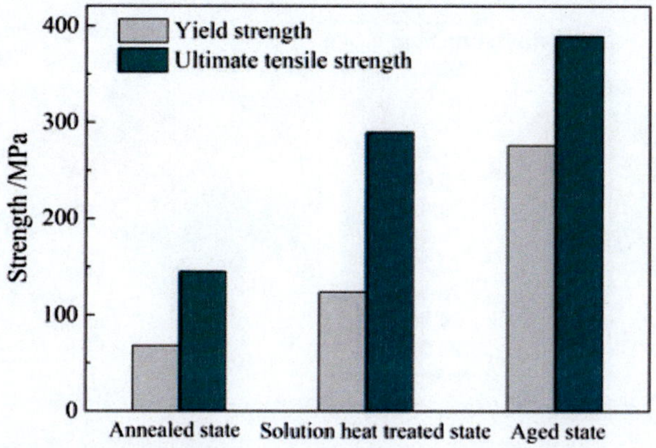

Fig. 7.42 The strength of 2219 aluminium alloy at different heat treatment conditions

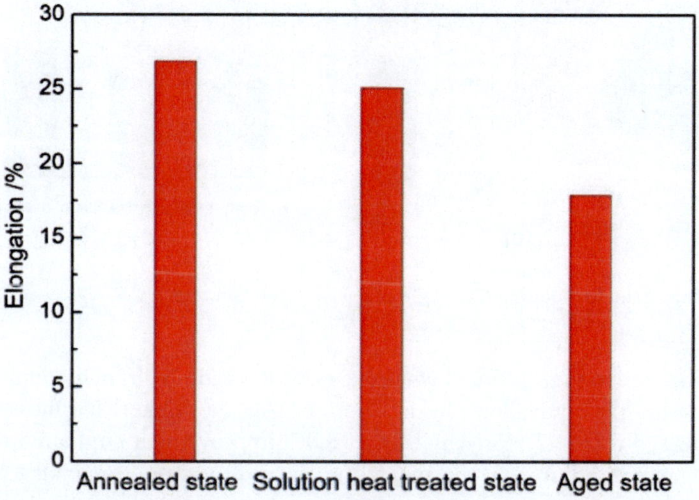

Fig. 7.43 The elongations at different heat treatment conditions

the experimentally determined maximum strain contours of the straight wall zone. As shown in the figure, the magnitude of strain varies obviously with variations of directions. The maximum principal strain of 0°, 45° and 90° is 0.28, 0.46 and 0.29. The deformation amounts are similar for the directions of 0° and 90°, and the degree of plastic deformation peaks at the direction of 45°. To further illustrate the effect of sheet hydroforming deformation on the strengthening of material, samples at different locations on the cylindrical part with a flat bottom were selected and used to conduct hardness and uniaxial tensile tests. The obtain strengthening results are shown in Fig. 7.45.

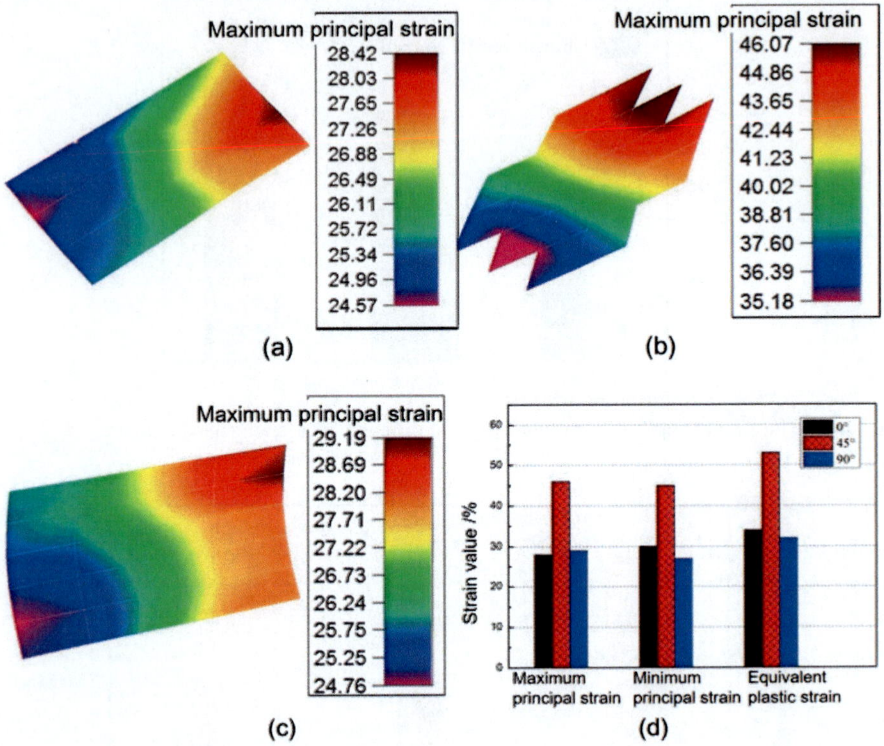

Fig. 7.44 The strain distribution of the cylindrical part with a flat bottom

As can be seen in Fig. 7.45a, the material exhibits obvious strain hardening effect. With increasing the equivalent plastic strain to 0.46, the Vickers hardness of material is increased to 151.2 HV, which is a 18.0% improvement compared to that of artificial ageing condition. In the meantime, according to the uniaxial tensile test results, the ultimate tensile strength of specimen at directions of 0°, 45° and 90° are 408.65, 415.23 and 438.92 MPa. Particularly, the true stress–strain curve of 90° specimen is shown in Fig. 7.45b. An obvious strengthening effect was obtained for the hydroformed part. The ultimate tensile strength increases 12.5% compared to that of artificial ageing condition.

The plastic deformation process has a certain effect on the microstructural evolution. Figure 7.46 shows the EBSD observations of T6 conditioned and deformed specimen. The microstructure results have shown that, the material occurred significant plastic deformation along the deep drawing direction. The initial material has an average grain size of 57.04 μm. In comparison, the average grain sizes of the straight wall are 52.48 μm, 37.83 and 49.96 μm for 0°, 45° and 90° directions, indicating a decreased grain size. Through the analysis of small angled grain boundaries at different zones, it can be seen that a large amount of small angled grain boundaries

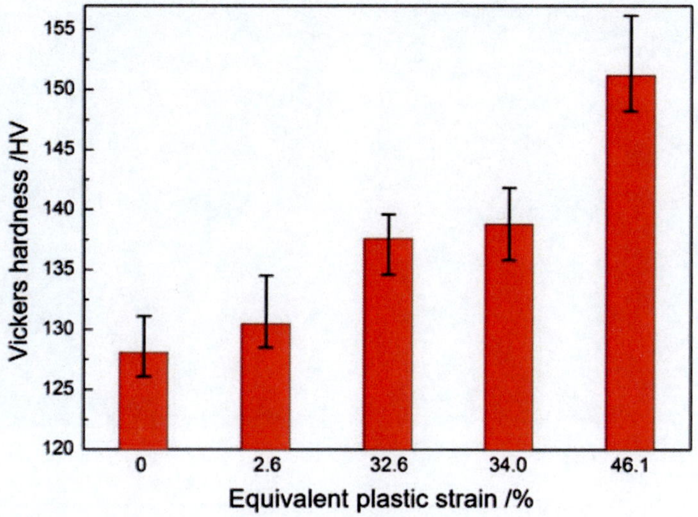

(a)

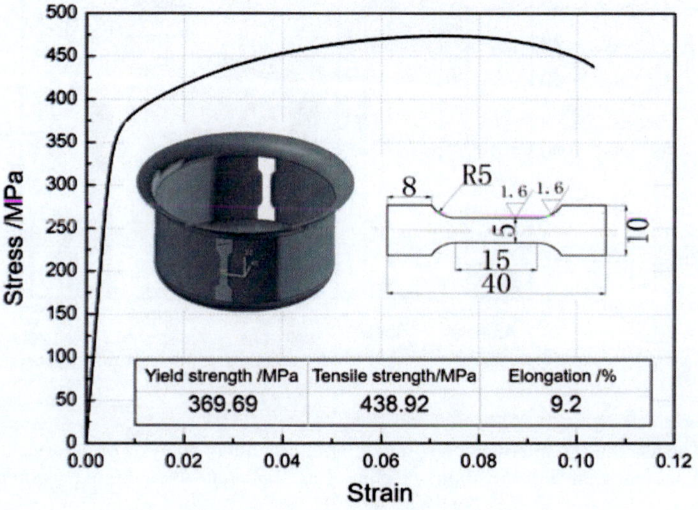

(b)

Fig. 7.45 The hardness **a** and strength **b** of the cylindrical component with a flat bottom

were generated at the straight wall zone of this cylindrical part, with a percentage of 75% approximately, which indicates the sheet occurred obvious plastic deformation.

2219 aluminium alloy is a typical Al–Cu alloy. After solution heat treatment and quenching, the precipitation sequence during ageing is:$\alpha \rightarrow \alpha + GP\ \text{I} \rightarrow \alpha + GP\ \text{II}(\theta'') \rightarrow \alpha + \theta' \rightarrow \alpha + \theta$. The alloy becomes unstable supersaturated solid solute, which can precipitate strengthening phases during the subsequent ageing

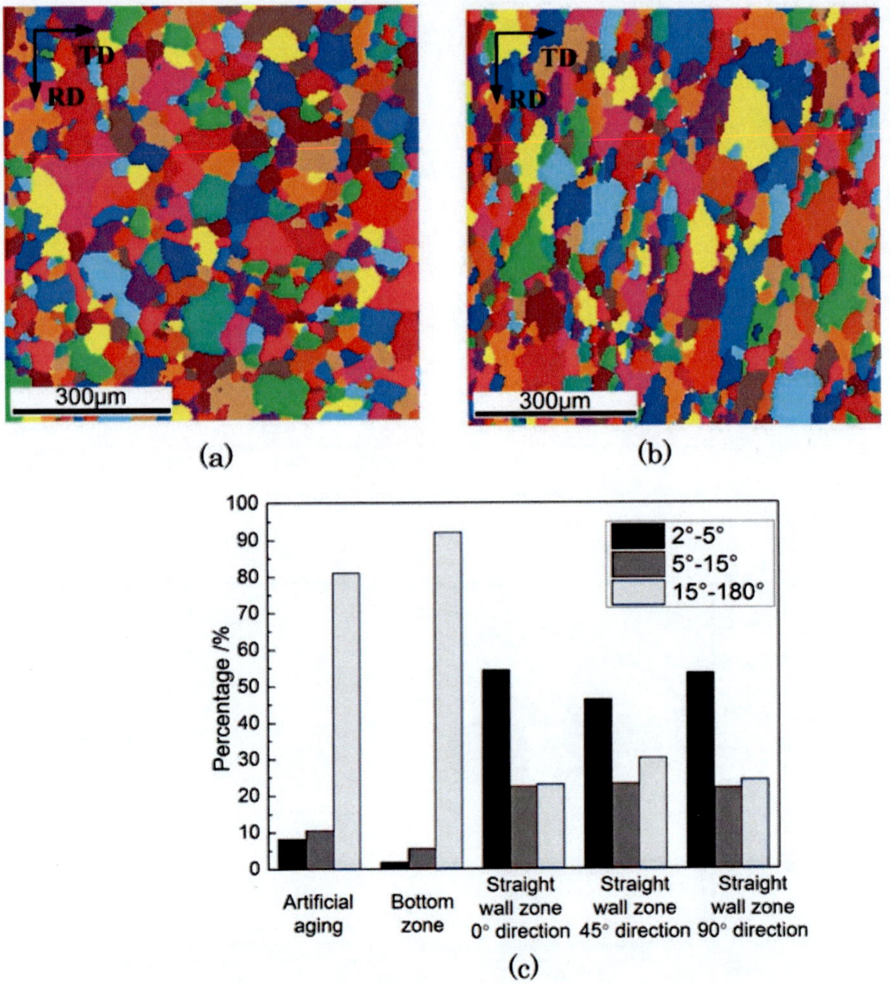

Fig. 7.46 The as-received microstructure and grain morphology analysis of deformed part. **a** T6 condition; **b** straight wall at 0° direction; **c** the percentage of small angle grain boundaries

process. Transmission Electron Microscopic observations were performed for the specimen selected from the straight wall of hydroformed sheet part. Figure 7.47 shows the obtained bright-field photo and diffraction pattern. A great amount needle-shaped precipitates and large dislocation density can be observed. The diameter and thickness of precipitates are roughly around 50–100 nm and 10–20 nm, respectively. Using the diffraction pattern, the dense distributed precipitate is identified as θ' phase. The results have shown that the part experienced an obvious enhanced strengthening using the sheet hydroforming.

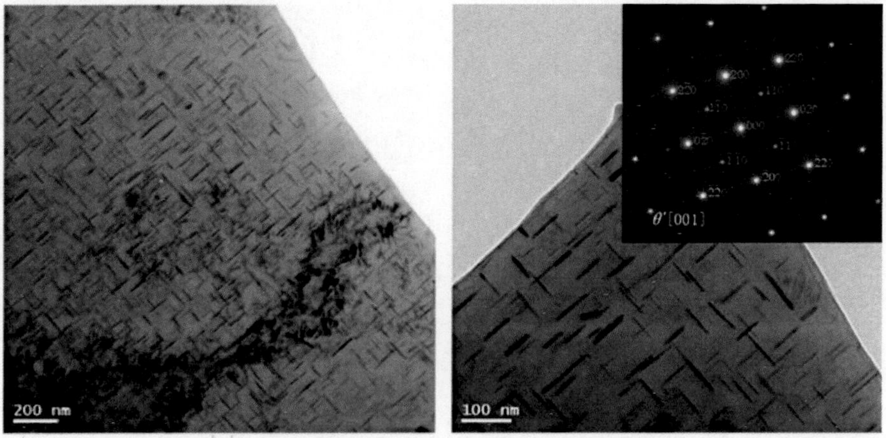

(a) (b)

Fig. 7.47 TEM observations of specimen sampled from the straight wall zone of a cylindrical part with flat bottom

Chapter 8
Die-Less Hydroforming of Shells

8.1 Structure of Shells and Manufacturing Process

Spherical vessels have two main advantages over cylindrical vessels. Firstly, the spherical vessel has uniform stress and larger capacity. With the same thickness and diameter, its capacity is one time higher than that of a cylindrical vessel. Secondly, under the same capacity the spherical vessel has a smaller surface area, so less material is used for the manufacturing of the spherical vessel, which has the characteristic of lighter weight and less heat loss. Therefore, the cost of manufacturing, storage and transportation is reduced. In the conventional manufacturing process of shells, the die pressing process is mainly used. For spherical shells with different diameters and thickness, the spherical shell petals are stamped using different dies and then welded into the whole sphere. The main disadvantages of this process are worktable press with large bed and large forming dies are required, and the pressing and secondary cutting process are complicated and costly. For the spherical shells with a smaller diameter, the deep drawing process can also be used to form the hemispheres or a number of spherical petals, which are then welded into the whole sphere. The disadvantage of deep drawing is the high cost of die, and it is prone to wrinkling and splitting during deep drawing.

Ellipsoidal vessels have the advantages of low gravity center, smaller wind area and more attractive appearance, which are especially suitable for manufacturing of the petrochemical vessels and large water towers. In addition, some of the ellipsoidal shells are often used for the head of a cylindrical vessel, the fairing of large carrier rocket. The die pressing technology for ellipsoidal shell is more complicated than that of spherical shell. This is because a spherical shell with a diameter only needs one set of dies, while the curvature of the ellipsoidal shell from the equator to the south and north poles is changing, which requires more than one set of dies. This is the main reason restricting the wide application of ellipsoidal shells. However, if the die-less hydroforming technology is used to manufacture the ellipsoidal vessels, ellipsoidal heads and other ellipsoidal shells, it has the advantages of simple process, low manufacturing cost and strong adaptability.

S. Yuan, *Modern Hydroforming Technology*,
https://doi.org/10.1007/978-981-19-5775-8_8

8.2 Die-Less Hydroforming for Spherical Vessels

8.2.1 Forming Principle and Advantages

The basic process of die-less hydroforming for spherical vessel is as follows: a closed polyhedral shell is firstly assembled and welded by flat plates or single curvature petals manufactured by roll bending. Then the closed polyhedral shell is filled with liquid medium (usually water), and internal pressure is applied to the closed polyhedral shell through a pressurization system. Under the internal pressure, plastic deformation occurs on the shell, which is bulged into to the desired spherical shell gradually. For single-curvature shell, the main process is that blanking → bending → assembly and welding → hydroforming. There are two fundamental basis in theory for the die-less hydroforming of spherical vessel. First, under the action of moment, the preformed polyhedral shell will gradually become a spherical shell with the increase of forming pressure. On one hand, in any position of the preformed shell, if the radius of curvature is relatively large, it firstly deforms during loading, and accordingly the radius of curvature becomes smaller and the deforming stops. On the other hand, for the position where the radius of curvature is previously relatively small, its curvature radius is now relatively large and begins to deform. In this way, the curvature radius everywhere are equal and it becomes a spherical shell. The other is the automatic adjustment of plastic deformation of metal materials. During the forming process, plastic deformation first begins where the yield conditions are satisfied first. With the increase of deformation, the material can be strengthened, and the plastic deformation is transferred to other relatively weak regions, while the previously relatively strong regions becomes relatively weak regions and plastic deformation occurs. With such cyclic adjustment, the thickness distribution of the final formed spherical shell is relatively uniform.

The main advantages of die-less hydroforming technology for spherical vessels is that: large die and press are not needed and the production cost can be reduced; the production cycle is shortened with no need of dies; the residual stress of welding is reduced effectively after overload bulging.

8.2.2 Forming Pressure

Supposing that the radius of the spherical shell is r, the thickness is t, and the bearing internal pressure is p, the relationship between stress and internal pressure of the shell is as follows:

$$\sigma_\theta = \sigma_r = \frac{pr}{2t} \tag{8.1}$$

For the die-less hydroforming of spherical shell, its stress state is$\sigma_1 = \sigma_2 = \sigma_\theta = \sigma_r$, $\sigma_3 = 0$. Therefore, according to the Tresca yield criterion $(\sigma_1 - \sigma_3 = \sigma_s)$, the internal pressure resulting in the initial plastic deformation of the shell, also known as the initial yielding pressure, can be obtained

$$p_s = \frac{2t}{r}\sigma_s \tag{8.2}$$

Supposing the spherical shell material obeys the hardening law $\sigma_i = K\varepsilon_i^n$, the internal pressure of the shell is

$$p = \frac{2t}{r}\sigma_i \exp[-\frac{3}{2}\varepsilon_i] \tag{8.3}$$

where σ_i—Equivalent stress;

ε_i—Equivalent strain;

n—Strain hardening exponent;

K—Strength coefficient.

According to the tensile instability (splitting) condition $dp = 0$, the splitting pressure of the shell is calculated as

$$p_{\max} = \frac{2t}{r}k\left(\frac{2n}{3e}\right)^n \tag{8.4}$$

The die-less hydroforming pressure p of the spherical vessel is between the initial yield pressure p_s and the splitting pressure $p_{\max}$. For example, after substituting the radius of spherical shell $r = 1000\,\text{mm}$, the thickness $t = 1.5\,\text{mm}$, $\sigma_s = 48\,\text{MPa}$, $n = 0.19$ and $k = 179\,\text{MPa}$ into Eqs. (8.2) and (8.4), it can be obtained that initial yielding pressure $p_s = 0.144\,\text{MPa}$ and cracking pressure $p_{\max} = 0.3\,\text{MPa}$. In actual hydroforming, the forming pressure is $p = 0.195\,\text{MPa}$.

8.2.3 Polyhedral Structure Before Hydroforming

There are two kinds of polyhedral structures before hydroforming. One kind is flat structure, in which the elements that compose the polyhedral shell are flat plates with different shapes, such as football shape, icosahedron, dodecahedron, and volleyball form, as shown in Fig. 8.1. The other is single-curvature structure, such as basketball and tennis, as shown in Fig. 8.1. The main parameters of these preformed polyhedral shell are compared in Table 8.1 (the diameter of spherical shell is 4 m). In Table 8.1, the minimum dihedral angle refers to the minimum value of the dihedral angle of the adjacent petal on the polyhedral shell. The smaller the value of the minimum dihedral angle, the larger the deformation of the weld after forming. Therefore, the weld is more prone to crack. Weld length is a major evaluation index of manufacturing cycle

Table 8.1 Comparison of main parameters for different preformed polyhedral shell (the diameter of spherical shell is 4 m)

Parameters	Football	Basketball (15 petals in equatorial zone)	Volleyball	Tennis
Minimum dihedral angle/(°)	138	156	150	90
Weld length/m	96	86.6	98.5	22
Volume of polyhedron/m^3	28.71	32.52	31.1	20.77
Rate of volume variation/%	16.7	3	7.7	61.3
Area of polyhedron/m^2	46.56	49.51	47.04	41.2
Average thinning ratio/%	7.36	1.49	6.37	18.0

and cost. The change rate of volume refers to the volume change from polyhedral preformed structure to spherical shell. If the change rate of volume is large, then on the one hand it needs more water injection from polyhedron to spherical shell, and the time for hydroforming is long; On the other hand, the surface area changes greatly and the shell thickness decreases greatly.

The choice of polyhedral shell structure before hydroforming is a key to the hydroforming of spherical vessel. As can be seen from Table 8.1, the average thinning ratio of tennis structure is the largest, reaching 18%, and the average thinning ratio of football and volleyball structure is the same. Moreover, the average thinning ratio of basketball structure is the smallest.

In industry, the commonly used polyhedral structure before hydroforming is in the form of football and basketball. The advantage of the football structure is that the polyhedral shell is only composed of regular pentagon and regular hexagon, which is simple for blanking and easy to assemble. However, by comparing with the basketball structure with 15 petals, the football structure has a smaller minimum dihedral angle, a larger change rate of volume, and a longer weld length, which is the main shortcoming of the football structure. Especially the large dihedral angle and the longer weld will easily cause the weld crack during hydroforming. Another drawback of the football structure is the low utilization of materials. For the basketball structure, it has the advantages of small dihedral angle and small thinning ratio. Moreover, the more petals in the equatorial zone, the smaller the dihedral angle and the smaller the weld deformation. However, the increase in the number of petals in the equatorial zone increases the weld length. The main parameters for the basketball shells with different petal numbers in the equatorial zone are shown in Table 8.2.

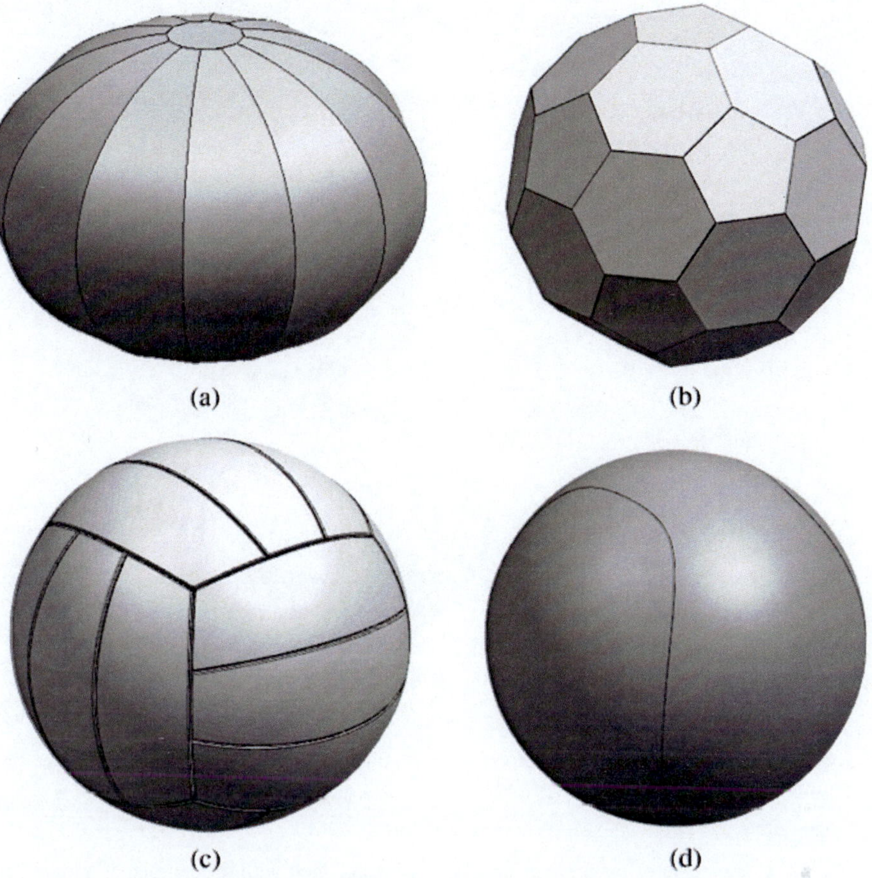

Fig. 8.1 Type of shell structures. **a** Basketball; **b** football; **c** volleyball; **d** tennis

8.3 Die-Less Hydroforming of Ellipsoidal Shells

8.3.1 Die-Less Hydroforming of Ellipsoidal Shell

The ellipsoidal vessels are suitable for the structures of large water tower, head of pressure vessel and tank dome of carrier rocket because of their advantages of low gravity center, large volume, bearing small wind area and attractive appearance. Because the curvature radius of the ellipsoidal shell gradually changes from the pole to the equator, the die press forming of an ellipsoidal shell needs more than one set of dies, which is the main reason to limit the wide application of ellipsoidal shells. In order to solve the problems existing in the conventional die pressing process, a new method for manufacturing of thin-walled ellipsoidal shell by die-less hydroforming was proposed. Figure 8.2 shows the principle of die-less hydroforming of ellipsoidal

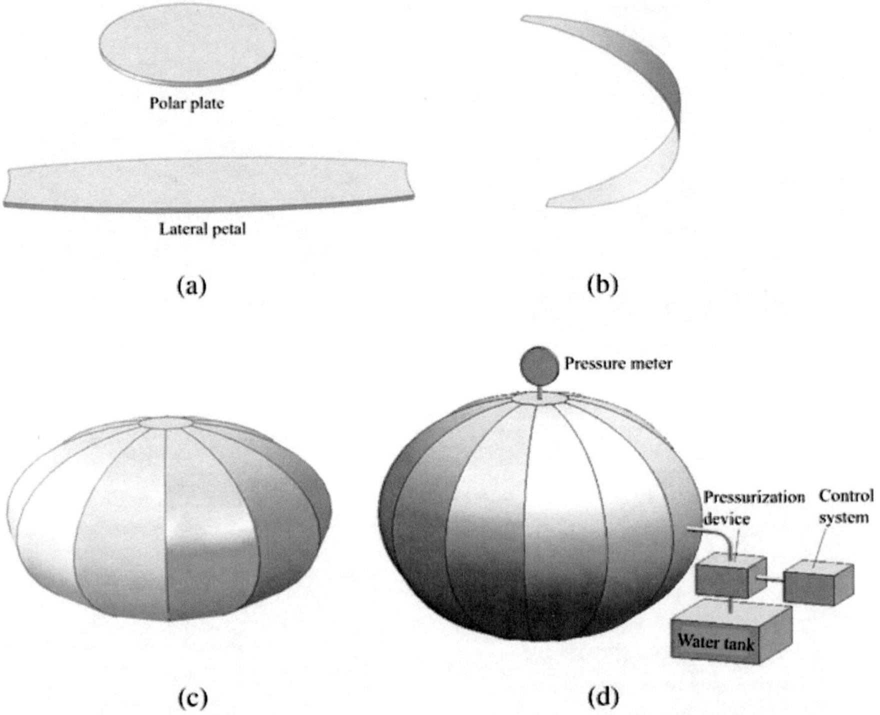

Fig. 8.2 Basic process of die-less hydroforming of ellipsoidal shells. **a** Blanking; **b** roll-bending; **c** assembling and welding; **d** hydroforming

Table 8.2 Comparison of main parameters for basketball shell with different petal number (the diameter of spherical shell is 4 m)

Petal number in equatorial zone	12	15	20	25
Minimum dihedral angle/(°)	150	156	162	165.6
Weld length/m	69.02	86.6	110.95	137.01
Volume of polyhedron/m^3	31.94	32.52	32.91	33.1
Change rate of volume/%	4.9	3	1.8	1.2
Area of polyhedron/m^2	49.1	49.51	49.82	49.97
Average thinning ratio/%	2.26	1.45	0.84	0.545

shell. The basic process is that first a number of lateral plates and two polar plates are blanked from the plate, in which the lateral plates are then roll-bent into the lateral petals with single curvature. These single curvature shell petals are then assembled and welded into closed polyhedral shell, after which the closed polyhedral shell is filled with water medium, and internal pressure is applied to the closed polyhedral shell through a pressurization system. Under the action of internal pressure, plastic deformation occurs on the shell, which is bulged into to the desired spherical shell gradually.

The main advantages of die-less hydroforming technology for ellipsoidal vessels is that die and press are not needed and the cost can be reduced. Moreover, the production cycle is shortened and product changes are easy as forming dies are not needed. The residual stress of welding is reduced after overload bulging calibration, and it has a high dimensional accuracy can be obtained.

8.3.2 *Relationship Between Stress and Axial Length Ratio of an Ellipsoidal Shell*

Under the action of internal pressure, the stress distribution of an ellipsoidal shell has a certain correspondence with the curvature radius. In order to obtain the stress distribution of the ellipsoidal shell under internal pressure, the geometry and stress state of an ideal ellipsoidal shell are analyzed, as shown in Fig. 8.3. In this figure, r_1 is the first principal curvature radius, r_2 is the second principal curvature radius, t is the thickness of shell, p is the internal pressure, φ is the angle between r_2 and the rotation axis (y axis), a is the equatorial radius (radius of the non-rotating axis), b is the radius of rotation axis.

The axial length ratio of the ellipsoidal is defined as $\lambda = a/b$. Figure 8.4 shows the relationship between the shape of the ellipsoidal shell and the axial length ratio. It can be found that when $\lambda < 1$, the shell is a prolate ellipsoidal shell, whose rotating axis b is the long axis, while non-rotating axis a is the short axis. When $\lambda = 1$, the shell is a spherical shell, whose radius of the non-rotating axis is equal to the radius of the rotating axis, i.e. $a = b$. When $\lambda > 1$, the shell is a oblate ellipsoidal shell, whose non-rotating axis a is the long axis, while rotating axis b is the short axis. In

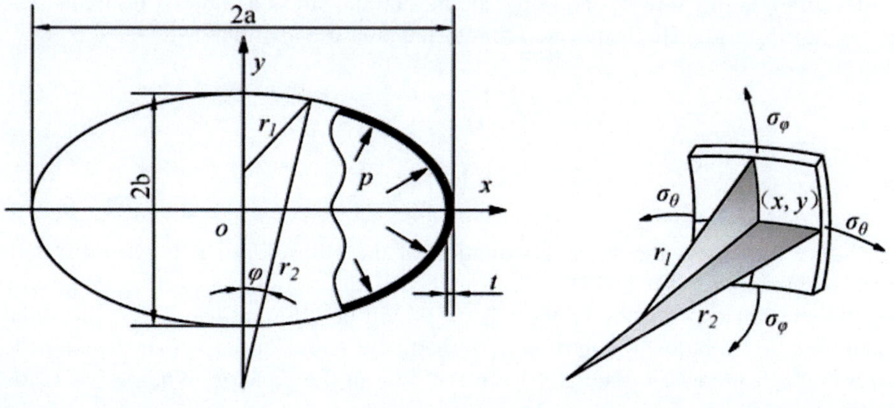

Fig. 8.3 The geometry and stress of an ellipsoidal shell

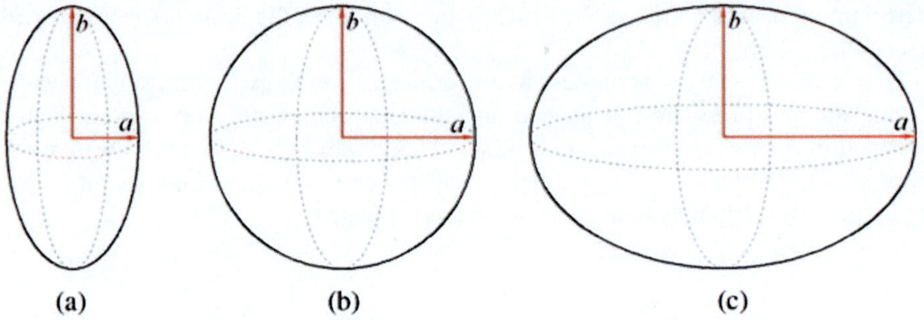

Fig. 8.4 Shape of ellipsoidal shell with different axial length ratios. **a** Prolate ellipsoidal shell; **b** spherical shell; **c** oblate ellipsoidal shell

the absence of a special reference, the ellipsoidal shell mentioned refers to a oblate ellipsoidal shell.

For the ideal ellipsoidal shell under internal pressure, its meridional stress σ_φ and latitudinal stress σ_θ are respectively:

$$\sigma_\varphi = \frac{p}{2t} r_2 \quad \sigma_\theta = \frac{pr_2}{2t}\left(2 - \frac{r_2}{r_1}\right), \tag{8.5}$$

where

$$\begin{aligned} r_1 &= \lambda a k^3 \\ r_2 &= \lambda a k \\ k &= \frac{1}{\sqrt{(\lambda^2 - 1)\sin^2\varphi + 1}} \end{aligned} \tag{8.6}$$

By arranging Eqs. (8.5) and (8.6), the meridional stress σ_φ and latitudinal stress σ_θ of the ellipsoidal shell subjected to internal pressure are obtained.

$$\begin{aligned} \sigma_\varphi &= \frac{pa}{2t} \lambda k \\ \sigma_\theta &= \sigma_\varphi \left(2 - \frac{1}{k^2}\right) \end{aligned} \tag{8.7}$$

Figure 8.5 shows the stress distribution of the ellipsoidal shells with different axial length ratios λ. For an ideal ellipsoidal shell, the meridional stress is always the tensile stress. However, the latitudinal stress will be different when the ellipsoidal shell with different axial length ratios. When $\lambda < 1$, the latitudinal stress is tensile stress, which increases gradually from the pole to the equator. When $\lambda > 1$, the latitudinal stress gradually decreases from the pole to the equator, and when $\lambda = 1$, the latitudinal stresses of the shell are equal everywhere. When $\lambda = \sqrt{2}$, the latitudinal stress is 0 at the equator. When, $\lambda > \sqrt{2}$ the latitudinal stress changes from tensile

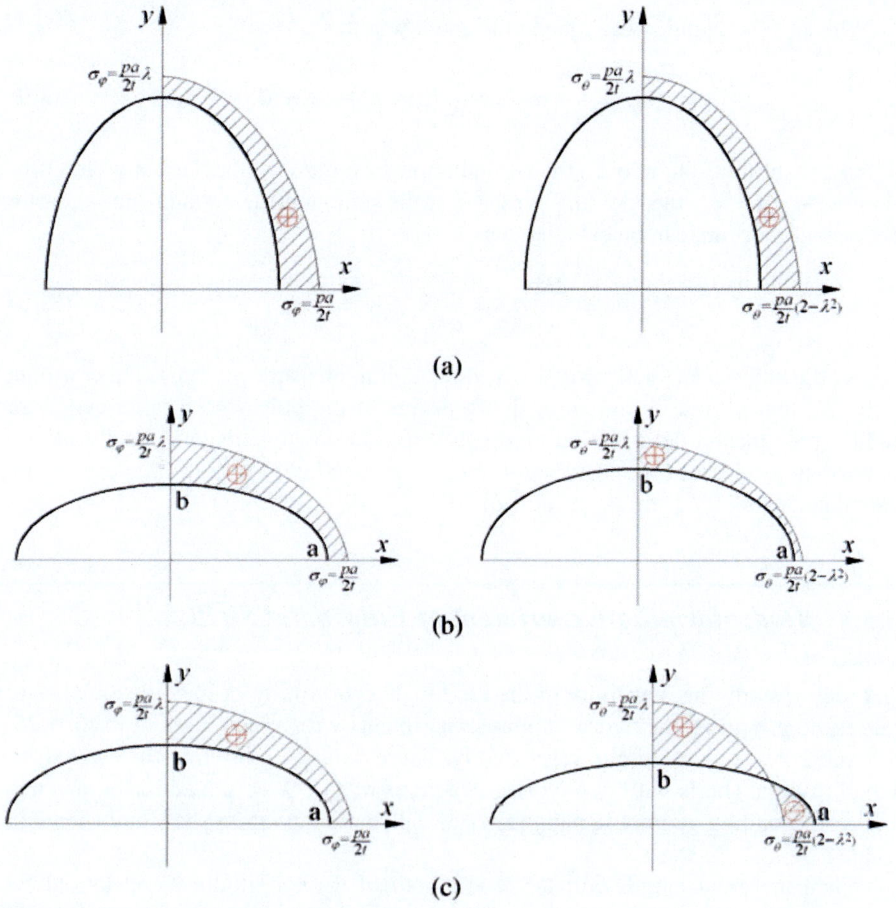

Fig. 8.5 Relationship between stress and axial length ratio of ellipsoidal shell under internal pressure. **a** $\lambda < 1$; **b** $1 < \lambda \leq \sqrt{2}$; **c** $\lambda > \sqrt{2}$

stress to compressive stress at a certain spherical angle, which is defined as the tension–compression stress dividing angle. At this point, the latitudinal stresses in the equatorial zone (the region near the equator) are all compressive stresses, which leads to instability and wrinkling in the hydroforming process of the ellipsoidal shell.

8.3.3 Hydroforming Pressure for Ellipsoidal Shells

The stress state at the pole is

$$\sigma_1 = \sigma_2 = \frac{pa^2}{2bt}, \ \sigma_3 = 0 \tag{8.8}$$

When $\lambda < \sqrt{2}$, the stress state at the equator is

$$\sigma_1 = \frac{pa}{2t}, \sigma_2 = \frac{pa}{2t}\left(2 - \lambda^2\right), \sigma_3 = 0 \tag{8.9}$$

By substituting the above stress calculation formula into the Tresca yield criterion $\sigma_1 - \sigma_3 = \sigma_s$, the yielding pressure at the pole and the equator when plastic deformation occurs can be calculated as

$$p_1 = \frac{2bt}{a^2}\sigma_s = \frac{2t}{a\lambda}\sigma_s, p_2 = \frac{2t}{a}\sigma_s \tag{8.10}$$

For the ellipsoidal shell with $\lambda > 1$, the yielding pressure $p_1 < p_2$. This formula indicates that plastic deformation firstly occurs at the pole with a large curvature radius, then plastic deformation gradually expands downwards, and finally plastic deformation occurs at the weld point in equator. The final forming pressure can be calculated using the formula $p_2 = \frac{2t}{a}\sigma_s$.

8.3.4 Hydroforming Experiment of Ellipsoidal Shell

In order to study the feasibility of the die-less hydroforming of the ellipsoidal shell and its deformation, the hydro-bulging experiments with axial length ratios of 1.25, 1.8 and 2.2 were carried out, respectively. These axial length ratios corresponds to the ellipsoidal shells with two typical structures of $1 < \lambda < \sqrt{2}$ and $\lambda > \sqrt{2}$, and the corresponding geometric parameters of the single-curvature shells are shown in Table 8.3.

When the axial length ratio $\lambda < \sqrt{2}$ both of the longitudinal and latitudinal stresses of the ellipsoidal shell are tensile stresses, thus compression instability and wrinkling will not occur in the bulging process of the ellipsoidal shell with an axial length ratio of $\lambda = 1.25$. It can be gradually expanded under the internal pressure until the required ellipsoidal shell is formed, as shown in Fig. 8.6a.

When the axial length ratio $\lambda > \sqrt{2}$, the latitudinal stress changes from tensile stress to compressive stress at a certain spherical angle, which is defined as the tension–compression stress dividing angle. At this point, the latitudinal stress near the equator is compressive stress, and when the compressive stress reaches a certain

Table 8.3 Geometric parameters of a single-curvature shell

Semi-major axis a/mm	Semi-minor axis b/mm	Axial length ratio λ	The initial thickness t_0/mm	Number of lateral petals N
250	200	1.25	1.5	11
225	125	1.8	0.9	12
500	225	2.2	1.0	12

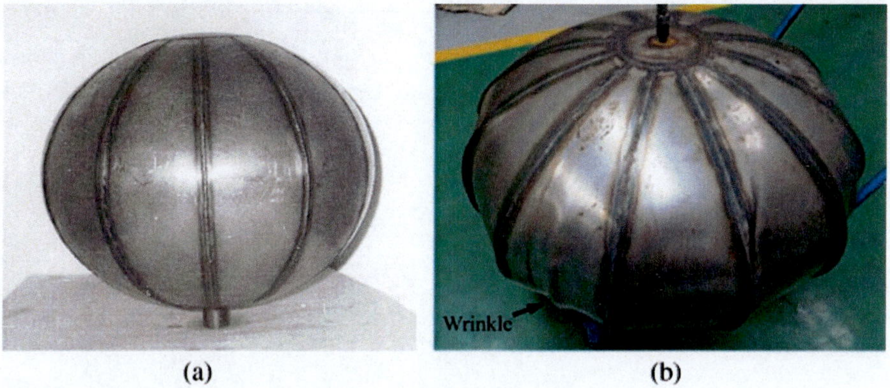

Fig. 8.6 Hydroforming experiment of ellipsoidal shell. **a** Axial length ratio $\lambda = 1.25$; **b** axial length ratio $\lambda = 2.2$

level, it leads to instability and wrinkling in the forming process of ellipsoidal shell. When the axial length ratio is relatively small, such as $\lambda = 1.8$, the long axis begins to contract with the increase of internal pressure in the shell hydroforming process, and wrinkles will appear at the equator at the initial stage of deformation. As the internal pressure continues to increase, the wrinkles are gradually flattened until the long axis no longer contracts, and gradually becomes elongated. The details will be introduced in the next section. With the further increase of the axial length ratio, such as $\lambda = 2.2$, the wrinkling could appear at the initial stage of shell hydroforming. The degree of wrinkling becomes severe as the internal pressure increases, until finally a dead wrinkle that cannot be flattened is formed, as shown in Fig. 8.6b.

8.3.5 Deformation and Wrinkling Behavior in Hydroforming of Ellipsoidal Shell

In order to investigate the variation of shell shape, axial length size and wrinkle shape in the whole bulging process of ellipsoidal shell, the die-less hydroforming experiment of ellipsoidal shell with axial length ratio of $\lambda > \sqrt{2}$ was carried out. The formation, development and disappearance of wrinkles in equatorial zone and the change of shell shape were observed under the condition that the short axis is unconstrained. Finally, the variation of axial length with internal pressure is obtained. Through the variation of axial length, the shell structure before hydroforming can be optimized, so as to guide the forming of ellipsoidal vessels with large diameter and proper axial length ratio.

Figure 8.7 shows the structure of the ellipsoidal shell before hydroforming, which is welded into a whole by 12 lateral petals and 2 polar plates. The long axis of the ellipsoidal shell is 225 mm and the short axis is 125 mm, which corresponds to an

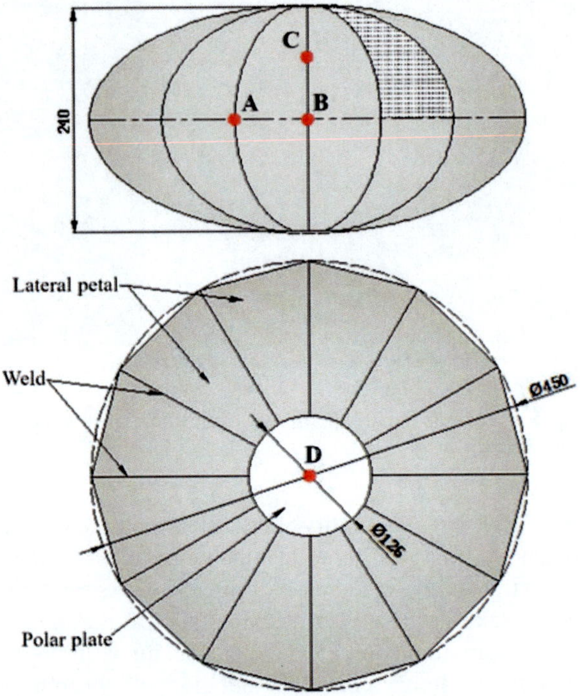

Fig. 8.7 Structure and dimension (mm) of ellipsoidal shell

initial axial length ratio of 1.8. The shell is manufactured with the 304 stainless steel, whose nominal thickness is 0.9 mm (actual thickness is 0.88 mm), yield strength is 313 MPa, tensile strength is 815 MPa, and elongation is 54%. In Fig. 8.9, points A, B, C and D are the locations for deformation measurement, where point A is the weld point on the equatorial surface, point B is the center point of the lateral petal in the equatorial zone, point C is the center point of the lateral petal in the temperate zone, and point D is the pole (Fig. 8.8).

Figure 8.8 shows the shape variation of the ellipsoidal shell with internal pressure during die-less hydroforming process. Figure 8.11 shows the profile changes of ellipsoidal shell under different internal pressures. It can be found that the deformation firstly occurs at the polar plate of ellipsoid shell with the increase of internal pressure, and the short axis begins to elongate. When the internal pressure reaches 1.0 MPa, the short axis reaches the initial design size. At the same time, small wrinkles begin to appear at the weld of equatorial zone due to the continuous increase of the latitudinal compressive stress value in the equatorial belt. When the internal pressure reaches 2.2 MPa, the wrinkling is the most serious. As the internal pressure continues to increase, the long axis begins to contract. When the internal pressure reaches 3.8 MPa, the wrinkles at the equatorial belt disappear, and the contour line of the equatorial plane is consistent with the initial state, showing a regular dodecagon. With the continuous increase of internal pressure, the 12 lateral petals that constitutes the ellipsoidal shell gradually become arcs under the action of internal pressure. When

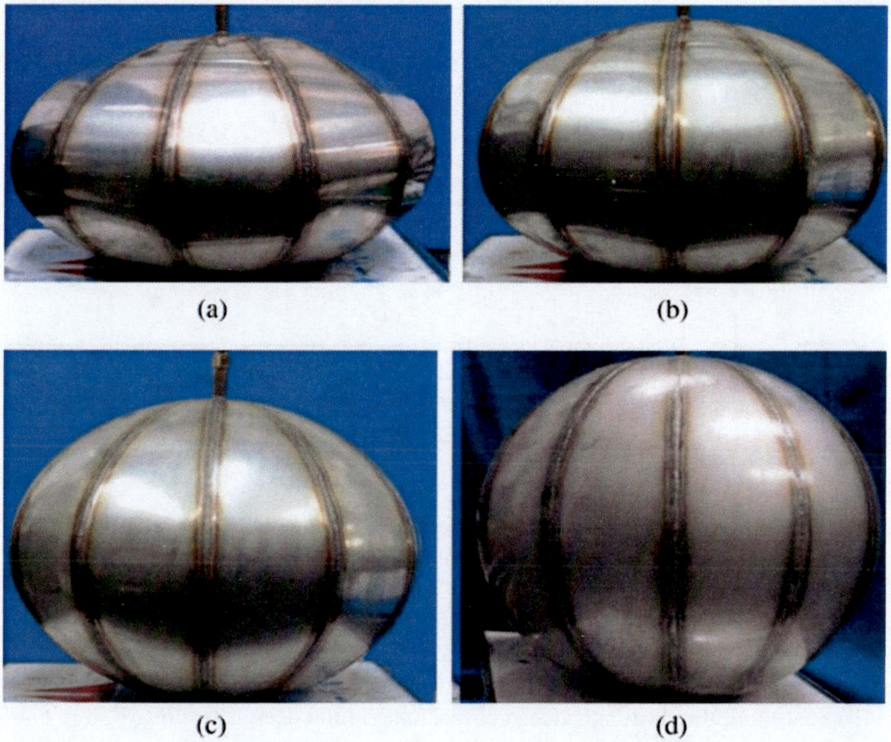

Fig. 8.8 The shape variation of the ellipsoidal shell during die-less hydroforming process. **a** Initial state; **b** 2.2 MPa; **c** 3.8 MPa; **d** 6.8 MPa

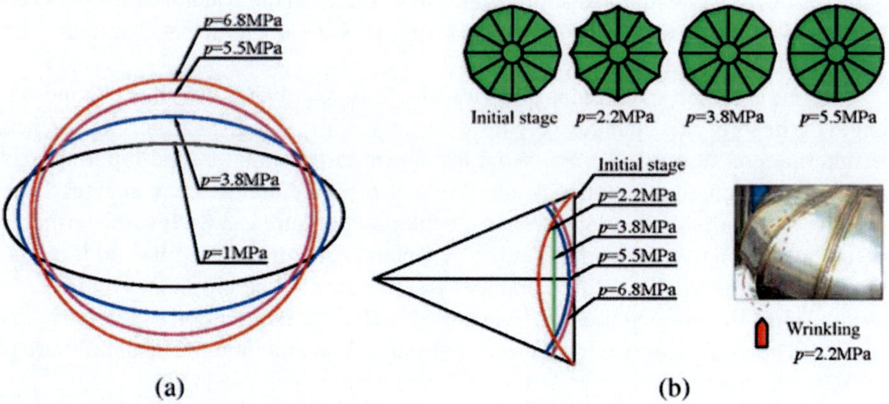

Fig. 8.9 Change of section profile in the die-less hydroforming process of ellipsoidal shell. **a** Meridional cross-section; **b** equatorial plane

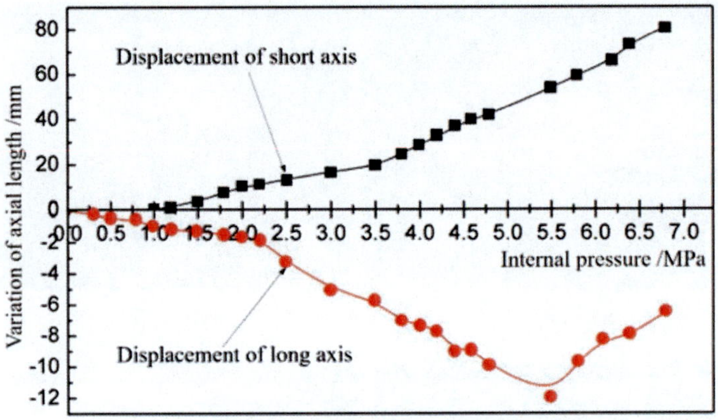

Fig. 8.10 The variation of axial length with internal pressure

the internal pressure reaches 5.5 MPa, the equatorial plane becomes completely arc-shaped, and the long axis changes from continuous contraction to elongation, while the short axis continues to elongate. When the internal pressure is 6.8 MPa, the final shape is approximately spherical, and the actual axial length ratio is 1.06.

Figure 8.10 shows the variation of axial length with internal pressure. Since the height of the initial polar plate is less than the short axis length of the circumscribed ellipsoidal shell, the short axis does not reach the initial design size of 250 mm until the internal pressure increases to 1.0 MPa. With the increase of internal pressure, the plastic deformation develops and the short axis continues to elongate. When the internal pressure is greater than 3.8 MPa, the elongation rate of the short axis increases obviously with the disappearance of wrinkles in the equatorial zone. When the final internal pressure is 6.8 MPa, the short axis is elongated by 80.8 mm and the corresponding elongation is 64.6%.

With the increase of internal pressure, the long axis begins to shorten, and the length of the long axis changes very little during the initial deformation stage. When the internal pressure is 2.2 MPa, wrinkles appear at the equator, and the long axis begins to shorten significantly. As the internal pressure continues to increase, the wrinkles are gradually flattened. When the internal pressure is 3.8 MPa, the wrinkles are completely flattened, and the equatorial plane is again in a regular dodecagon. After that, the internal pressure continues to increase and the long axis continues to shrink. When the internal pressure reaches 5.5 MPa, the contraction of the long axis stops, and gradually turns into an elongated state. When the final internal pressure is 6.8 MPa, the long axis is shortened by 6.5 mm with a length reduction of 2.9%.

Figure 8.11 shows the relationship between volume and internal pressure during the die-less hydroforming process of ellipsoidal shell. At the early stage of deformation, only the polar plate has plastic deformation, and the variation of volume is small. When the internal pressure is within the range of 2.2–3.5 MPa, the volume reduction caused by the contraction of the long axis is almost equal to the volume increment caused by the elongation of the short axis. Therefore, the variation of volume is very

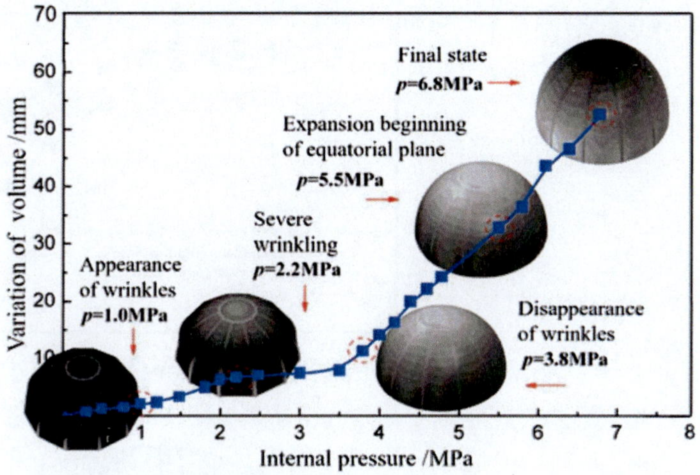

Fig. 8.11 Relationship between volume of ellipsoidal shell and internal pressure

small, which only increases from 6.8 to 8.0%. As the internal pressure continues to increase, the elongation of the short axis is faster even though the long axis is still contracting. Thus the volume of the ellipsoidal shell gradually increases. When the internal pressure reaches 3.8 MPa, the variation of volume is 11.4%, and the increase of volume directly leads to the flattening of the wrinkles.

With the continuous increase of internal pressure, the elongation of the short axis becomes larger, and the pole and temperate zone gradually change into a spherical shell, while the contraction of the equatorial zone is relatively small. In this case, the variation of volume is very significant. When the internal pressure is up to 5.5 MPa, the variation of volume is 32.8%. At this point, the lateral petals become smooth until a perfect circular arc. The subsequent increase in volume and surface area lead to the expansion of the equatorial zone from contraction to expansion. When the final internal pressure is 6.8 MPa, the shell volume increases by 52.7%, which is more than half of the initial volume.

Figure 8.12 shows the variation of thickness with an internal pressure in the hydroforming process of ellipsoidal shell. It is seen clearly that thinning occurs first at the poles and last in the equatorial zone. At the end of the forming, the thinning ratio at the vicinity of weld (point A) and the center of the lateral petal (point B) in the equatorial plane is 13.2% and 14.6%, respectively. Moreover, the thinning ratio at the center of the lateral petal in temperate zone (point C) and at the pole (point D) is 19.8% and 29.4%, respectively. In addition, the contour line of shell thickness distribution when an internal pressure is 6.8 MPa is given in the numerical simulation. It can be seen from the overall thickness distribution that the maximum thinning is located at the pole and the minimum thinning is located near the weld in the equatorial zone. From the pole to the center of the lateral petal in equatorial zone, the maximum thinning ratio decreases from 31.8 to 17%. Moreover, the maximum

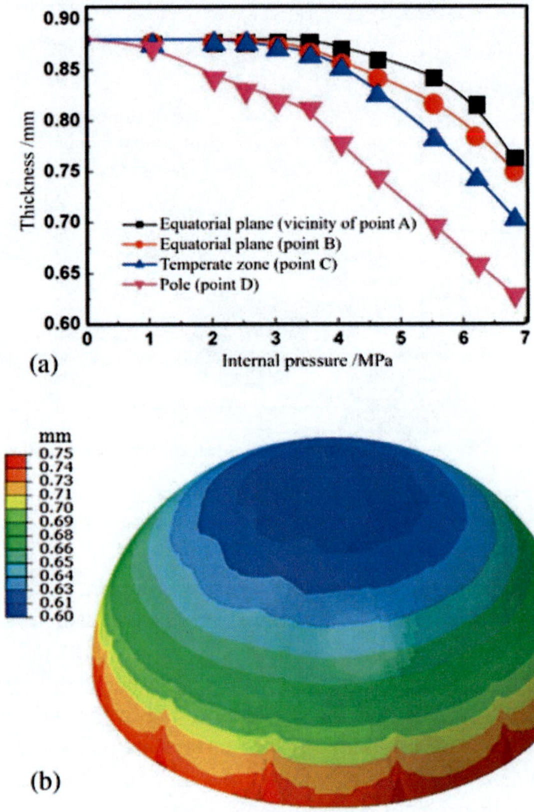

Fig. 8.12 Thickness distribution during hydroforming process of ellipsoidal shell. **a** Experiment; **b** numerical simulation

thinning ratio increases from 14 to 17% from the vicinity of weld in the equatorial zone to the center of lateral petal. The difference of thickness is small.

8.4 Die-Less Hydroforming of Ellipsoidal Shells with Double Generating Lines

8.4.1 Principle of Hydroforming of Ellipsoidal Shell with Double Generating Lines

Through theoretical analysis and a large number of experimental studies, it is proved that in the hydroforming process of ellipsoidal shell, there is a latitudinal compressive stress in the equatorial zone when the axis length ratio $\lambda > \sqrt{2}$, which will cause wrinkling of the shell and failure of forming. However, the ellipsoidal shell can be formed successfully when the axis length ratio is $1 < \lambda < \sqrt{2}$. In order to solve this

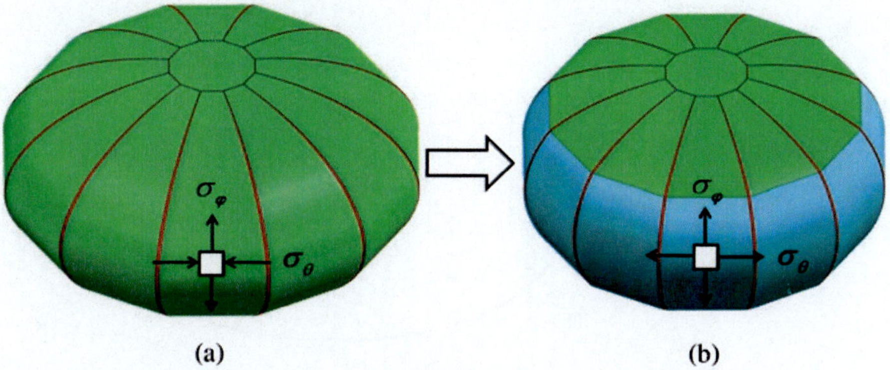

Fig. 8.13 Principle of hydroforming of ellipsoidal shell with double generating line. **a** Ellipsoidal shell with single generating line; **b** ellipsoidal shell with double generating line

problem, researchers have proposed a method that the elongation of short axis is restricted using press and center tube. For the method that using press to restrict the elongation of short axis, only small size ellipsoidal shells can be produced because of the limitation of press table size. For the method that using center tube to restrict the elongation of short axis, the center tube needs to be removed and the polar plates need to be added for secondary hydro-bulging. Moreover, the ability of the center tube to withstand external pressure instability is limited. Therefore, it is also not suitable for making large-diameter ellipsoidal shells.

To resolve the problem that wrinkling occur at the equator of an oblate ellipsoidal shell whose axis length ratio is $\lambda > \sqrt{2}$ during hydroforming, a hydroforming method using an ellipsoidal shell structure with double generating lines as the prefabricated shell has been proposed. The basic idea is that: in order to make the whole shell in the biaxial tensile stress state, the equatorial zone with the latitudinal compressive stress is replaced by a segment of ellipsoidal shell whose axis length ratio is $1 < \lambda < \sqrt{2}$, so the ellipsoidal shell structure with double generating lines can be obtained. The shell has two axis length ratios, that is, a segment of ellipsoidal shell whose axis length ratio is $1 < \lambda < \sqrt{2}$ in the equatorial zone and another segment of ellipsoidal shell whose axis length ratio is $\lambda > \sqrt{2}$, as shown in Fig. 8.13.

8.4.2 *Structural Design of Ellipsoidal Shell with Double Generating Lines*

For the ellipsoidal shell with axis length ratio $\lambda > \sqrt{2}$, there exists a tension–compression stress dividing angle of latitudinal stress α, and $\tan\alpha = y/x$, as shown in Fig. 8.14. And the latitudinal stress at this point is that $\sigma_\theta = 0$. Combining Eqs. (8.6) and (8.7), the relationship between the axis length ratio λ and the rotation angle φ can be derived as Eq. (8.11).

Fig. 8.14 Tension–compression stress dividing angle of latitudinal stress of ellipsoidal shell

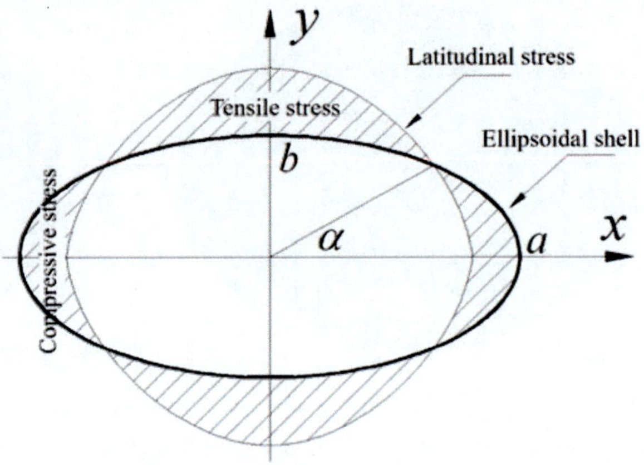

$$\sin^2 \varphi = \frac{1}{\lambda^2 - 1} \tag{8.11}$$

It is known that $\sin \varphi = x/r_2 = x/\lambda ak$. Hence, by substituting it into Eq. (8.11), Eq. (8.12) can be obtained.

$$\frac{x^2}{a^2} = \frac{\lambda^2}{2(\lambda^2 - 1)} \tag{8.12}$$

Any point with the coordinate (*x*, *y*) on the ellipsoid is satisfied the geometric equation for ellipsoidal shells as Eq. (8.13).

$$\frac{x^2}{a^2} + \frac{y^2}{b^2} = 1 \tag{8.13}$$

Based on Eqs. (8.12) and (8.13), the relationship between tension–compression dividing globe angle α corresponding to the latitudinal stress and axis length ratio λ can be obtained as Eq. (8.14).

$$\alpha = arc \tan \frac{\sqrt{\lambda^2 - 2}}{\lambda^2} \tag{8.14}$$

According to Eq. (8.14), it can be seen that, with different initial axis length ratios λ, tension–compression stress dividing angle α_0 of latitudinal stress are also different. Figure 8.15 shows the relationship between tension–compression stress dividing angle α_0 and axis length ratio λ. It can be seen that with the increase of axis length ratio λ, the tension–compression stress dividing angle α_0 increases first and then decreases.

Derivative operation on Eq. (8.14), the derivative function is expressed in Eq. (8.15).

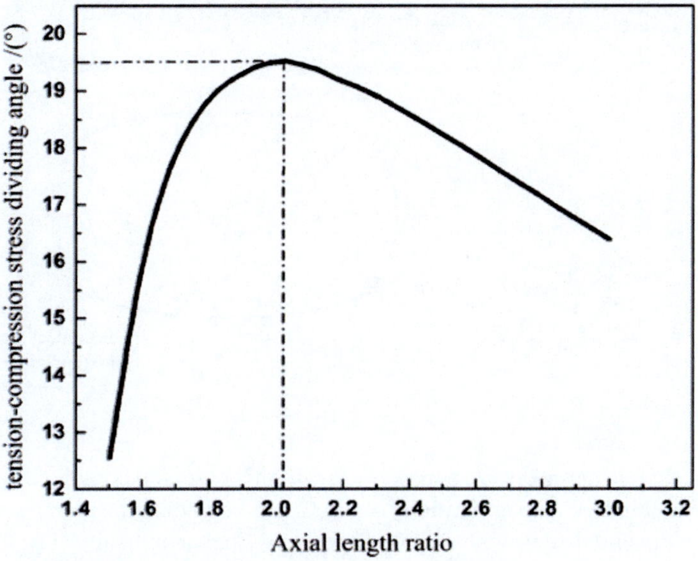

Fig. 8.15 The relationship between tension–compression stress dividing angle and axis length ratio

$$\alpha_0' = f'(\lambda) = \frac{\lambda(4 - \lambda^2)}{(\lambda^4 + \lambda^2 - 2)\sqrt{\lambda^2 - 2}} \tag{8.15}$$

According to the derived function $\alpha_0' = 0$, it is shown that when the initial axial length ratio is $\lambda = 2.0$, there is a maximum value of 19.5° for the tension–compression stress dividing angle. In other words, for the oblate ellipsoid shells with arbitrary axial length ratio of $\lambda > \sqrt{2}$, their maximum tension–compression stress dividing angle of the latitudinal stress will not exceed 19.5°.

In order to avoid the wrinkling in the equatorial zone, the shell segment between the tension–compression stress dividing angle and the equator line that under latitudinal compressive stress is replaced by a segment of ellipsoidal shell whose initial axis length ratio is $1 < \lambda < \sqrt{2}$, so that an ellipsoidal shell with double generating lines is obtained. The schematic diagram of its structural design is shown in the Fig. 8.16. The combined ellipsoidal shell with double generating lines refers to the ellipsoidal shell composed of two segments with different axial length ratios, in which the first ellipsoidal shell with an axis length ratio of $\lambda > \sqrt{2}$ constitutes the polar zone and the temperate zone while the second ellipsoidal shell with an axial length ratio of $\lambda < \sqrt{2}$ constitutes the equator zone. In order to ensure that the latitudinal stress of the first ellipsoidal shell is fully tensile stress, the actual dividing point (line) of the designed two-segment lateral petals should be higher than the theoretical dividing point α with 5°. It is assumed that the globe angle corresponding to the dividing point of the two segments of the lateral petals is θ, then $\theta = \alpha + 5°$. The length of the long axis of the ellipsoidal shell with double generating lines is a_2 while the length of the short axis is b_1, thus the effective axis ratio of the combined

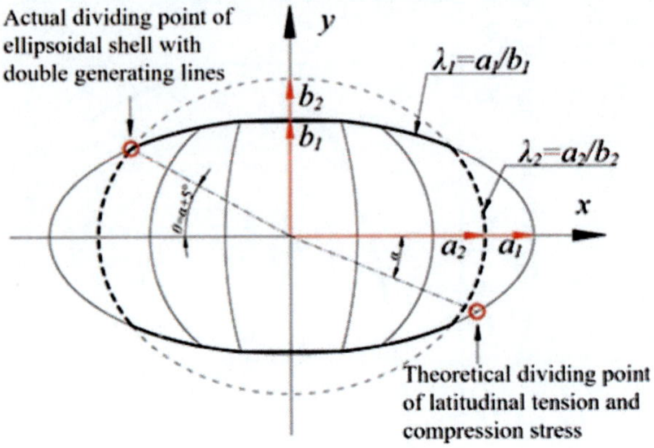

Fig. 8.16 Structure design of the ellipsoidal shell with double generating lines

ellipsoidal shell is defined as $\overline{\lambda}$, and $\overline{\lambda}=a_2/b_1$. In this case, the combined ellipsoidal shell with an initial axis length ratio of $\overline{\lambda} > \sqrt{2}$ can be obtained, so as to ensure that the whole ellipsoidal shell with double generating lines is not subject to latitudinal compression stress under the action of internal pressure. Further, the buckling and wrinkling during forming can be avoided.

8.4.3 Hydroforming Experiment of Ellipsoidal Shell with Double Generating Lines

The structure design of the ellipsoidal shell with double generating lines is shown in Fig. 8.17, which is a closed shell welded by 12 lateral petals and 2 polar plates. The globe angle corresponding to dividing point is $\theta = 22°$. An ellipsoidal shell with an axis length ratio of 1.6 is used from the pole to the dividing point, and an ellipsoidal shell with an axis length ratio of 1.2 is used from the dividing point to the equator zone. The main dimension parameters of the ellipsoidal shell are shown in Table 8.4, in which the diameter of equatorial plane and the polar plate is 420 mm and 104 mm, respectively. The shell material is 304 stainless steel sheet with a thickness of 1.1 mm, the yield strength is 325 MPa, tensile strength is 810 MPa, and the elongation is 53%. In Fig. 8.17, A, B, C, D and E are the locations of deformation measurement points, in which point A is the pole, point B and point C are the points near the dividing line, point D is the center of the lateral petal on the equatorial plane, and point E is the point on the weld of the equatorial plane.

Figure 8.18 shows the shape variation of the ellipsoidal shell with double generating lines with internal pressure in the experiment. Figure 8.19 shows the cross-sectional profile variation of the ellipsoidal shell with double generating lines under different internal pressures in experiment. When the internal pressure is applied, the deformation firstly occurs on the polar plate and the short axis was elongated. When the internal pressure is less than 1.0 MPa, deformation does not occur in all

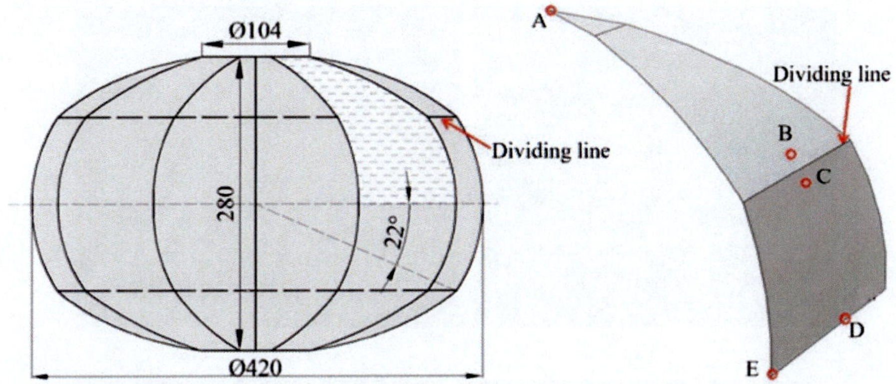

Fig. 8.17 Structure size (mm) and measurement points of the ellipsoidal shell with double generating lines

Table 8.4 Structural parameters of ellipsoidal shell with double generating lines

a_1	b_1	λ_1	a_2	b_2	λ_2	$\overline{\lambda}$
230 mm	143 mm	1.60	210 mm	175 mm	1.20	1.47

areas except the plate area. When the internal pressure is 1.5 MPa, plastic deformation would be expanded to the dividing line and its vicinity, where the shells with different curvatures is gradually flattened. When the internal pressure is greater than 2.5 MPa, the plastic deformation is expanded to the equatorial zone, and the cross-sectional shape of the equatorial zone is changed from polygon to a perfect circle gradually. After that the dividing line of the combined ellipsoidal shell with double generating lines disappear completely as the internal pressure is increased to 3.0 MPa, and the whole shell has been formed as an ellipsoidal shell. As the internal pressure continues to increase to 3.5 MPa, the equatorial zone is completely arc-formed under the action of biaxial tensile stress, and the required ellipsoidal shell is obtained. In the hydroforming process, there are no wrinkles in the equatorial zone, which proves that the ellipsoidal shell with double generating lines can effectively suppress the occurrence of wrinkling.

Figure 8.20 shows the variation of axis length during the hydroforming process of ellipsoidal shell with double generating lines. When the internal pressure is greater than 1.0 MPa, the short axis begins to elongate. When the internal pressure is greater than 2.5 MPa, the length of the long axis shortens significantly. When the final internal pressure reaches 3.5 MPa, the final length of the long axis is 208.9 mm which corresponds to a shrinking percentage of 0.5%, while the final size of the short axis is 157.4 mm which corresponds to an elongation of 10.1%. Moreover, the out-of-roundness of the equatorial plane is 1% after hydroforming.

Figure 8.21 shows the thickness variation of the ellipsoidal shell with double generating lines. It can be found from this figure that the thickness at the pole (point

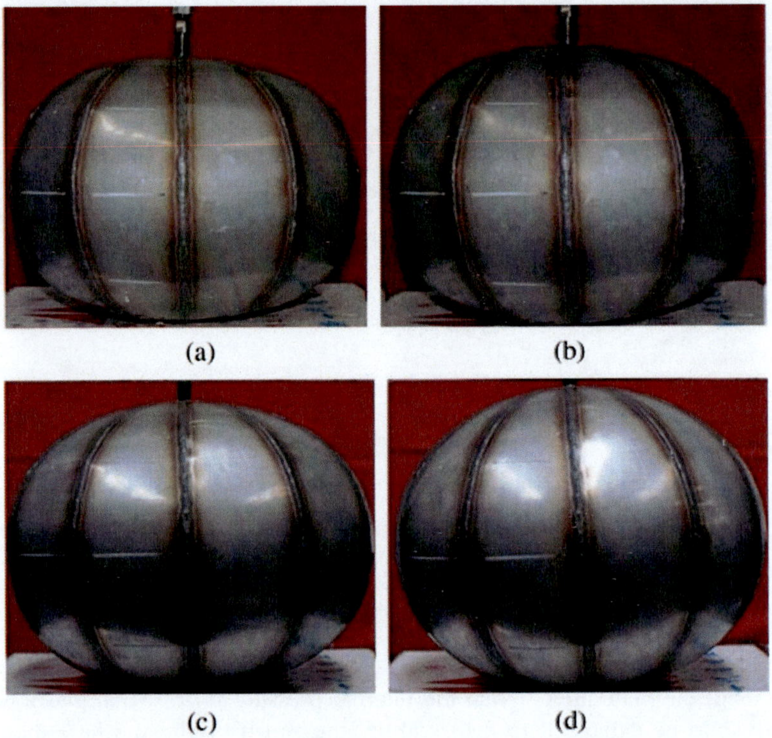

Fig. 8.18 Shape variation of the ellipsoidal shell with double generating lines in hydroforming process. **a** Initial state; **b** 1.5 MPa; **c** 2.5 MPa; **d** 3.5 MPa

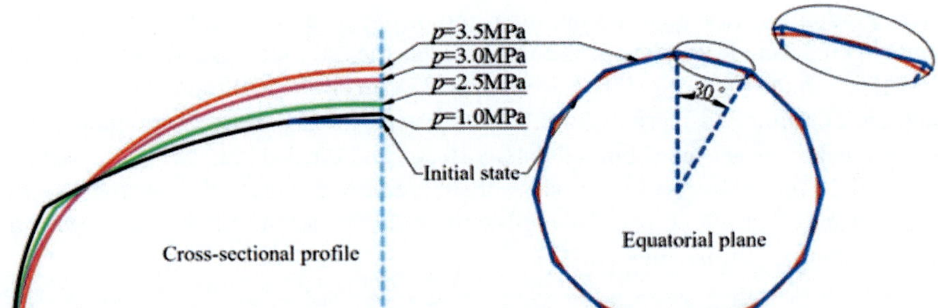

Fig. 8.19 Schematic diagram of cross-sectional profile variation of the ellipsoidal shell with double generating lines

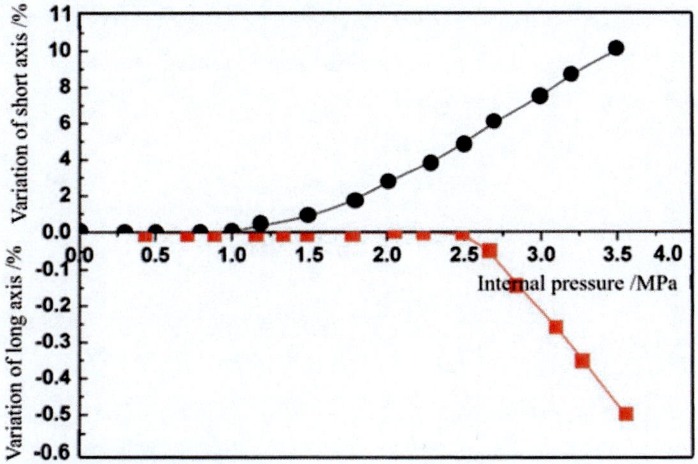

Fig. 8.20 Variation of axis length of the ellipsoidal shell with double generating lines

A) is 1.049 mm, and the thinning ratio is 4.6%, the thinning ratio at the vicinity of dividing line of lateral petals (points B and C) is 2.2%, the thinning ratio at the equatorial plane (point D) is 1%。In general, the thickness distribution is uniform after the hydroforming of ellipsoidal shell with double generating lines. Moreover, the thickness distribution contour is given by numerical simulation. For the ellipsoidal shell between the dividing line and the equator zone, the thickness near the weld line is basically unchanged, and the thickness at the vicinity of dividing line and center of the lateral petals is relatively large, but is only 1.8%. For the ellipsoidal shell between the pole and the dividing line, the maximum thinning of 7.3% occurs at the pole.

8.4.4 Variation of Curvature Radius During Hydroforming of Ellipsoidal Shell with Double Generating Lines

As mentioned above, the forming pressure for the ellipsoidal shell with double generating lines cannot be lower than the yielding pressure of the equatorial zone. In addition, under the action of internal pressure for the forming, the final shell shall be an ellipsoidal shell with single generating line.

According to the parametric Eq. (8.6) of the ellipsoidal shell, the curvature radius for both of the first and the second ellipsoidal shell on dividing angle can be expressed as Eq. (8.16):

$$
\begin{aligned}
R_1 &= \lambda_1 a_1 / \sqrt{(\lambda_1^2 - 1)\sin^2\phi_1 + 1} \\
R_2 &= \lambda_2 a_2 / \sqrt{(\lambda_2^2 - 1)\sin^2\phi_2 + 1}
\end{aligned}
\tag{8.16}
$$

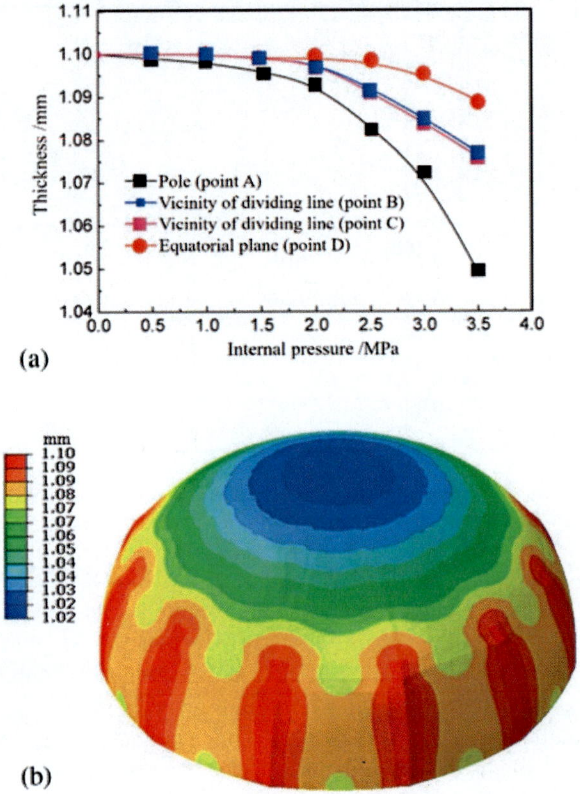

Fig. 8.21 Thickness variation of the ellipsoidal shell with double generating lines. **a** Experiment; **b** numerical simulation

For the point (x_0, y_0) on the dividing angle, $\sin\phi_1 = x_0/R_1$ and $\sin\phi_2 = x_0/R_2$, hence the relationship between curvature radius at the dividing angle and axis lengths can be written as follows.

$$R_1^2 = \frac{a_1^4}{b_1^2} - \frac{a_1^2 x_0^2}{b_1^2} + x_0^2$$
$$R_2^2 = \frac{a_2^4}{b_2^2} - \frac{a_2^2 x_0^2}{b_2^2} + x_0^2 \tag{8.17}$$

During the hydroforming of ellipsoidal shell with double generating lines, the short axis b_1 of the first ellipsoidal shell, the long axis a_2 of the second ellipsoidal shell, and the coordinate (x_0, y_0) of the dividing angle are easy to be recorded with the increasing of internal pressure. Hence, the curvature radius at the dividing angle can be transformed as follows

$$R_1 = \sqrt{\frac{x_0^4 y_0^2 + x_0^2 (b_1^2 - y_0^2)^2}{(b_1^2 - y_0^2)^2}}$$

$$R_2 = \sqrt{\frac{(a_2^2 - x_0^2)^2 + x_0^2 y_0^2}{y_0^2}} \tag{8.18}$$

During hydroforming process, the theoretical geometric criterion of ellipsoidal shell with double generating lines changing into single generating line can be written as follows: curvature radius on the dividing angle $R_1 = R_2$. Figure 8.22 shows the curvature radius variation on dividing angle with internal pressure during hydroforming of ellipsoidal shell with double generating lines. It can be seen that for the ellipsoidal shell with double generating lines and initial axis length ratio of $\lambda = 1.5$, the theoretical criterion of ellipsoidal shell with double generating lines changing into single generating line is satisfied when internal pressure P_s is applied. In the subsequent shell deformation process, the curvature radius of the first and second ellipsoidal shell are exactly the same at the dividing angle.

Figure 8.23 shows a schematic diagram of the curvature radius variation with the internal pressure at the dividing angle during the hydroforming process of the ellipsoidal shell with double generating lines. At the early stage of deformation, there are obvious differences in the curvature radius at the dividing angle of the ellipsoidal shell with double generating lines. This is because that the curvature radius of the first ellipsoidal shell at the dividing angle is much larger than that of the second ellipsoidal shell. With the increase of internal pressure, the curvature radius of the first ellipsoidal shell at the dividing angle decreases significantly, while the curvature radius of the second ellipsoidal shell first increases and then decreases. This is because that during the deformation of the first ellipsoidal shell, the second ellipsoidal shell is slightly bent inwards, which leads to the first increase of the curvature radius. In the middle stage of deformation, the curvature radius of the first ellipsoidal shell at the dividing angle continues to decrease, while the curvature radius of the second ellipsoidal shell at the dividing angle begins to decrease. But the reduction rate is less than the former,

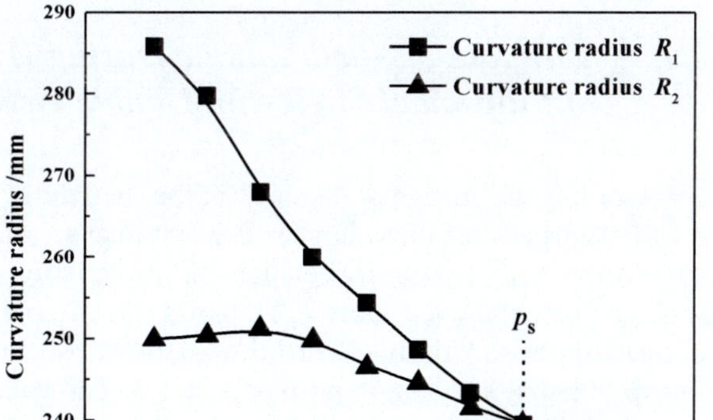

Fig. 8.22 Curvature radius variation on dividing angle during hydroforming of ellipsoidal shell with double generating lines

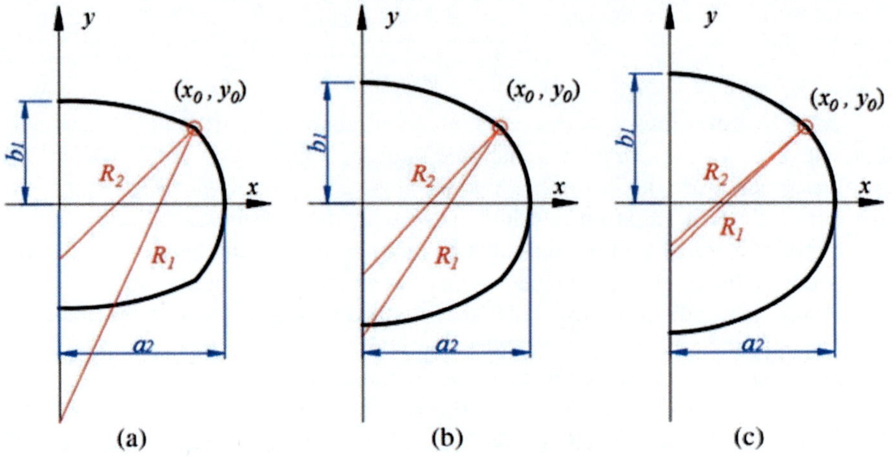

Fig. 8.23 Curvature radius variation at the dividing angle of the ellipsoidal shell with double generating lines. **a** Early stage; **b** middle stage; **c** last stage

and the two curvature radii are close to each other. At the last deformation stage, the two curvature radii at the dividing angle merge into one, and the whole shell changes from the initial ellipsoidal shell with double generating lines to an ellipsoidal shell with single generating line. At this time, different target ellipsoidal shells can be obtained by adjusting the forming internal pressure.

In conclusion, when the internal pressure is not less than $1.1p_s$, not only the initial yield occurs at the equator can be satisfied during the hydroforming of the ellipsoidal shell with double generating lines, but also the target ellipsoidal shell with single generating line can be obtained.

8.4.5 Variation of Shell Volume During Hydroforming of Ellipsoidal Shell with Double Generating Lines

By recording the change of the mass of the shell during the deformation process, the relationship between the volume and the internal pressure can be obtained, and consequently the relationship between the volume variation and the axis length ratio of the shell can be obtained. Figure 8.24 shows the volume variation with the increasing of internal pressure during hydroforming of ellipsoidal shell with double generating lines and initial axis length ratio of $\lambda = 1.5$. The volume variation is exponentially increased with the increase of internal pressure. At the early stage of deformation ($p < 0.5ps$), the volume variation is not obvious, but there is a rapid increase of internal pressure. At the last stage of deformation ($ps < p < 1.5ps$), as the whole ellipsoidal shell enters the yield state, the plastic deformation of the shell increases, and the volume increases significantly. In this stage, the volume variation is much

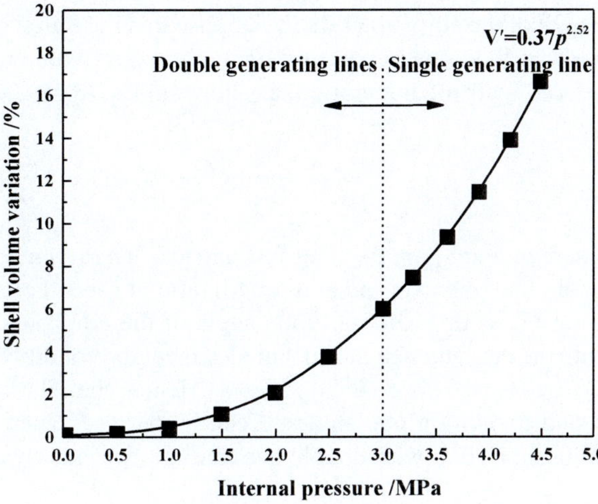

Fig. 8.24 Volume variation with internal pressure during hydroforming of ellipsoidal shell with double generating lines

higher than the linear growth. With each increase of 0.1 MPa in internal pressure, the liquid filling volume will be larger and the filling time will be longer. After further analysis of the second derivative of the volume variation, it can be found that in the pressure range shown in Fig. 8.24, the larger the initial axis length ratio is, the larger the volume variation is under the same pressure, the more obvious the shell volume variation is, and the larger amount of liquid needed during the hydroforming process. When the forming pressure is $1.5ps$, the volume variation for the three shells with initial axis length ratios of 1.5, 1.7 and 2.2 are 16.6%, 26.1% and 41.0%, respectively.

By studying the volume variation with internal pressure during the hydroforming process of ellipsoidal shell with double generating lines, the volume variation required for the forming of shells with different target sizes can be obtained. In this case, the forming accuracy of shell in the experiment process can be effectively controlled by controlling the filling amount of liquid in the hydroforming process. However, if the ellipsoidal shell with the same size is obtained by designing different prefabricated shell structures before hydroforming, the required forming pressure and the volume of the filling liquid are also different. When the initial axis length ratio of the prefabricated shell before hydroforming is 1.5, 1.7 and 2.2, respectively, and the final axis length ratio of the target shell is 1.30, the required forming pressure is $1.1Ps$, $1.3Ps$ and $1.4Ps$, respectively. The corresponding volume variations are 8.5%, 18.5% and 37.5%, respectively. The larger the axis length ratio of the target shell, the larger the corresponding initial axis length ratio of the prefabricated shell before hydroforming.

The shell structure before hydroforming is assumed to be an ideal ellipsoidal shell, then the volume can be expressed by the following functional relation: $V = f(a, \lambda)$. Moreover, it can be seen from the axis length variation in the experimental process that the shrinkage of long axis in hydroforming is very small, and can be considered unchanged. Therefore, the expression of volume variation can be written as $V' = f(\lambda)$, which indicates that the volume variation V' is only related to the axis length ratio λ of the ellipsoidal shell. Suppose that the initial structure is an ideal ellipsoidal shell with double generating lines, the liquid volume before hydroforming can be derived by triple integration, as shown in Eq. (8.19).

$$V_0 = \frac{4\pi a_1^3}{3\lambda_1}(1 - \cos\phi_1) + \frac{4\pi a_2^3}{3\lambda_2}\cos\phi_2 \tag{8.19}$$

where a_1 and λ_1 are the long axis and axis length ratio of first ellipsoidal shell, a_2 and λ_2 are the long axis and axis length ratio of second ellipsoidal shell. For an arbitrary point (x, y) on dividing globe angle of the ellipsoidal shell, it is not only satisfied with the curvature equation, but also meet the conditions that $\tan\alpha_0 = y/x$, $\sin\phi_1 = x_0/\lambda_1 a_1 k_1$ and $\sin\phi_2 = x_0/\lambda_2 a_2 k_2$. Hence, the relationship between rotation angle ϕ and dividing globe angle α_0 can be derived. Then the volume expression of an ellipsoidal shell with double generating lines is derived.

$$V_0 = \frac{4\pi a_1^3}{3\lambda_1}\left(1 - \sqrt{\frac{\lambda_1^4}{\cot^2\alpha_0 + \lambda_1^4}}\right) + \frac{4\pi a_2^3}{3\lambda_2}\sqrt{\frac{\lambda_2^4}{\cot^2\alpha_0 + \lambda_2^4}} \tag{8.20}$$

In the hydroforming process, as the generatrix of the first ellipsoidal shell and the second ellipsoidal shell become consistent gradually, a target ellipsoidal shell with single generating line is obtained, and the volume can be written as follows.

$$V = \frac{4\pi a_2^3}{3\lambda_F} \tag{8.21}$$

where, λ_F is the final axis length ratio.

According to the Eqs. (8.20) and (8.21), the volume variation can be obtained as Eq. (8.22).

$$V' = \frac{V - V_0}{V_0} = \frac{1}{\frac{\lambda_F}{\lambda_1}\left(1 - \sqrt{\frac{\lambda_1^4}{\cot^2\alpha_0 + \lambda_1^4}}\right)\left(\frac{2\lambda_1^4 - 2\lambda_1^2}{\lambda_1^4 + \lambda_1^2\lambda_2^2 - 2\lambda_2^2}\right)^{3/2} + \frac{\lambda_F}{\lambda_2}\sqrt{\frac{\lambda_2^4}{\cot^2\alpha_0 + \lambda_2^4}}} - 1 \tag{8.22}$$

It is obvious that the volume variation in the hydroforming process of the shell depends on the axis length ratio of the shell before and after deformation. According to Eqs. (8.19) and (8.21), the relationship between the volume variation of the ideal ellipsoidal shell and the initial and final axis length ratio can be calculated, as shown in Fig. 8.25. The larger the initial axis length ratio, the larger the volume variation. The smaller the final axis length ratio, the greater the volume variation. The volume variation obtained by theoretical calculation has a good agreement with the experimental results.

The theoretical calculation and experimental results of the shell volume variation show that when the shell structure before hydroforming is fixed, the shell volume variation during hydroforming has a one-to-one corresponding relation with the axis

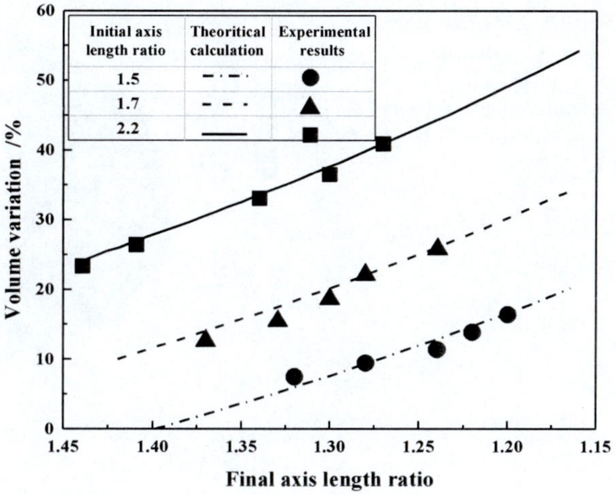

Fig. 8.25 Volume variation with axis length ratio of ellipsoidal shell with double generating lines

length ratio of shell. That is to say, by studying the volume variation with internal pressure during the hydroforming process of ellipsoidal shell with double generating lines, the volume variation required for the forming of shells with different target sizes can be obtained. Based on this, the shape of the final ellipsoidal shell can be controlled by controlling the filling amount of liquid in the hydroforming process.

8.4.6 Variation of Stress During Hydroforming of Ellipsoidal Shell with Double Generating Lines

With the increase of the internal pressure, plastic deformation takes place in the dividing line of the double generating lines and gradually extends to the poles and the equatorial zone. At the same time, the plastic deformation zone at the polar plate develops toward to the dividing line of the double generating lines along the direction of the weld. Then the equatorial belt enters the plastic deformation zone, and the equatorial plane gradually arcs under the action of internal pressure. As the internal pressure continues to increase, the plastic deformation spread throughout the entire ellipsoidal shell, and the dividing line of double generating lines disappears. At this moment, the ellipsoidal shell reaches the design size, as shown in Fig. 8.26b.

Although the dividing line area enters the plastic deformation zone earlier, the meridional stresses of ellipsoidal shells on both sides of the dividing line generated under the same internal pressure are different because of their different curvature radii. And the dividing line that has entered the plastic deformation zone is a strong constraint relative to other regions. Because the elongation of the short axis is slow at the initial stage of deformation, hence the pulling force on the meridional stress is small. As a result, there is a compression effect in a certain area on both sides of

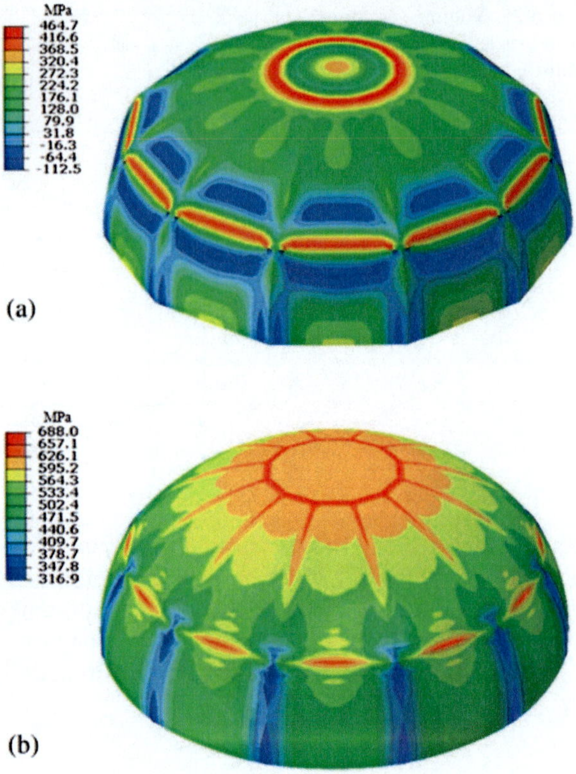

Fig. 8.26 Meridional stress distribution for the ellipsoidal shell with double generating lines. **a** Early stage (1.5 MPa); **b** last stage (3.5 MPa)

the dividing line. Figure 8.26 shows the meridional stress distribution in the hydroforming process of the ellipsoidal shell with double generating lines. In the early stage of deformation, this compression effect will increase, but will not cause instability. With the increase of the internal pressure, the elongation of the short axis exerts a bigger pulling force on the meridional stress. The size of plastic deformation zone is continuously enlarged, and the stress around the dividing line gradually turns into tensile stress. As the deformation continues, the dividing line of the double generating lines gradually disappears under the pull of the longitudinal stress, and the ellipsoidal shell with designed single axis length ratio is obtained.

The latitudinal stresses of ellipsoidal shells on both sides of the dividing line generated under the same internal pressure are also different because of their different curvature radii. And the dividing line that has entered the plastic deformation zone is a strong constraint relative to other regions. The deformation on both sides of the dividing line in latitudinal direction needs to be coordinated and eventually converged. In the case of gradual elongation of the short axis, the plane contraction along the latitude direction of the ellipsoidal shell is driven. However, it is difficult to be driven to contract because the dividing line enters yield hardening early. Therefore, in the early stage of deformation, it can cause the area near the dividing line to shrink and produce compressive stress, as shown in Fig. 8.26a. But this compressive stress cannot cause instability. As the plastic deformation zone is enlarged toward to both sides of the dividing line, the compressive stress value gradually decreases and changes to the tensile stress, and the lateral petals are driven to arc gradually under the action of internal pressure. Finally, an ellipsoidal shell with single axis length ratio of design size is obtained, as shown in Fig. 8.26b.

In general, in the hydroforming process of ellipsoidal shell with double generating lines, although both meridional stress and latitudinal stress have existed as compressive states, the value of compressive stress is small, which is not enough to cause instability and wrinkling of the shell. Therefore, no wrinkles appear in the deformation process of the ellipsoidal shell with double generating lines. At the same time, an ellipsoidal shell with single axis length ratio of design size can be obtained.

8.5 Die-Less Hydroforming of Prolate Ellipsoidal Shells with Double Generating Lines

For the shell subjected to internal pressure, deformation always occurs first in the position and direction with large curvature radius, and the curvature radius is automatically decreased. If the curvature radius is small, it will be increased passively, and the deformation occurs later or even does not occur in this position. The deformation characteristics of prolate ellipsoidal shells with axis length ratio less than 1 are significantly different from those oblate ellipsoidal shells with axis length ratio greater than 1. In the hydroforming of prolate ellipsoidal shell, the polar zone is difficult to form. For this reason, using the method of overcoming the wrinkling at the equator for reference, the prolate ellipsoid shell with double generating lines is proposed as a prefabricated shell before hydroforming to overcome the difficulty of deformation in the polar zone.

According to the stress distribution characteristics of the above ellipsoidal shells, in order to decrease the initial yielding pressure at the polar zone and achieve uniform deformation everywhere, a prolate ellipsoidal shell structure with double generating lines is proposed, as shown in Fig. 8.27. The polar zone and its vicinity that is difficult to deform are composed of a prolate ellipsoidal shell with an initial axis length ratio of $\lambda = 0.9$, which is defined as the first prolate ellipsoidal shell. The rest is composed of another prolate ellipsoidal shell with the initial axis length ratio of $\lambda = 0.5$, which is defined as the second prolate ellipsoidal shell. In this case, a prolate ellipsoidal shell with initial axis length ratio of $\lambda = 0.7$ can be obtained. For this prolate ellipsoidal shell with double generating lines, its short axis and long axis are a_2 and b_1 respectively, which corresponds to an initial axis length ratio of $\lambda = a_2/b_1$.

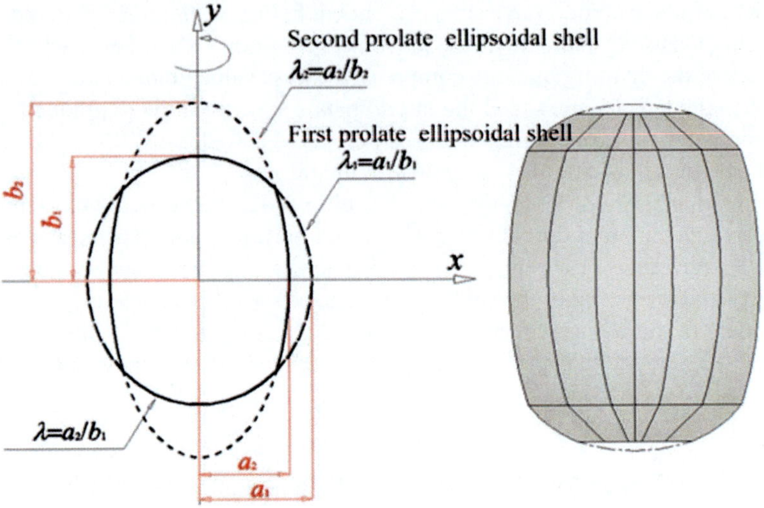

Fig. 8.27 Structure design of the prolate ellipsoidal shell with double generating lines

8.5.1 Hydroforming Experiment of Prolate Ellipsoidal Shell with Double Generating Lines

The prefabricated shell structure before hydroforming is composed of 12 lateral petals and two polar plates, and the polyhedral shell is inscribed in the ideal prolate ellipsoidal shell with double generating lines.

Table 8.5 shows the structural parameters of a prolate ellipsoidal shell with double generating lines and an initial axis length ratio of 0.71. The short axis length and long axis length are 125 mm and 175 mm, respectively. The diameter of the polar plate is 90 mm, which corresponds to a globe angle of 75° and a height of 167.5 mm. The shell is made of stainless steel SUS304 with an initial thickness of 1.1 mm. According to Eq. (8.26), theoretically, the yielding pressures of the equatorial zone and the polar zone are 3.5 MPa and 4.5 MPa, respectively.

Figure 8.28 shows the shape variation with internal pressure for the ellipsoidal shell with double with generating lines. Figure 8.28 shows the cross-sectional profile variation of the prolate ellipsoidal shell under different internal pressures. In the early stage of deformation ($p \leq 4$ MPa), the deformation first occurs in the equatorial zone. The lateral petals are hydro-bulged outwards under the action of internal pressure, and the dihedral angle has been spread out soon to a smooth arc. The polar plate is

Table 8.5 Dimensions of prolate ellipsoidal shell structure with double generating lines

a_1/mm	b_1/mm	λ_1	a_2/mm	b_2/mm	λ_2	$\lambda = a_2/b_1$	H/mm
160	175	0.92	125	250	0.50	0.71	167.5

convex under the action of internal pressure, and the initial dihedral angle along the meridional direction is slightly spread out. When the internal pressure is 3.5 MPa, the equatorial belt begins to arc gradually and the out-of-roundness decreases. At the middle stage of deformation (4 MPa $< p \leq$ 5 MPa), the plastic deformation starts to occur in the polar plate zone and extends to the dividing angle of the prolate ellipsoidal shell with double generating lines. Under the action of internal pressure, the shell segment in the polar zone is gradually deformed toward a smooth arc, and the dividing angle is gradually spread out. At the same time, the equatorial plane expands as a whole. When the internal pressure is 4.5 MPa, the long axis reaches the initial design size, and the polar zone and its adjacent areas become into smooth arcs in both meridional and latitudinal directions. At the final stage of deformation ($p >$ 5 MPa), the initial dihedral angle along the meridional direction has been completely spread out to a smooth arc, which indicates that the prolate ellipsoidal shell with double generating lines has been changed into a prolate ellipsoidal sheel with single generating line. In this process, both the long axis and the short axis are elongated compared with the middle stage of deformation. With a forming pressure of 5.5 MPa, an ellipsoidal shell with a shape conforming to the design requirements is obtained (Fig. 8.29).

Figure 8.30 shows the length variation of short axis on the equatorial plane and the height variation of the polar plates for the prolate ellipsoidal shell with double generating lines. In the initial stage of deformation ($p \leq$ 4 MPa), the ellipsoidal shell above the dividing angle does not deform when the internal pressure is less than 2.0 MPa, that is, the height of the polar plate does not change. With the increase of internal pressure, the polar plate slightly protrudes outward under the action of internal pressure. When the internal pressure is greater than 3.0 MPa, the height of the polar plate increases significantly. In the middle stage of deformation (4 MPa $< p \leq$ 5 MPa), the height of the polar plate reaches the initial design size of the long axis

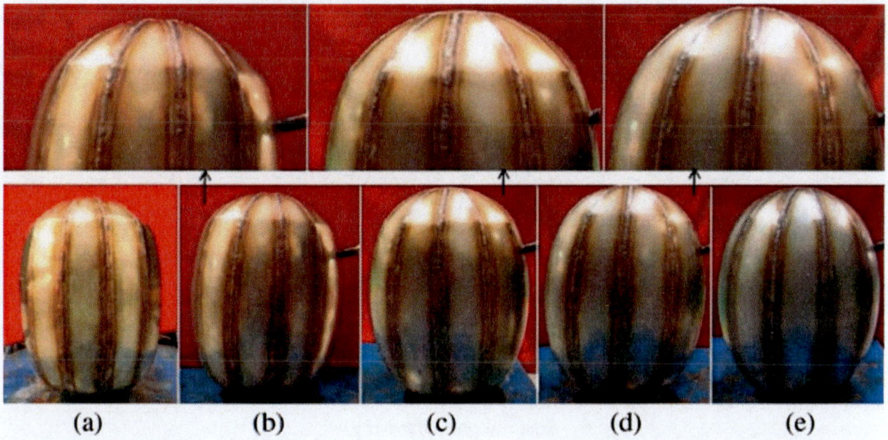

Fig. 8.28 Shape variation of the prolate ellipsoidal shell. **a** Initial polyhedral shell; **b** 3.5 MPa; **c** 4.5 MPa; **d** 5.0 MPa; **e** 5.5 MPa

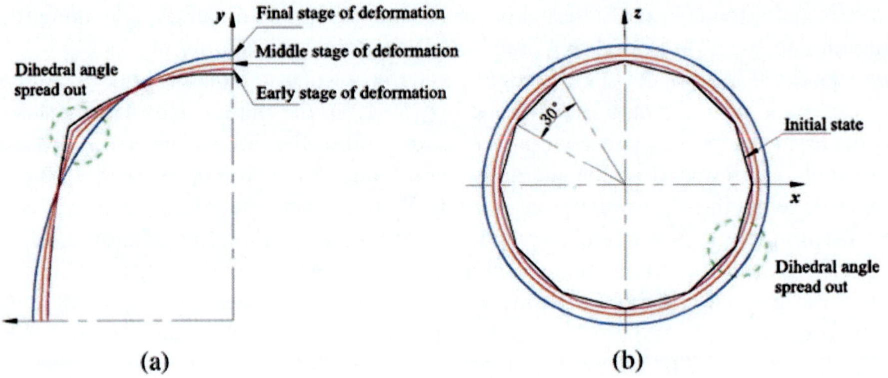

Fig. 8.29 Cross-sectional profile variation of the prolate ellipsoidal shell. **a** Meridional cross-section; **b** equatorial plane

when the internal pressure is 4.5 MPa. In the last deformation stage ($p > 5$ MPa), when the internal pressure is 5.5 MPa, the length of the long axis is 178.7 mm, which is 2.1% longer than the original design value. In the height variation process of the polar plate, the deformation rate gradually decreases. This is because with the increase of internal pressure, the curvature radius of the polar plate and its surrounding ellipsoidal shell gradually decreases, which needs a higher subsequent yielding pressure. Therefore, the deformation rate also decreases accordingly.

In the initial stage of deformation (p ≤ 4 MPa), the length of the short axis on the equatorial plane does not change when the internal pressure is less than 3.0 MPa, but the out-of-roundness of the equatorial plane decreases gradually. With the internal

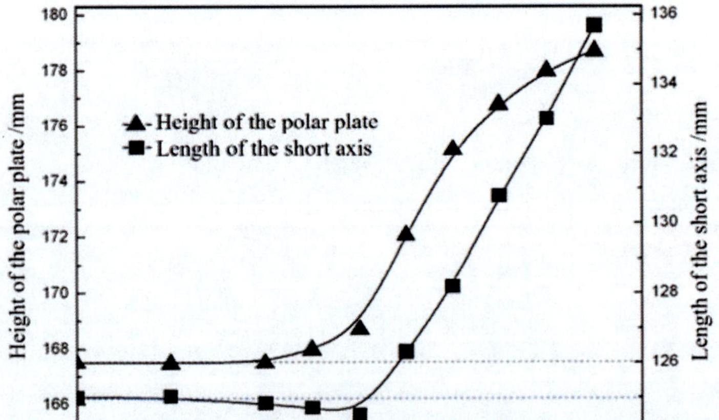

Fig. 8.30 The length variation of short axis and height variation of polar plates of the prolate ellipsoidal shell

pressure exceeding 3.5 MPa, the equatorial plane begins to expand as a whole, and the length of the short axis gradually increased. In middle or late stage of deformation (p > 5 MPa), with the increase of internal pressure, the elongation rate of short axis increases gradually. When the internal pressure is 5.5 MPa, the length of the short axis is 135.7 mm, which is an elongation of 8.7% from the initial design value. At this time, the axis length ratio is 0.76.

8.5.2 Volume Variation of Prolate Ellipsoidal Shell with Double Generating Lines in Hydroforming

Figure 8.31 shows the volume variation with the internal pressure in the hydroforming of prolate ellipsoidal shell with double generating lines. In the initial stage of deformation ($p \leq 4$ MPa), the polyhedral shell only protrudes outwards under the action of internal pressure before hydroforming, and the plate shell has not yet entered into plastic deformation. Hence the volume change is very small. As the internal pressure is increased to 3.0 MPa, the volume variation is only 1.5%. When the internal pressure reaches 3.5 MPa, the equatorial zone will undergo plastic deformation, and there will be a significant variation in volume.

In the middle stage of deformation (4 MPa $< p \leq$ 5 MPa), the volume variation rate increases gradually with the development of plastic deformation from the equatorial zone to the polar zone. At the late deformation stage ($p > 5$ MPa), once the whole ellipsoidal shell enters the yield state, the plastic deformation of the shell increases and the volume significantly increases. At this moment, the internal pressure changes relatively slow. This is because that more liquid must be filled into the shell in a longer time to fulfill each increase of 0.1 MPa in internal pressure. When the internal pressure is 5.5 MPa, the volume variation is 7.3%.

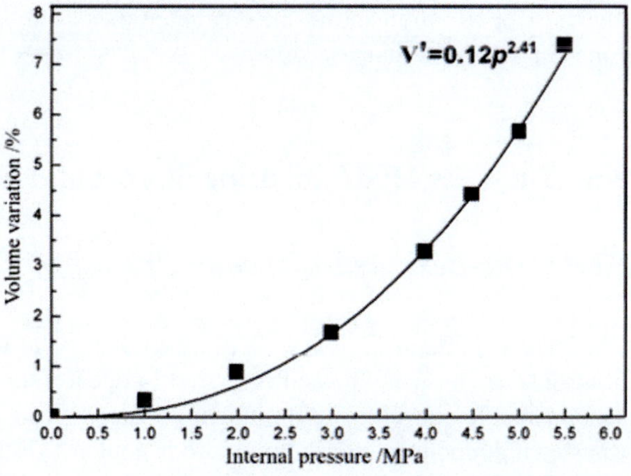

Fig. 8.31 Volume variation with the internal pressure in the hydroforming of prolate ellipsoidal shell

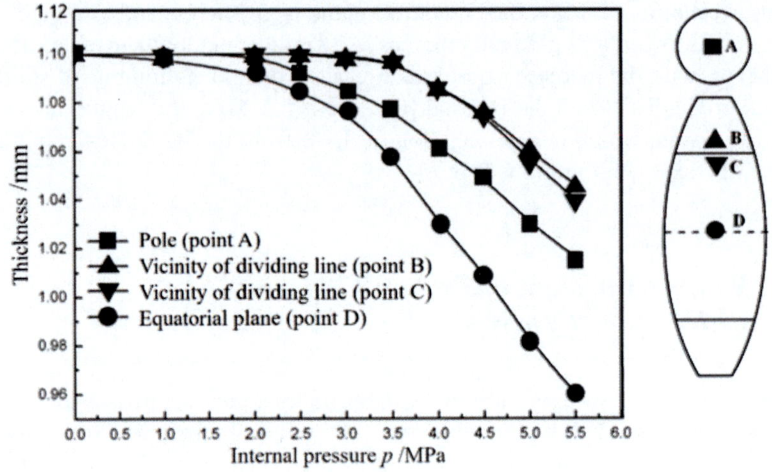

Fig. 8.32 Thickness variation of prolate ellipsoidal shell in hydroforming

8.5.3 Thickness Variation of Prolate Ellipsoidal Shell with Double Generating Lines in Hydroforming

Figure 8.32 shows the thickness variation at different typical points on the ellipsoidal shell with internal pressure during the experiment. The thickness thinning begins as the different regions of the shell enter the plastic deformation zone. After hydroforming, the maximum thinning is located in the center point of the equatorial lateral petals, where the finial thickness is 0.960 mm; and accordingly, the thinning ratio is 12.7%. The minimum thinning is located at the vicinity of the dividing line, where the thickness thinning ratio is only 4.5%. Moreover, the wall thickness at the pole is 1.016 mm, and the corresponding thinning ratio is 7.6%. In conclusion, the overall thickness distribution shows the characteristics of thinner polar zone and equatorial zone and thicker temperate zone.

8.6 Die-Less Hydroforming of Toroidal Shells

8.6.1 Die-Less Hydroforming Process of Toroidal Shells

The basic manufacturing process of elbow using die-less hydroforming technology of toroidal shell is as follows: first, a multi-edge toroidal shell with a polygonal cross-section should be welded. After the multi-faced toroidal shell is filled with pressure-transferring medium, internal pressure is applied. Under the action of internal pressure, the cross-section gradually turns into a circle from a polygon, and finally a

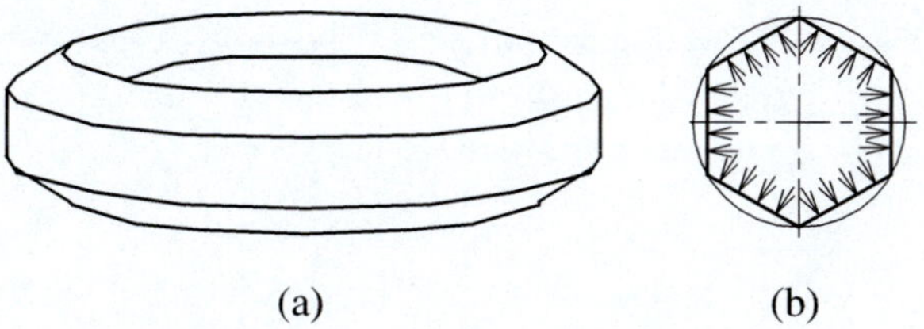

Fig. 8.33 Initial polyhedral toroidal shell structure. **a** Polyhedral toroidal shell; **b** cross-section

circular toroidal shell can be formed, as shown in Fig. 8.33. A circular toroidal shell can be cut into four 90° elbows or six 60° elbows or other specifications as required. Compared with the original elbow manufacturing method, the main advantages of this process are:

(1) Tube material is not needed, which can save the cost of tube manufacturing equipment and die, and can get the elbows with an arbitrarily large diameter and relatively thin wall thickness.
(2) The blank is flat or developable surface, which is easy to cut, assemble and weld. Moreover, its accuracy can be easily guaranteed.
(3) Short manufacturing cycle and low production cost. Because it does not need any special equipment, it is especially suitable for manufacturing large elbow on site.

8.6.2 *Stress Distribution and Forming Pressure of Toroidal Shells*

As shown in Fig. 8.34, in an ideal toroidal shell under uniform internal pressure, its tangential stress σ_φ and circumferential stress σ_θ can be derived by the internal forces equilibrium condition in the vertical direction of portion of the shell defined by the angle φ and the general equilibrium equation of the shell as follows.

$$\sigma_\varphi = \frac{pr(r' + R_b)}{2r't} \tag{8.23}$$

$$\sigma_\theta = \frac{pr}{2t}$$

where r—Radius of the tube shell (mm);
R_b—Middle radius of bending (mm);
r'—The distance from any point on the toroidal shell to the symmetry axis (mm);
t—Thickness (mm).

It can be seen from Eq. (8.23) that the circumferential stress σ_θ of an ideal toroidal shell subjected to internal pressure is constant. However, the tangential stress σ_φ varies with r, and reaches the minimum value of $\frac{pr}{2t}\left(1 + \frac{R_b}{R_b+r}\right)$ at the point c on the

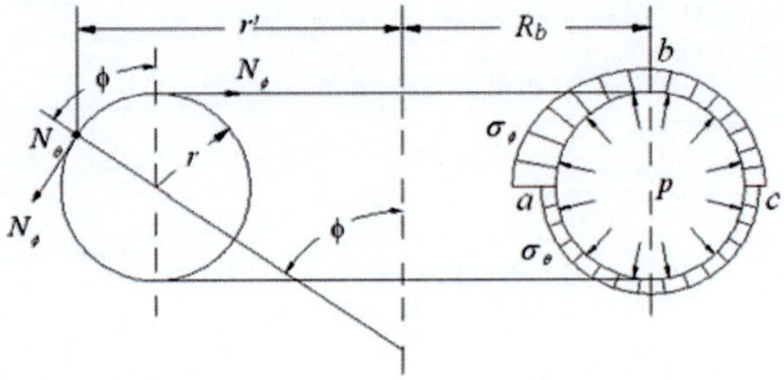

Fig. 8.34 Stress distribution of toroidal shell

outside of the toroid, and the maximum value of $\frac{pr}{2t}\left(1+\frac{R_b}{R_b-r}\right)$ at the point *a* on the inside of the toroid. Moreover, at the point *b*, the tangential stress is equal to $\frac{pr}{t}$.

According to Eq. (8.23), it is known that $\sigma_\varphi > \sigma_\theta$. By ordering $\sigma_1 = \sigma_\varphi, \sigma_2 = \sigma_\theta$, $\sigma_3 = 0$, the internal pressure producing plastic deformation in different positions of the toroidal shell can be calculated based on the Tresca yield criterion $\sigma_1 - \sigma_3 = \sigma_s$ as follows

$$\text{At point } a \quad p_a = \frac{2t}{r}\sigma_s \frac{1}{1+\frac{R_b}{R_b-r}} \tag{8.24}$$

$$\text{At point } b \quad p_b = \frac{t}{r}\sigma_s \tag{8.25}$$

$$\text{At point } c \quad p_c = \frac{2t}{r}\sigma_s \frac{1}{1+\frac{R_b}{R_b+r}} \tag{8.26}$$

where σ_s—Yield stress of the material (MPa).

From Eqs. (8.24) to (8.26), there is a relationship that $p_a < p_b < p_c$, which indicates that the plastic yielding occurs first at the point *a* followed by point *b*, and the point *c* yields at last. That is, plastic deformation occurs first in the inner side of the toroidal shell and then gradually extends to the outer side of the toroidal shell. The forming pressure at point *c* is the final forming pressure of the whole toroidal shell.

8.6.3 Die-Less Hydroforming Experiment of Toroidal Shells

The structure of the multi-edge toroidal shell before hydroforming is shown in Fig. 8.35. The design dimensions of the toroidal shell are: middle radius of bending $R_b = 150$ mm, radius of the tube shell $r = 50$ mm, and the thickness $t = 1.5$ mm. The cross-section of the multi-edge toroidal shell before hydroforming is a regular hexagon. In Fig. 8.35, the toroidal shell is composed of four shell elements, namely cylindrical shell A and cone shell B under external pressure, and cone shell C and cylindrical shell D under internal pressure. In addition, the Q235A low carbon steel was used for the shell, whose yield limit is that $\sigma_s = 208$ MPa.

According to the structure size and material mechanical properties of the toroidal shell, the forming pressure of the shell at different points can be calculated from Eqs. (8.24) to (8.26). At the points *a*, *b* and *c*, their forming pressure are $p_a =$ 4.99 MPa, $p_b = 6.24$ MPa and $p_c = 7.13$ MPa, respectively. In experiment, the final forming pressure is 7.35 MPa, which is basically agreement with the theoretical calculation of forming pressure p_c.

Figure 8.36 shows the normal displacement variation of each sub-shell element with internal pressure. It can be seen from the figure that the normal displacement in each sub-shell element increases with the increase of internal pressure. After hydroforming, the cross-section of the toroidal shell changes from a regular hexagon to a circle, as shown in Fig. 8.37.

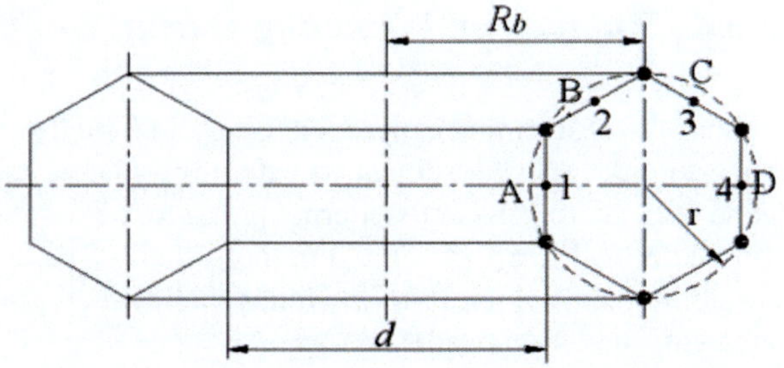

Fig. 8.35 Toroidal shell structure and the arrangement of measuring point

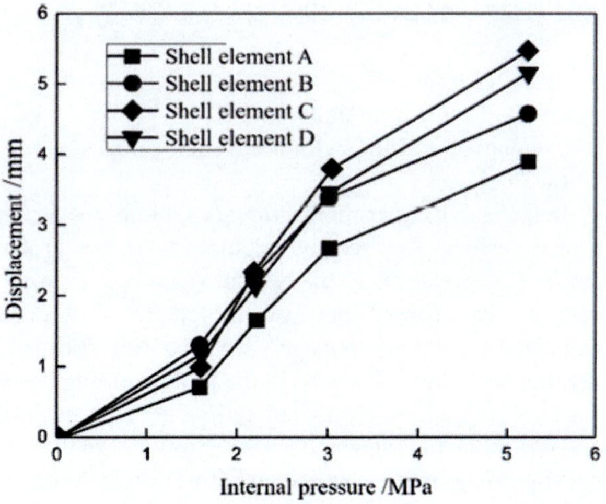

Fig. 8.36 Displacement variation of the shell with the internal pressure

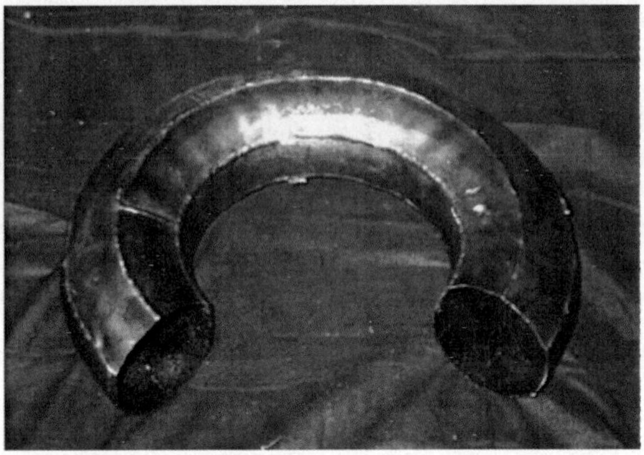

Fig. 8.37 A toroidal shell

8.6.4 Analysis on Wrinkling During Toroidal Shell Forming

In above hydroforming experiment of toroidal shell, since the inner part of the multi-edge toroidal shell is subject to external pressure, that is, cylindrical shell A and cone shell B are subject to external pressure, compressive stress and large compression deformation are generated in these shells. The compression stress could cause instability and wrinkling of the inner shell, which also has been found in the finite element simulation results.

For the cylindrical shell A under external pressure, its critical instability stress is

$$\sigma_{cr} = 1.1\frac{Et^2}{d^2} \tag{8.27}$$

where E—Young's modulus (MPa);

d—Diameter of the cylindrical shell (mm);

t—Thickness (mm).

In the above experiments, $d = 213.4$ mm, $t = 1.5$ mm, $E = 210{,}000$ MPa. Therefore, the critical stress can be calculated as $\sigma_{cr} = 11.4$ MPa. When the circumferential stress σ_θ is greater than the critical stress σ_{cr}, wrinkle will occurs on the shell.

In the experiment, because the lengths of cylindrical shell A and cone shell B subjected to external pressure are relatively shorter, and the welding seam can be regarded as a rigid structure in the hydroforming process, the actual value of critical stress is larger than the calculated value. Moreover, the phenomenon of instability and wrinkling first appears on the cylindrical shell A in the hydroforming experiment because of its smaller critical instability stress. With the increase of internal pressure, wrinkling also occurs on the cone shell B.

In the hydroforming of toroidal shell, for the inner part of the multi-edge toroidal shell subjected to external pressure, although wrinkle will occurs on the shell when the circumferential stress σ_θ is greater than the critical stress σ_{cr}, it can be eliminated by local heating and cooling. Moreover, it is also possible to prevent wrinkling by changing the structure of polyhedral toroidal shell before hydroforming or by some other technological methods.

8.6.5 Effect of Initial Structures of Toroidal Shell on Hydroforming

In the previous investigation about toroidal shell hydroforming, a large compression deformation occurs in these sub-shells subjected to external pressure, which will result in instability and wrinkling of the inner sub-shells. In order to reduce the compression deformation of the inner sub-shells of the toroidal shell, the effect of the initial structure of the toroidal shell on the hydroforming was studied by using the hexagonal and octagonal polygonal shell structures.

Because the deformation in toroidal shell hydroforming process is complicated, it is difficult to give an analytical solution in theory. In view of this situation, it can be assumed that the points on each sub-shell of the toroidal shell in hydroforming process move to the outer circle along the line between the point and the center of the circle (that is, the direction of radius). Then the strain value of each point can be obtained using the method of geometric analysis. Because the wrinkling occurs on the inner shell under external pressure, only the circumferential deformation of this inner sub -shell is analyzed.

Figure 8.38 shows the initial structures of toroidal shell. For the toroidal shell with hexagonal cross-section, it is composed of three sub-shell elements, namely cone shell 1 under external pressure, plate 2 and cone shell 3 under internal pressure. For the toroidal shell with octagonal cross-section, the shell elements subjected to external pressure are inner cone shell 1 and inner cone shell 2. In hydroforming, the initial line segment element *AB* sandwiched by the angle $d\phi$ of inside of the toroidal shell moves to the circle, and becomes into the arc element *CD*. Therefore, the circumferential strain ε_θ can be obtained as follows

$$\varepsilon_\theta = \ln \frac{\frac{R_b}{r} - \cos(\frac{\pi}{6} + \phi)}{\frac{R_b}{r} - \frac{\cos\frac{\pi}{6}\cdot\cos(\frac{\pi}{6}+\phi)}{\cos\phi}} \tag{8.28}$$

where R_b—Middle radius of bending;

r—Radius of the tube shell;

ϕ—The angle between the point on the subshell under external pressure and its neutral surface.

Figure 8.39a shows the circumferential strain distribution of cone shell 1 under external pressure of hexagonal toroidal shell. As can be seen from Fig. 8.39a, with the increase of the ratio between the middle radius of bending and the radius of the tube shell R_b/r, the absolute value of the circumferential compressive strain of the inner shell gradually decreases. This indicates that with the increase of R_b/r, it will

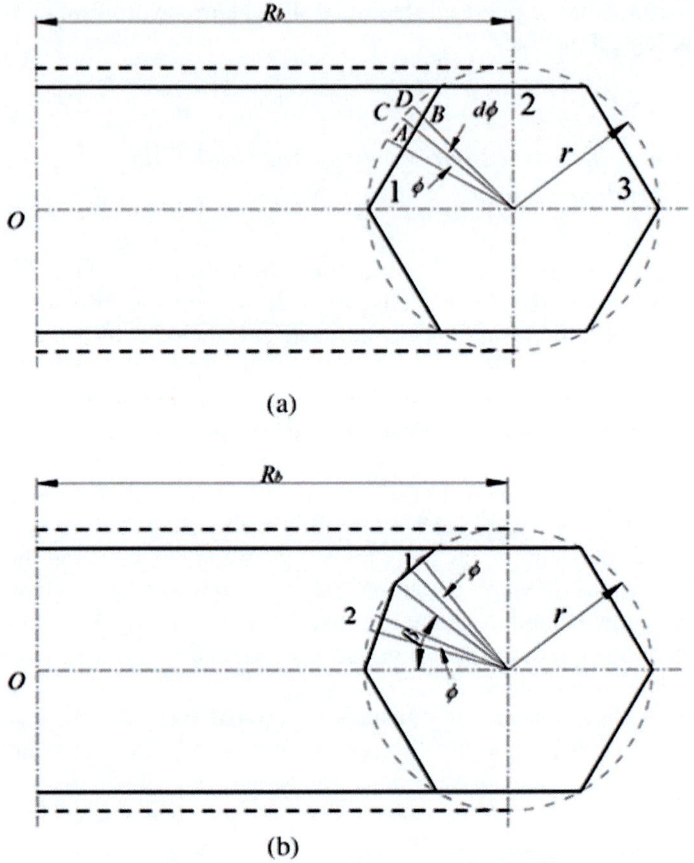

Fig. 8.38 Initial structures of toroidal shell. **a** Hexagonal toroidal shell; **b** octagonal toroidal shell

reduce the wrinkling tendency on the shell and be beneficial to the hydroforming of the toroidal shell.

In the hydroforming of toroidal shell, the inner side wrinkling is mainly caused by the large compression deformation of the inner side. Therefore, in order to ensure that the absolute value of circumferential compressive strain is minimum for the octagonal toroidal shell, the lengths of inner cone shell 1 and 2 under external pressure should be reasonably distributed, that is, the angle β should be rationally selected. Figure 8.38b shows the circumferential strain distribution of cone shells 1 and 2 of octagonal toroidal shells under different angles β. As can be seen from Fig. 8.38b, it can be found that with the increase of β, the circumferential strain value of inner cone shell 1 increases, while the circumferential strain value of inner cone shell 2 decreases. Considering the circumferential strain of both inner cone shell 1 and 2, it can be seen that the absolute value of circumferential compressive strain is the minimum

Fig. 8.39 Circumferential strain distribution of toroidal shell subjected to external pressure. **a** Hexagonal toroidal shell; **b** octagonal toroidal shell

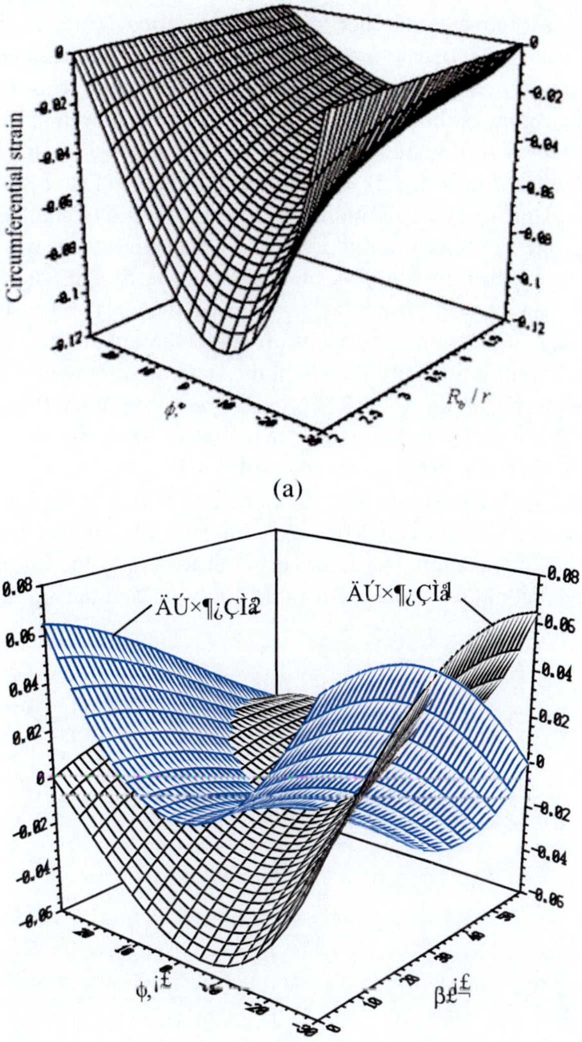

(a)

(b)

for the whole inner shell when $\beta = 30°$. For the toroidal shell with hexagonal cross-section, the maximum circumferential compressive strain of the inner shell can be calculated as $\varepsilon_\theta = -0.055$. However, for the octagonal toroidal shell with $\beta = 30°$, the maximum circumferential compressive strain of the inner shell is $\varepsilon_\theta = -0.016$, whose absolute value is much less than that of the hexagonal toroidal shell. It can be seen that the increase of the number of inner shells under external pressure can effectively improve the compression deformation of these shells, thus reducing their tendency of wrinkling.

Moreover, the upper-lower asymmetrical structure was adopted to study the effect of the initial structure of the shell on the toroidal shell hydroforming. For upper-lower asymmetrical toroidal shell, its top half part is octagon and bottom half part is hexagon, as shown in Fig. 8.40. The material used in experiment is the stainless steel sheet with thickness of 1.5 mm. The dimensions of the toroidal shell are: middle radius of bending $R_b = 225$ mm, the radius of tube shell $r = 75$ mm.

During the experiment, the strain variation of shell elements A, B and C, which are prone to wrinkling under external pressure, were measured. Figure 8.41 shows the strain variation with internal pressure for the shell elements under external pressure. It can be seen from the figure that during the whole hydroforming process, the tangential strain ε_ϕ of each sub-shell element is in the state of tensile strain, while the circumferential strain ε_θ is in the state of compressive strain. However, the absolute value of circumferential compressive strain of shell element B and C of the octagon structure is much smaller than that of shell element A of the hexagon structure. When the internal pressure $p = 1.8$ MPa, obvious wrinkling phenomenon appears on the shell element A, but shell element B and C are not wrinkled. When the internal pressure reaches 8.0 MPa, the whole toroidal shell is basically formed. It can be found that the octagonal structure is better for the hydroforming of toroidal shell, while the hexagonal structure is unstable and wrinkled during the hydroforming process.

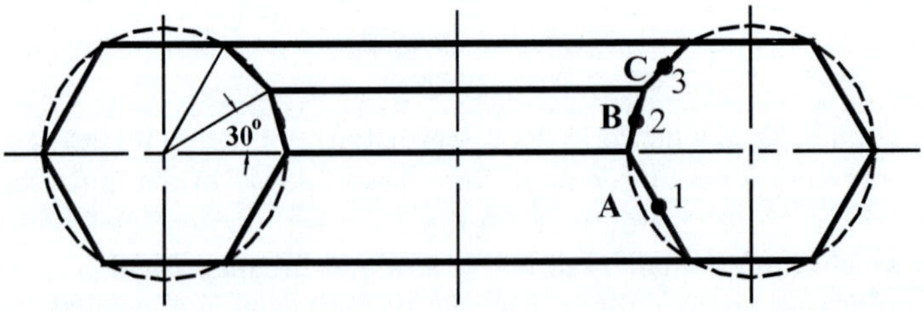

Fig. 8.40 Structure of toroidal shell and arrangement of strain measuring point before hydroforming

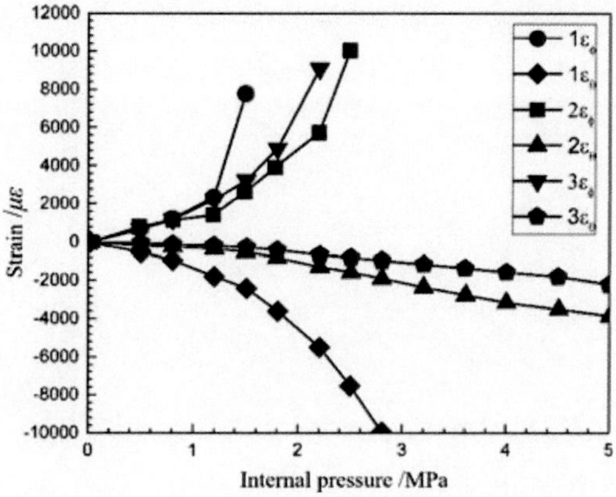

Fig. 8.41 The strain variation with internal pressure

References

1. Yuan, S.J., and Z.R. Wang. 2002. Progress on Hydroforming Applications. *China Mechanical Engineering* 13 (9): 783–786.
2. Dohmann, F., Bohm, A., and K. Dudziak. The Shaping of Hollow Shaped Workpieces by Liquid Bulge Forming. In *Proceedings of 4th ICTP, Beijing, China, 1993*, 447–452.
3. Amino, H., K. Nakamura, and T. Nakagawa. 1990. Counter-pressure Deep Drawing and Its Application in the Forming of Automobile Parts. *Journal of Materials Processing Technology* 23: 243–265.
4. Wang, Z.R., T. Wang, and D.C. Kang. 1989. The Technology of the Hydro-bulging of Whole Spherical Vessels and Experimental Analysis. *Journal of Mechanical Working Technology* 18: 85–94.
5. Dohmann, F., and Ch. Hartl. 1996. Hydroforming-a Method to Manufacture Lightweight Parts. *Journal of Materials Processing Technology*, 60 (1–4): 669–676.
6. Dohmann, F., and H.A.R.T.L. Ch. 1997. Tube Hydroforming. *Journal of Materials Processing Technology* 71: 174–186.
7. Dohmann, F., and H.A.R.T.L. Ch. 2004. Hydroforming-Applications of Coherent FE-Simulations to the Development of Products and Processes. *Journal of Materials Processing Technology* 150: 18–24.
8. Koc, M., and T. Altan. 2001. An Overall Review of the Tube Hydroforming (THF) Technology. *Journal of Materials Processing Technology* 108: 384–393.
9. Hartl, Ch. 2005. Research and Advances in Fundamentals and Industrial Applications of Hydroforming. *Journal of Materials Processing Technology*, 167 (2–3): 383–392.
10. Horton, F. 2003. Hydroforming Application Trends in Automotive Structures. In *Proceedings of 3rd International Conference on Hydroforming, Stuttgart, Germany, 2003,* 7–13.
11. Volerstern, F., Prange, T., and M. Sander. 1999. Hydroforming: Needs, Developments and Perspective. In *Proceedings of 6th ICTP, Nuremberg, Germany, 1999*, 1197–1210.
12. Sigert, K. 2003. Research in Hydroforming at the Institute of Metal Forming of University of Stuttgart. In *Proceedings of 3rd International Conference on Hydroforming, Stuttgart, Germany, 2003*, 245–258.
13. Liewald, M. 2007. State-of-the-Art of Hydroforming Tubes and Sheets in Europe. In *Proceedings of TUBEHYDRO 2007, Harbin, China*, 19–26.
14. Hennig, K.P. 2008. Economical Aspects and Trends of the Hydroforming Technology. In *Proceedings of TUBEHYDRO 2007, Harbin, China*, 11–18.
15. Asnafi, N., T. Nilsson, and G. Lassl. 2003. Tubular Hydroforming of Automotive Side Members with Extruded Aluminium Profile. *Journal of Materials Processing Technology* 142: 93–101.
16. Fuchizawa, S. 1984. Influence of Strain-Hardening Exponent on the Deformation of Thin-walled Tube of Finite Length Subjected to Hydrostatic Internal Pressure. In *Proceedings of 1st ICTP, Tokyo, Japan, 1984*, 297–302.

S. Yuan, *Modern Hydroforming Technology*,
https://doi.org/10.1007/978-981-19-5775-8

17. Fuchizawa, S. 1987. Influence of Plastic Anisotropy on Deformation of Thin-walled Tube in Bulge Forming. In *Proceedings of 2nd ICTP, Stuttgart, Germany, 1987*, 727–732.
18. Fuchizawa, S., and M. Narazaki. 1993. Bulge Test for Determining Stress-strain Characteristics of Thin Tubes. In *Proceeding of 4th ICTP, Beijing, China, 1993*, 488–493.
19. Fuchizawa, S. 2007. Recent Developments in Tube Hydroforming Technology in Japan. In *Proceedings of TUBEHYDRO 2007, Harbin, China*, 1–10.
20. Manabe, K., Suetake, M., H. Koyama, et al. 2004. Hydroforming Process Optimization of Aluminum Alloy Tube Using Intelligent Control Technique. In *Proceedings of 1st ICNFT, Harbin, China, 2004*, 490–495.
21. Manabe, K., Suzuki, K., and S. Mori, et al. 1984. Bulge Forming of Thin Walled Tubes by Micro-computer Controlled Hydraulic Press. In *Proceedings of 1st ICTP, Tokyo, Japan, 1984*, 279–284.
22. Huang, Y.-M., and W.-C. Chen. 2005. Analysis of Tube Hydroforming in a Square Cross-sectional Die. *International Journal of Plasticity* 21: 1815–1833.
23. Nefussi, G., and A. Combescure. 2002. Coupled Buckling and Plastic Instability for Tube Hydroforming. *International Journal of Mechanical Science* 44: 899–914.
24. Liu, S.-D., and D. Meuleman. 1998. Analytical and Experimental Examination of Tubular Hydroforming Limits. SAE Technical Paper No: 980449, 139–150.
25. Smith, L.M., Ganeshmurthy, N., P. Murty, et al. 2004. Finite Element Modeling of the Tubular Hydroforming Process: Part 1. Strain Rate-independent Material Model Assumption. *Journal of Materials Processing Technology* 147 (1): 121–130.
26. Ray, P., and B.J. Mac Donald. 2004. Determination of the Optimal Load Path for Tube Hydroforming Processes Using a Fuzzy Load Control Algorithm and Finite Element Analysis. *Finite Elements in Analysis and Design* 8: 173–192.
27. Jirathearanat, S., Hartl, Ch., and T. Altan. 2004. Hydroforming of Y-shapes-Product and Process Design Simulation and Experiments. *Journal of Materials Processing Technology*, 146: 124–129.
28. Uchida, M., and M. Kojima. 2002. Hydropiercing of Tube Wall in Hydroforming. In *Proceedings of 7th ICTP, Yokohama, Japan, 2002*, 2, 1483–1488.
29. Nakamura, K., T. Nakagawa, and H. Amino. 1997. Various Application of Hydraulic Counter Pressure Deep Drawing. *Journal of Materials Processing Technology* 71: 160–167.
30. Amino, H., Makita, K., and T. MAKI. 2000. Sheet Fluid Forming and Sheet Dieless NC Forming. In *International Conference on New Developments in Sheet Metal Forming, Stuttgart, Germany, 2000*, 39–67.
31. Kleiner, M., Homberg, W., and A. Brosius. 1999. Process and Control of Sheet Metal Hydroforming. In *Proceedings of 6th ICTP, Nuremberg, Germany, 1999*, 1243–1252.
32. Danket, J., and K.B. Nielsen. 2000. Hydromechanical Deep Drawing with Uniform Pressure on the Flange. *Annuals of the CIRP* 49 (1): 217–220.
33. Khandeparker, T., and M. Liewald. 2008. Hydromechanical Deep Drawing of Cups with Stepped Geometries. *Journal of Materials Processing Technology* 202: 246–254.
34. Kleiner, M., and W. Homberg. New 100,000KN Press for Sheet Metal Hydroforming. In *Proceedings of 2nd International Conference on Hydroforming, Stuttgart, Germany, 2001*, 351–362.
35. Groche, P., and C. Metz. 2004. Investigation of Active-Elastic Blank Holder Systems for High-Pressure Forming of Metal Sheets. In *Proceedings of 1st ICNFT, Harbin, China*, 447–452.
36. Yuan, S.J. 2006. Technologies and Development Perspective of Material Plasticity Forming. *Advanced Manufacture and Material Application Technologies* 4: 1–14.
37. Yuan, S.J. 2003. Development and Trends in Internal High Pressure Forming. *Metal Forming Technology* 21 (3): 1–3.
38. Yuan, S.J., Liu, G., and S. Wang, et al. 2007. Development and Applications of Tube Hydroforming in China. In *Proceedings of TUBEHYDRO 2007, Harbin, China*, 27–38.
39. Yuan, S.J., G. Liu, and X.R. Huang. 2004. Hydroforming of Typical Hollow Components. *Journal of Materials Processing Technology* 151: 203–207.

40. Yuan, S.J., and X.S. Wang. 2002. Fundamentals and Application of Tube Hydro-forming. *Chinese Journal of Mechanical Engineering* 38: 12–15.
41. Yuan, S.J., X.S. Wang, and G. Liu. 2007. Control and Use of Wrinkles in Tube Hydroforming. *Journal of Materials Processing Technology* 182: 6–12.
42. Yuan, S.J., W.J. Yuan, and X.S. Wang. 2006. Effect of Wrinkling Behavior on Formability and Thickness Distribution in Tube Hydroforming. *Journal of Materials Processing Technology* 177: 668–671.
43. Yuan, S.J. 2007. Innovative Processes in Tube Hydroforming and Applications. In *Proceedings of 2nd ICNFT, Bremen, Germany, 2007*, 1–13.
44. Liu, G., S.J. Yuan, and X.S. Wang. 2005. Effects of Loading Paths on thickness distribution of A Hydroforming Component. *Materials Science & Technology* 13 (2): 162–165.
45. Yuan, W.J., X.S. Wang, and S.J. Yuan. 2008. Thickness Dividing Circle in Hydroforming of Variable Diameter Tubes. *Materials Science & Technology* 16 (2): 196–199.
46. Liu, G., W.C. Xie, S.J. Yuan, et al. 2004. Internal High Pressure Forming of Hollow Part with a Big Section Difference. *Journal of Material Science and Technology* 12 (4): 398–401.
47. Yuan, S.J., Lang, L.H., S. X.S. Wang, et al. 2001. Experiment and Numerical Simulation of Aluminum Tube Hydroforming. In *Proceedings of 2nd International Conference on Hydroforming, Stuttgart, Germany, 2001*, 339–349
48. Wang, X.S., S.J. Yuan, X.R. Huang, et al. 2008. Research on Hydroforming Tubular Part with Large Perimeter Difference. *Acta Metallurgica Sinica (English Letters)* 21 (2): 133–138.
49. He, Z.B., S.J. Yuan, W.W. Zha, et al. 2008. Force and Deformation Analysis of Tube Ring Specimen During Hoop Tensile Test. *Acta Metallurgica Sinica* 44 (4): 423–427.
50. Wang, X.S. 2005. *Research on Wrinkling Behavior in Tube Hydroforming*. Harbin: Harbin Institute of Technology.
51. Lin, J.F. 2007. *Mechanism of Hydroforming of a Hollow Crankshaft*. Harbin: Harbin Institute of Technology.
52. Yuan, W.J. *Research on Hydroforming of Unsymmetrical Tubes with Larger Expansion Ratio*. Harbin: Harbin Institute of Technology.
53. Yuan, S.J., C. Han, and X.S. Wang. 2006. Hydroforming of Automotive Structural Components with Rectangular-sections. *International Journal of Machine Tools and Manufacture* 46 (11): 1201–1206.
54. Han, C. 2006. *Fundamentals on Hydroforming Curved Tubular Components with Polygonal—Sections*. Harbin: Harbin Institute of Technology.
55. Yuan, S.J., G. Liu, X.W. Tian, et al. 2003. Numerical Simulation and Experiments of Hydroforming of Rectangular-section Tubular Component. *Journal of Materials Science & Technology* 19 (Suppl. 1): 35–37.
56. Yuan, S.J., Wang, X.S., G. Liu, et al. 2002. Research on Hydroforming of Tubular Components with Changeable Cross-sections. In *Proceedings of 7th ICTP, Yokohama, Japan, 2002*, 11, 1495–1500.
57. Han, C., Wnag, X.S., S.J. Yuan. 2007. Preform Section Design and Application in Hydroforming of an Engine Cradle. In *Proceedings of TUBEHYDRO2007, Harbin, China, 2007*, 141–146.
58. Liu, G., S.J. Yuan, and B.G. Teng. 2006. Analysis of Thinning at the Transition Corner in Tube Hydroforming. *Journal of Materials Processing Technology* 177: 688–691.
59. Wang, X.S., S.Q. Zhu, and S.J. Yuan. 2007. Internal High Pressure Forming of Aluminum Tubular Part with Special Shape Sections. *Aeronautical Manufacturing Technology* 291: 519–522.
60. Yuan, S.J., Liu, G., and Z.R. Wang. 2003. Hydroforming of Rectangular-Section Structural Components with Relatively Lower Pressure. In *Proceedings of TUBEHYDRO 2003 Nagoya Japan 2003*, 13–17.
61. Han, C., and S.J. Yuan. 2008. Reduction of Friction and Calibration Pressure by Section Preform during Hydroforming of Tubular Automotive Structural Components. *Advanced Materials Research* 44–46: 143–150.

62. Liu, G., S.J. Yuan, and B.G. Teng. 2006. Stress Analysis on Corner of Rectangular Section in Tube Hydroforming. *Chinese Journal of Mechanical Engineering* 42 (6): 150–155.
63. Yuan, S.J., G. Liu, and C. Han. 2006. Mechanism Analysis on Reducing Pressure of Tube Hydroforming through reform. *Journal of Aeronautical Materials* 26 (4): 46–50.
64. Teng, B.G., G. Liu, S.J. Yuan, et al. 2007. Internal High Pressure Forming Technology Used for Manufacturing Exhaust Pipes of Automotive Engine. *Journal of Plasticity Engineering* 14 (3): 88–92.
65. Cheng, D.M., B.G. Teng, B. Guo, et al. 2008. Analysis of Deformation and Defects in Hydroforming of Y-shaped Tubes. *Journal of Harbin Institute of Technology* 15 (2): 206–210.
66. Cheng, D.M. 2008. *Research on Hydroforming Mechanism of a Y-Shaped Thin Wall Tube.* Harbin: Harbin Institute of Technology.
67. Li, H.Y., X.S. Wang, S.J. Yuan, et al. 2004. Typical Stress States of Tube Hydroforming and Their Distribution on the Yield Ellipse. *Journal of Materials Processing Technology* 151: 345–349.
68. Li, H.Y. 2002. *Stress and Strain Analysis of Internal High-Pressure Forming (IHF) and the Study of IHF Multidiameter Shaft.* Harbin: Harbin Institute of Technology.
69. Wu, H.F., S.J. Yuan, and Z.R. Wang. 2002. Derivation of General Equations for Elasto-plastic Stability of Cylindrical Shell under Axial Compression. *Journal of Harbin Institute of Technology* 34 (1): 35–39.
70. Yuan, S.J., Q. Liu, and G. Liu. 2007. Effect of Internal Pressure on Hydro-joining Strength of Steel Elements. *Journal of Plasticity Engineering* 14 (3): 84–87.
71. Liu, Q., G. Liu, S.J. Yuan, et al. 2007. A Hydro-joining Assembled Hollow Camshaft and Its Joining Strength Measurement. *Chinese Internal Combustion Engine Engineering* 28 (5): 68–70.
72. Liu, Q., J.G. Ge, G. Liu, et al. 2007. Analysis of Stress and Strain of Hydro-joining for the P/M Cam and Tube. *Material Science & Technology* 15 (5): 614–618.
73. Han, C., Yuan, S.J., S.U.H. Bous, et al. 2006. Numerical Simulation on Hydro-piercing. *The Chinese Journal of Nonferrous Metals*, 1: 22–28.
74. Liu, G., S.J. Yuan, and G.N. Chu. 2007. FEA on Deformation Behavior of Tailor-welded Tube in Hydroforming. *Journal of Materials Processing Technology* 187–188: 287–291.
75. Liu, G., X.L. Yin, and S.J. Yuan. 2006. External Pressure Forming and Buckling Analysis of a Tubular Part with Ribs. *Journal of Material Science & Technology* 22 (5): 708–712.
76. Kang, D.C., Y.C. Xu, and Y. Chen. 2005. Hydromechanical Deep Drawing of Superalloy Cups. *Journal of Materials Processing Technology* 166: 243–246.
77. Xu, Y.C., and D.C. Kang. 2004. Investigation of SUS304 Stainless Steel with Warm Hydro-mechanical Deep Drawing. *Journal of Material Science & Technology* 20 (1): 92–93.
78. Xu, Y.C. 2003. *Sheet-Metal Hydroforming and the Numerical Simulation of the Forming Process.* Harbin: Harbin Institute of Technology.
79. Xu, Y.C., S.J. Yuan, Y. Chen, et al. 2006. Investigation on Hydroforming of SUS304 Stainless Steel Dome. *Chinese Journal of Aeronautics* 19: 261–264.
80. Liu, X.J., Y.C. Xu, and S.J. Yuan. 2008. Effects of Loading Paths on Hydrodynamic Deep Drawing with Independent Radial Hydraulic Pressure of Aluminum Alloy Based on Numerical Simulation. *Journal of Material Science and Technology* 24 (3): 395–399.
81. Liu, X.J. 2008. *Research on Hydroforming Process of 5A06 Aluminium Alloy Sheet With Controllable Radial Pressure.* Harbin: Harbin Institute of Technology.
82. Wang, Z.R., S.J. Yuan, and Y.S. Zeng. 1999. Principle and Research Progress of the Integrated Dieless Hydro-Building Spherical Vessels. *Chinese Journal of Mechanical Engineering* 35 (4): 64–66.
83. Yuan, S.J., Teng, B.G., X.Y. Dong, et al. 2004. Progress in Large Vessel Forming: Introduction of some Innovations of Prof. Z. R. Wang. *Journal of Materials Processing Technology* 151: 12–17.
84. Wang, Z.R., G. Liu, and S.J. Yuan. 2005. Progress in Shell Hydroforming. *Journal of Materials Processing Technology* 167 (2–3): 230–236.

85. Yuan, S.J., and Z.R. Wang. 1997. Safety Analysis of 200m^3 LPG Tank Manufactured by the Dieless Hydro-bulging Process. *Journal of Materials Processing Technology* 70: 215–219.
86. Yuan, S.J., and Z.R. Wang. 1996. Effect of the Hydro-Bulging Process on Mechanical Properties of Spherical Pressure Vessels. *International Journal of Machine Tools and Manufacture* 36: 829–834.
87. Yuan, S.J., and Z.R. Wang. 1994. Research on the thick wall shell bulging using wide sheet tension tests. *China Mechanical Engineering* 5 (3): 9–10.
88. Yuan, S.J., and Z.R. Wang. 1996. An Investigation into Formability of Welded Joints. *Journal of Materials Processing Technology* 55: 33–36.
89. Yuan, S.J., Zeng, Y.S., Z.R. Wang. 1996. The Integral Hydro-bulge Forming of Elliptical Shells. In *Proceedings of 5th ICTP, Columbus, USA, 1996,* 943–946.
90. Zeng, Y.S., S.J. Yuan, and Z.R. Wang. 1997. Research on the Integral Hydrobulge Forming of Ellipsoidal Shells. *Journal of Materials Processing Technology* 72: 28–31.
91. Yuan, S.J., Z.R. Wang, and Q. He. 1999. Finite Element Analysis of Hydroforming Process of Toriodal Shells. *International Journal of Machine Tools and Manufacture* 39: 1439–1450.
92. Yuan, S.J., J. Qi, and Z.B. He. 2006. An Experimental Investigation into the Formability of Hydroforming 5A02 Al-tubes at Elevated Temperature. *Journal of Materials Processing Technology* 177: 680–683.
93. He, Z.B., X.S. Wang, S.J. Yuan, et al. 2007. Formability of AZ31B Magnesium Alloy Extruded Tube Under Internal High Pressure. *Acta Metallurgica Sinica* 43 (5): 534–538.
94. He, Z.B., G. Liu, S.J. Yuan, et al. 2007. Formability Evaluation of AZ31B Extruded Tube for Hydroforming Process. *RARE METALS* 26 (S1): 83–87.
95. Liu, G., He, Z.B., S.J. Yuan, et al. 2007. Warm Hydroforming Device for Mg-Alloy and Property of Tube Forming. *Aeronautical Manufacturing Technology*, 2007 (Supplement): 470–477.
96. He, Z.B., J. Qi, and S.J. Yuan. 2007. Warm Hydroforming of Aluminum Alloy Tubular Parts. *Aeronautical Manufacturing Technology* 291: 542–545.
97. QI, J. 2008. *Hydroforming of 5A02 Aluminium Alloy Tubes at Elevated Temperatures.* Harbin: Harbin Institute of Technology.
98. Tang, Z.J., Liu, G., S.J. Yuan, et al. 2008. Warm Hydroforming of an AZ61A Tubular Component with Multiple Cross Section Shapes. In *The 5th International Conference on Advanced Materials and Processing (ICAMP-5), Harbin, China, 2008.*
99. He, Z.B., Liu, G., J. Wu, et al. 2008. Mechanical Properties Testing and Formability Evaluation of AZ31B Extruded Tube at Elevated Temperature. In *Proceedings of the 3rd International Materials Research Conference, Chongqing, China 2008.*
100. Preytag, P., Neubert, J., and S. Kluge. Hydroforming in High-Volume Production. In *Proceedings of International Conference on New Development in Sheet Metal Forming and Hydroforming, Stuttgart, Germany, 2016,* 364–372.
101. Gericke, D. 2016. Hydroform Intensive Body Structure (HIBS) with Advanced and Ultra High Strength Steel. In *Proceedings of International Conference on New Development in Sheet Metal Forming and Hydroforming, Stuttgart, Germany, 2016,* 353–363.
102. Merten, C., Riedel, M., and M. Knape. EXFREE-Expansion Free Hydroforming a Process Option with New Possibilities. In *Proceedings of International Conference on Hydroforming of Sheets, Tubes and Profiles, Stuttgart, Germany, 2010,* 53–67.
103. Kuwabara, T., K. Yoshida, K. Narihara, and S. Takahashi. 2005. Anisotropic Plastic Deformation of Extruded Aluminum Alloy Tube under Axial Forces and Internal Pressure. *International Journal of Plasticity* 21: 101–117.
104. Kuwabara, T., and F. Sugawara. 2013. Multiaxial Tube Expansion Test Method for Measurement of Sheet Metal Deformation Behavior under Biaxial Tension for a Large Strain Range. *International Journal of Plasticity* 45: 103–118.
105. Manabe, K., X. Chen, D. Kobayashi, and K. Tada. 2014. Development of in-Process Fuzzy Control System for T-Shape Tube Hydroforming. *Procedia Engineering* 81: 2518–2523.
106. Sato, H., K. Manabe, K. Ito, D. Wei, and Z. Jiang. 2015. Development of Servo-type Micro-hydromechanical Deep-drawing Apparatus and Micro Deep-drawing Experiments of Circular Cups. *Journal of Materials Processing Technology* 224: 233–239.

107. Hartl, C., Anyasodor, G., and J. Lungershausen. 2011. Formability of Micro-Tubes in Hydroforming. In *2011 Proceedings of 14th International Conference on Material Forming Esaform*, 1353: 529–534.
108. Hartl, C., and G. Anyasodor. 2010. Experimental and Numerical Investigations into Microhydroforming Processes and Machine Design. *Steel Research International* 81: 1193–1196.
109. Liewald, M. 2010. State-of-the-Art and Recent Developments in Hydroforming in Europe—A Market Survey of Research and Industry. In *Proceedings of International Conference on Hydroforming of Sheets, Tubes and Profiles, Stuttgart, Germany, 2010*, 1–29.
110. Landgrebe, D., Albert, A., and A. Paul. 20 Years of Hydroforming Experience at the Fraunhofer IWU-Innovative Process Variants. In *Proceedings of International Conference on New Development in Sheet Metal Forming and Hydroforming, Stuttgart, Germany, 2016*, 403–423.
111. He, Z., S. Yuan, Y. Lin, X. Wang, and W. Hu. 2014. Analytical Model for Tube Hydro-bulging Tests, Part II: Linear Model for Pole Thickness and Its Application. *International Journal of Mechanical Sciences* 87: 297–306.
112. Yuan, S.J., X.L. Cui, and X.S. Wang. 2015. Investigation into Wrinkling Behavior of Thin-walled 5A02 Aluminum Alloy Tubes under Internal and External Pressure. *International Journal of Mechanical Sciences* 92: 245–258.
113. Cui, X.L., X.S. Wang, and S.J. Yuan. 2014. Deformation Analysis of Double-sided Tube Hydroforming in Square-Section Die. *Journal of Materials Processing Technology* 214: 1341–1351.
114. Cui, X.L., X.S. Wang, and S.J. Yuan. 2014. Experimental Verification of the Influence of Normal Stress on the Formability of Thin-walled 5A02 Aluminum Alloy Tubes. *International Journal of Mechanical Sciences* 88: 232–243.
115. Xie, W.C., C. Han, G.N. Chu, and S.J. Yuan. 2015. Research on Hydro-pressing Process of Closed Section Tubular Parts. *International Journal of Advanced Manufacturing Technology* 80: 1149–1157.
116. Wang, K.H., G. Liu, J. Zhao, J.L. Wang, and S.J. Yuan. 2015. Formability and Microstructure Evolution for Hot Gas Forming of Laser-Welded TA15 Titanium Alloy Tubes. *Materials & Design* 91: 269–277.
117. Liu, G., Wang, J., Dang, K., and Z. Tang. 2014. High Pressure Pneumatic Forming of Ti-3Al-2.5V Titanium Tubes in a Square Cross-Sectional Die. *Materials* 7: 5992–6009.
118. Liewald, M., and S. Wagner. 2012. Current Research Work into Sheet Metal Forming at the Institute for Metal Forming Technology (IFU) at the University of Stuttgart. In *Proceedings of International Conference on New Development in Sheet Metal Forming and Hydroforming, Stuttgart, Germany, 2012*, 241–278.
119. Maki, T. 2012. Sheet Hydroforming of Aluminum Body Panels. In *Proceedings of International Conference on Hydroforming of Sheets, Tubes and Profiles, Stuttgart, Germany, 2012*, 41–56.
120. Chen, Y.Z., W. Liu, Y.C. Xu, and S.J. Yuan. 2015. Analysis and Experiment on Wrinkling Suppression for Hydroforming of Curved Surface Shell. *International Journal of Mechanical Sciences* 104: 112–125.
121. Chen, Y.Z., W. Liu, and S.J. Yuan. 2015. Strength and Formability Improvement of Al-Cu-Mn Aluminum Alloy Complex Parts by Thermomechanical Treatment with Sheet Hydroforming. *JOM Journal of the Minerals Metals and Materials Society* 67: 938–947.
122. Yuan, S.J., W.W. Zhang, and B.G. Teng. 2015. Research on Hydro-forming of Combined Ellipsoidal Shells with Two Axis Length Ratios. *Journal of Materials Processing Technology* 219: 124–132.
123. Zhang, W.W., and S.J. Yuan. 2015. Pre-form Design for Hydro-forming Process of Combined Ellipsoidal Shells by Response Surface Methodology. *International Journal of Advanced Manufacturing Technology* 81: 1–10.
124. Yuan, S.J., and W.W. Zhang. 2016. Analysis of Shape Variation During Hydro-forming of Ellipsoidal Shells with Double Generating Lines. *International Journal of Mechanical Sciences* 107: 180–187.

125. Zhang, W.W., and S.J. Yuan. 2016. Research on Hydro-forming of Combined Prolate Ellipsoidal Shell with Double Generating Lines. *International Journal of Advanced Manufacturing Technology* 82: 595–603.
126. Limam, A., L.H. Lee, E. Corona, et al. 2010. Inelastic Wrinkling and Collapse of Tubes Under Combined Bending and Internal Pressure. *International Journal of Mechanical Sciences* 52 (5): 637–647.